DISCARDED

On Her Own Terms

*The publisher gratefully acknowledges
the generous contribution to this book
provided by the*
General Endowment Fund
of the
University of California Press Associates.

On Her Own Terms

*Annie Montague Alexander
and the Rise of Science
in the American West*

Barbara R. Stein

University of California Press
Berkeley   Los Angeles   London

University of California Press
Berkeley and Los Angeles, California

University of California Press, Ltd.
London, England

© 2001 by the Regents of the University of California

Chapters 8 and 17 are based on the following article: "Annie Montague Alexander: Extraordinary Patron," *Journal of the History of Biology* 30 (1997): 243–66. These revised versions of this article appear here with permission from Kluwer Academic Publishers.

Library of Congress Cataloging-in-Publication Data

Stein, Barbara R., 1955–
   On her own terms : Annie Montague Alexander and the rise of science in the American West / Barbara R. Stein.
      p.   cm.
   Includes bibliographical references and index.
   ISBN 0-520-22726-3 (cloth : alk. paper)
   1. Alexander, Annie Montague, 1867–1950.   2. Zoologists—United States—Biography.   I. Title.

QL31.A555 S74   2001
570'.92—dc21
[B]   2001027083

Manufactured in Canada

10 09 08 07 06 05 04 03 02 01
10 9 8 7 6 5 4 3 2 1

The paper used in this publication is both acid-free and totally chlorine-free (TFC). It meets the minimum requirements of ANSI/NISO Z39.48-1992 (R 1997) (*Permanence of Paper*).♾

Contents

	List of Illustrations	vii
	Acknowledgments	ix
	Introduction	xiii
1.	SAMUEL ALEXANDER AND HENRY BALDWIN	3
2.	LIFE IN OAKLAND	13
3.	A PASSION FOR PALEONTOLOGY	22
4.	AFRICA, 1904	35
5.	MEETING C. HART MERRIAM	48
6.	ALASKA, 1906	58
7.	MEETING JOSEPH GRINNELL	63
8.	FOUNDING A MUSEUM OF VERTEBRATE ZOOLOGY	76
9.	AN UNUSUAL COLLABORATION	88
10.	LOUISE AND PRINCE WILLIAM SOUND	97
11.	SUPPORT FOR PALEONTOLOGY	107
12.	HEARST, SATHER, FLOOD	114
13.	INNISFAIL RANCH	120
14.	VANCOUVER ISLAND AND THE TRINITY ALPS	138
15.	THE TEAM OF ALEXANDER AND KELLOGG	148
16.	FROM "A FRIEND OF THE UNIVERSITY"	155
17.	FOUNDING A MUSEUM OF PALEONTOLOGY	165

18.	A RESTLESS DECADE	181
19.	EUROPE, 1923	190
20.	THE TEMPLE TOUR	203
21.	THE "AMOEBA TREATMENT"	214
22.	FIELDWORK—THE LATER YEARS	224
23.	SALINE VALLEY	244
24.	THE END OF AN ERA	253
25.	HAWAII—"MY ONLY REAL HOME"	261
26.	THE SWITCH TO BOTANY	274
27.	BAJA CALIFORNIA—*TRES MUJERES SIN MIEDO*	290
28.	INVESTING IN THE FUTURE	299
29.	AN ENDURING LEGACY	308
	Epilogue	*315*
	Appendix	*317*
	Notes	*321*
	Index	*359*

Illustrations

Plates follow pages 110 and 174

FIGURES

1. Annie Montague Alexander's family tree   8
2. Chart highlighting the periods that relate to Alexander's paleontological fieldwork   25

MAPS

1. Route followed by Alexander and her father, British East Africa, 1904   36
2. Alexander expedition to southeastern Alaska, 1907   67
3. Innisfail Ranch   124

Acknowledgments

The veracity of any nonfiction work depends largely upon the extent and availability of primary source material. I am therefore deeply indebted to the following libraries, archives, museums, and their staffs who have allowed me access to, and have assisted my use of, their correspondence, books, photographs, and other research documents: Alaska State Library, Juneau (Kathryn Shelton, India Spartz, Gladi Kulp); Alexander & Baldwin, Inc., Honolulu (Howard Daniel, Michael J. Marks and staff); Alexander & Baldwin Sugar Museum, Puunene, Maui (Gaylord Kubota); American Museum of Natural History Special Collections, New York (Paula Willey); Bancroft Library and University Archives, University of California, Berkeley (William Roberts and staff); Bernice P. Bishop Museum of Natural History, Honolulu: Department of Vertebrate Zoology (Carla Kishinami), Bishop Museum Archives (Betty Lou Kam), Bishop Museum Library (Linda Laurence), Registrar's Office (Janet Ness); Hawaiian Mission Children's Society Library, Honolulu (Marilyn Reppun, Jodie Kakazu); Brennan Library, Lasell College, Newton, Massachusetts (James Boudreau); Maui Historical Society Archives, Wailuku (Kathy Riley, Holly Formolo); Punahou School, Honolulu (Bonnie Judd, Mary Judd); Smithsonian Institution Archives, Washington, D.C. (Bruce Kirby); Solano County Courthouse, Fairfield, California (staff); Solano Historical Society, Fairfield, California (Bertram Hughes, Susan Lemon); University of California Museum of Vertebrate Zoology, Berkeley; University of California Museum of Paleontology, Berkeley (Chris Bell, Mark Goodwin); University of California and Jepson Herbaria, Berkeley (Barbara Ertter); the University of Kansas Natural History Museum, Lawrence (Robert Timm); the University of Kansas Spencer Research Library, Lawrence (Ned Kehde); Vassar College Special Collections, Poughkeepsie, New York (Gita Nadas).

Correspondence, receipts, maps, and other paper documents provide a solid foundation around which to construct a biography, but the firsthand accounts of friends, relatives, lovers, business associates, and enemies create a much-needed third dimension. From that perspective, I am deeply grateful to Alice Q. Howard, who generously loaned me photographs of Alexander and Kellogg, their personal diaries, and relevant correspondence in her possession, entrusting me to interpret their contents fairly. Her support of this project and her willingness to invite me into her home and to share her knowledge of Kellogg family history with me were crucial to telling this story.

I am also grateful to members of Alexander's family who were kind enough to provide me with their recollections, in particular Martha (Pattie) Hurd Kreuter, Maryanna Stockholm, Barbara Toschi, and Alexander Waterhouse. Special thanks to the Wallace A. Gerbode Foundation (Thomas Layton, Maryanna Stockholm) for its generous support of my research outside the state of California.

Kellar Autumn and Valeurie Friedman brought my words and passion to the attention of the University of California Press. Without their effort, this project would have remained little more than a series of academic talks and papers.

Several years before this biography took form, discussions with Elihu Gerson, James Griesemer, and Leigh Star about the role of the Museum of Vertebrate Zoology (MVZ) in the development of vertebrate natural history and evolutionary biology stimulated my interest in the museum's history. When I finally undertook this project, Elihu kindly loaned me much valuable material from his personal research files.

Many individuals, including several whom I never met, provided personal recollections and contextual information that did not reside in any library, museum archive, or special collection. For their generosity in speaking or corresponding with me, I particularly acknowledge Richard Beidleman, Bryan Brewer, Daniel Cheatham, Lincoln Constance, Barbara Blanchard DeWolfe, Henry Fitch, Joseph Gregory, June Grinnell, Hubert Hall, Helen McDonald Harris, Steven Herman, Robert Hobdy, Donald Hoffmeister, John Howe, William Kaiwa, James Reid Macdonald, Leslie Marcus, Charles Mason, David Mason, Martha Hall Niccoll, Robert Orndurff, Anita Pearson, Oliver (Paney) Pearson, Frank Pitelka, Virginia Miller Russell, Ward Russell, Jack Schafer, Robert Stebbins, Keir Sterling, Carol Terry, Richard Wilson, and the late Warren Herb Wagner and Samuel Welles. Paney Pearson also made available to me the transcripts of his taped interviews with former MVZ staff members.

In addition to preparing preliminary drafts of the maps and figures in this volume, Karen Klitz volunteered me to speak in the symposium on the history of women at Cal, which took place on the Berkeley campus in the spring of 1994. I mark that event as the first of many steps culminating in this biography and thank her for her faith in me.

Chris Donlay served as the first "outside member of my review committee," and his kind words and constructive criticism in the early stages of this project were invaluable. Many months later, Robert and Sally Hoffmann graciously read the manuscript's entire first draft. Not only are they consummate reviewers but, as former MVZ inhabitants, they provided reflection that only practitioners of a discipline can supply. Robert Kohler offered most useful comments on the second draft of the manuscript.

Jessica Bolker, Margaret Carney, Anna Graybeal, Elizabeth Jockusch, Judith and Joseph Kelly, and Margo Rosenfeld provided inspiration and bolstering at various stages of this undertaking. Their words and encouragement meant more than they can know. I also extend my sincere thanks to David Wake, James Patton, and Harry Greene for their ongoing support.

I owe my deepest debt of gratitude to Karen Borell for her patience, understanding, love, and encouragement. She quickly realized that research for this book held out the potential for travel and eagerly embarked on each adventure, be it to the eastern Sierra or the Solano County Courthouse. While I worked, she willingly played and, at the end of each day, listened patiently as I recounted the results of my investigations. She was my chief critic and head cheerleader. I could not wish for a better partner.

This book is dedicated to the memory of my parents, who would have experienced tremendous joy in its publication.

Introduction

Among the many accomplished women during the last century and a half—adventurous Victorian ladies, pioneering women scientists who are finally being given the recognition they deserve, wealthy and often flamboyant patrons of the arts—few excelled in more than one category. Annie Montague Alexander did. She was an intrepid explorer, world traveler, amateur naturalist, farmer, philanthropist, and founder and patron of two natural history museums on the University of California's Berkeley campus, all at a time when women did not have the right to vote and few had any involvement in the world outside their homes. Alexander took part in many of the major endeavors that were changing the country, California in particular, as the twentieth century began—exploration, research, women's education, conservation, agriculture, business, and philanthropy.[1]

Alexander is remarkable not only for the diversity of her accomplishments but also for the contradictions that accompanied them. Fiercely independent and committed to her ideals, she battled the University of California Board of Regents for more than forty years as she strove to develop research programs in vertebrate natural history and paleontology that are now nationally and internationally known. Her efforts stand out all the more in light of the tremendous intellectual insecurities she harbored and her acute disdain for publicity.

Alexander's desire to establish a museum of vertebrate zoology on the Berkeley campus stemmed, in part, from her awareness of the rapidity with which the fauna in California was disappearing, succumbing to the state's spiraling population growth, rampant agricultural development, and increasing urbanization. At the same time, Alexander shared in this environmental perturbation by reclaiming a tract of land in the Sacramento and San Joaquin delta and establishing a farm there.

I was drawn to Alexander because of who she was as much as what she accomplished. Annie was an articulate and keen observer of human nature. She was a woman of vision who loved women and believed firmly in their capabilities. She was also a warm person who held friendships closely and cared deeply about her family. Alexander was not afraid to express herself or take charge, but she preferred to operate in the background rather than attract notice of any sort. She lived passionately and, perhaps most important, she dealt with the world on her own terms.

Observers usually described Alexander as quiet, certainly not outgoing, although in the same breath they acclaimed her as a friendly, caring, and generous individual who loved the natural world. Equally obvious was the fact that there was no stopping her when an idea captured her imagination. If a new discipline intrigued her, she read up on it and became familiar with its issues. She was captivated by new technology, making an airplane flight at the California State Fair in Sacramento in 1919 and a hydroplane flight while on vacation in Florida three years later. She quickly ascertained the advantages that the automobile would bring to fieldwork and she mastered the minute details of geologic ages and developed the ability to identify fragments of the vertebrate skeleton. She learned the techniques of farming and raised both crops and dairy cattle.

Alexander's appearance belied her physical strength and stamina. She was short and slender, her shoes were tiny, and she was almost always referred to as frail looking. She wore her long hair in a bun at the back of her head until 1929 when she impulsively bobbed it (the experience was perhaps more dramatic than traumatic but it was one that she commemorated in verse).[2] Her lifestyle was spare but she loved pretty things. She was well read and enjoyed the theater, opera, and movies.

As a young woman, Alexander experienced unspecified problems with her eyes and difficulty focusing at close range. In her early twenties, she began studying painting and drawing in Europe but ruefully abandoned these subjects at the onset of migraines and the threat of blindness. Freed from many of the detailed, domestic tasks that placed a strain on her eyes and chafed her restless nature, she struggled to find a place for herself and give her life meaning.

Much like her father, Samuel, in spirit and temperament, Alexander yearned for adventure. She traveled more and more often and, presumably under Samuel's tutelage, became an excellent shot (her distance vision remained unimpaired). A camera soon replaced brushwork and sketching. These activities coalesced with Alexander's introduction to paleontological fieldwork in 1900 and grew in importance following her decision in 1906 to create and develop a vertebrate natural history museum on the Berkeley campus.

Alexander's handwriting deteriorated badly in later years and it must have been with some relief for both writer and recipient that she began typing her correspondence and field notes, adding a typewriter to the standard retinue of field equipment that she and her partner of forty years, Louise Kellogg, carried with them on collecting expeditions. Alexander carried on a voluminous correspondence that offers a valuable and detailed record of her activities. Few today appreciate how truly pervasive the art of letter writing was for those who lived before the invention of the telephone. It was not until 1915 that phone service became available on the farm that Alexander and Kellogg maintained in the Sacramento River delta, their primary residence when not in the field. Alexander was forty-eight by then and would hardly have considered foregoing the correspondence that gave her the means to express ideas and feelings in a manner that the telephone would never duplicate. The letters document Alexander's intentions with respect to her many business agreements as well as her unedited sentiments and impressions about the people, places, and events in her life.

Despite the wealth of Alexander's correspondence, truly personal letters are rare. Those to her close friend Martha Beckwith convey the intensity of Alexander's feelings.[3] The presence of those letters makes it all the more unfortunate that, with only one or two exceptions, no letters between Alexander and Kellogg survive. Even Alexander's diaries, begun only after she and Kellogg lived together, do not shed any light on her feelings for Kellogg or the personal nature of their relationship. The diaries consist of "line-a-day" entries that served as a log of daily activities, letters written, and weather conditions and temperature on the farm. Kellogg apparently began the account, and her early entries cover the women's first trip together in 1908. Any reference to feelings, for example, "night with Annie—sweet!" soon disappears and, after the first year or so, there are no personal revelations, even so slight as this one.[4] The women took turns keeping the diary, trading off every three or four years, so that only a single set of diaries exists for the duration of their life together.

Close relationships between educated young women who belonged to the middle and upper classes were common at the end of the nineteenth century. These women sought out and discovered others like themselves—women who aspired to lives and occupations outside the confines of marriage and family. "Boston marriages" or "romantic friendships" allowed women to assume all or part of the professional, financial, and decision-making responsibilities held by men in traditional marriages.[5]

There has been much academic debate about the true nature of these late nineteenth-century romantic friendships. Adult women freely enjoyed intimacies such as sleeping in the same bed, holding hands, pledging eternal love, and penning romantic letters, deep personal expressions that are increasingly self-conscious and rare after the turn of the century. Their contemporaries simply described such proper single women as "close friends" or "devoted companions." And with or without sex, romantic friendships generally involved lives of love, shared values and experiences and, often, hard work. Alexander and Kellogg maintained such a relationship.[6]

By the end of World War I, tolerance for romantic friendships had virtually disappeared in this country. Several persons who were at the University of California shortly before or after World War II and who were aware of Alexander and her legacy of benefaction to the university referred to the friendship. One former Museum of Vertebrate Zoology employee said simply, "[Alexander] was a lesbian. We didn't talk about such things in those days. Her friend was Louise Kellogg and they did everything together to the end of their lives. They were both interested in natural history. It was a very happy relationship I think."[7]

Although creations of the Museum of Vertebrate Zoology (MVZ) and the University of California Museum of Paleontology (UCMP) are major facets of Alexander's life, this book is not a history of those institutions and I do not paint a complete picture of any of their colorful and influential inhabitants. The men whom Alexander hired and whose research she funded, those whose careers were built on the specimens she collected, are well known within the disciplines of vertebrate zoology and paleontology, for example, Joseph Grinnell, E. Raymond Hall, Charles Camp, John C. Merriam. Therefore, I merely included brief biographical sketches of these and other relevant figures in the history of evolutionary biology when I felt that such material helped develop the story about Alexander. Alternatively, I expounded on the growth and development of the corporation founded by Alexander's father and her uncle, because I believe that Alexander's business ventures and farming success owe much to her familiarity with, and understanding of, that enterprise.[8]

Throughout this story, I have used current place-name spellings when they differ from those that appear in Alexander and Kellogg's field notes, diaries, and letters. The women relied primarily on U.S. Geological Survey topographic maps and AAA maps available at the time and often pasted relevant sections of these maps into their field notebooks. When differences are striking and might be cause for confusion, the older spellings follow in parentheses.

This book actually began as a presentation in a symposium on the history of women at the University of California, Berkeley, during its first 125 years of existence.[9] Alexander was virtually unknown to the organizers of that symposium before her name was put forth for inclusion in the program. When I agreed to speak about her, my casual interest in her life, and my admiration for her many and varied accomplishments, soon became a passionate search to separate fact from the museum lore and apocrypha with which I had been indoctrinated since my arrival at the MVZ in 1985.

As I began to prepare my talk, I became engrossed in the many letters Alexander wrote to her family, her close friends—in particular a paleontological colleague, Edna Wemple McDonald, and the anthropologist Martha Beckwith—and to the faculty and administrators at the university, most frequently Joseph Grinnell, the biology instructor she hired to develop and direct the Museum of Vertebrate Zoology.[10] Accordingly, I chose to weave excerpts from these letters into the text of this biography to convey the extraordinary qualities that compelled me to undertake this work. I hope that by using Alexander's words I succeed, in some small measure, in giving life to a woman who came alive to me through her words.

Even the suggestion of a biography would probably have dismayed Alexander. She avoided publicity. She made it clear that she did not wish new species of plants, animals, or fossils to be named in her honor, although some were (see the appendix). Her numerous donations to the university were made with the stipulation that no publicity be associated with them, although announcements of her gifts occasionally appeared in the university's magazines and newspapers.

In the fall of 1935, when Alexander was sixty-eight years old, the MVZ's director wrote to her requesting that she have a photograph taken of herself for the museum's archives, feeling that he would be seriously remiss if such a record were not made for posterity. In an exchange of letters indicative of their unusual relationship, Alexander wrote that she would comply with Grinnell's request but added, "May I ask that no picture of myself be on exhibition as [sic] the museum during my life time." Grinnell replied that he would, of course, honor her request, "Even so, and all the more, do I personally appreciate the pains you have been to, to provide the permanent likeness that the future will demand."[11] Because of her extraordinary nature and accomplishments, I believe that the future also demands this biography.

On Her Own Terms

1

Samuel Alexander and Henry Baldwin

The death of her father in Africa in 1904 served as the catalyst for Annie Montague Alexander to found the Museum of Vertebrate Zoology on the Berkeley campus. At the age of thirty-seven, she felt the need to give meaning to her life and the idea of creating a natural history museum gradually took shape in her mind.

More than any other of Samuel Alexander's children, his second daughter, Annie, embodied her father's striking characteristics—his intensity and entrepreneurial spirit, as well as his generosity and an unshakable commitment to the causes in which he believed. Her life emulated his in extraordinary ways. From an early age, Alexander even seemed to display her father's head for business and his interest in farming. As a young girl, her uncle purportedly offered to pay her a quarter apiece for each avocado seedling she could bring him. She is said to have appeared in his yard some time later with an oxcart laden with small plants, each in its own tin, and to have presented the surprised man with a bill for $75.00![1]

Among all of Samuel's children, only Annie seemed to share her father's love of adventure, his passion for travel, and his deep appreciation for nature. Like him, she was drawn to challenge yet reluctant to take center stage. Annie's older sister, Juliette, became a writer and poet. Her brother, Wallace, followed in the family business.[2] Her younger sister, Martha, married and raised a family. Family chronicles never mention the youngest child, Clarence, who died before his fourth birthday.

The striking characteristics that Alexander displayed had previously been passed from father to son. Samuel Thomas Alexander was the third son and child born to the Reverend William Patterson Alexander and his wife, Mary Ann McKinney. Married barely a month when they set sail from New Bedford, Massachusetts in 1831, the twenty-six-year-old minister and his twenty-one-year-old bride were one of only nine couples comprising the fifth company of missionaries sent by the American Board of Commissioners of Foreign Missionaries (ABCFM) to proselytize among the natives on the Sandwich Islands (later, the Hawaiian Islands).³ King Kamehameha I had died in 1819 and the kingdom's new constitutional monarch quickly renounced the ancient religion that had long been a centerpiece of island life. Missionaries felt drawn there by Providence. Other missionary families whose names would become famous in Hawaiian history preceded the Alexanders to this tropical paradise. The Bishops had been a part of the second ABCFM company, the Gulicks had arrived with the third, and the Baldwins with the fourth. Amos Star Cooke and his wife, Samuel's future in-laws, would be in the eighth company.

Reverend Alexander's first posting was to a small congregation in Waoili on the island of Kauai, but in 1843 he was transferred to the town of Lahaina on the western coast of Maui, having been assigned to take charge of Lahainaluna Seminary, a missionary school for native Hawaiian boys. In the mid-nineteenth century, Lahaina was the principal town on Maui, a picturesque community and bustling whaling port, second only to Honolulu as an important commercial center on the islands.⁴ A dense growth of tropical foliage ran to the water's edge. Enormous kou trees, whose trunks measured up to 7 feet in diameter, grew to heights of 35 or 40 feet. To the east, the mountains of western Maui rose from the low-lying plains, their intermittent valley streams providing a valuable source of fresh water for the town's inhabitants. As it emerged from the dense foliage, the sparkling water cascaded to the lowlands in splendid falls. Clear streams rushed and tumbled from the hillsides, tripping over rocks and scouring streambeds, ultimately forming deep pools that were ideal for bathing. Only a few miles from the ocean's shore, one could experience the cool, bracing air of the valleys and escape the heat and humidity of town. The Lahainaluna Seminary, two miles away, stood in bold relief against the hills behind it.

For Samuel, growing up in Maui during the nineteenth century was in many ways idyllic. Life in general was lived out-of-doors—by the sea, in the mountains, under the sun. For children, tree climbing, horseback riding, and collecting land snails were favorite pastimes. The snails were astounding for the variety of colors, shapes, and sizes of their shells. Hours and even days might be spent turning over logs and rocks or peering under damp, decaying leaves amidst the lush foliage along the coast and up into the mountains in search of these fascinating gastropods.

Through activities as simple as gathering land snails and ferns, the Hawaiian missionaries and their children contributed significantly to what was known about the islands' natural history during the nineteenth century.[5] Many became passionate collectors who meticulously recorded extensive observations on all aspects of the natural world around them, describing unfamiliar botanical specimens and seemingly alien geologic structures that had resulted from volcanic activities not understood at the time.

Reverend Alexander transmitted an interest in natural history to his offspring.[6] Samuel, in particular, inherited his father's love of trees. Both men planted hundreds of saplings on Maui, young shoots that, in time, grew to provide their families with sweet fruit, delightful fragrances, and soothing shade. Years later, Samuel was to impart this love to his daughter Annie.

Samuel was seven when his family moved to Lahaina. In its carefree and gentle atmosphere, the Reverend Alexander raised his children, his sons in particular, to make the most of their lives and be of use in the world.[7] He based his own life on integrity and hard work and expected his children to live similarly. His financial resources were meager, but he stressed the merits of education and of learning, and he encouraged his children to think freely.

Even as a child, Samuel expressed a cheerful and outgoing personality. He interacted easily with people, and his popularity seemed based on his genuine good nature. His father reflected, "He has more energy & enterprise than any of his brothers. He will be more likely to make himself felt in the world than they. . . . He has more *go ahead* than any of my sons & a good deal more native talent."[8]

William's words would prove prophetic. While still a student at the Punahou School for missionary children in Honolulu, Samuel accepted a one-year position as bookkeeper and merchant on a sugar plantation on Kauai.[9] Gainful employment on the islands was a perpetual problem for all missionary children, and management positions that required a secondary education were coveted.

The conclusion of Samuel's internship on Kauai coincided with an economic downturn on the islands. The government was bankrupt and few schools had the financial resources to remain open. Unable to find a satisfactory position in Hawaii, Samuel set off in search of gold, "finding it preferable to risk the seduction of California, than to rot at the Islands," he explained to his older brother James.[10]

Before leaving for California, Samuel became engaged. His bride-to-be, Martha (Pattie) Eliza Cooke, was the daughter of missionaries Amos Star Cooke and Juliette Montague. Upon learning of Samuel's good fortune, James wrote home, "I should think, from what I have heard, that he had gained the affection of as accomplished, attractive, and noble-hearted a lady as there is in the Islands or anywhere. I hope he may thereby be clarified & ennobled himself."[11]

Samuel's search for gold proved much less romantic and far less profitable than he had hoped. Long days of back-breaking labor brought him little remuneration and he met few honest men or trustworthy companions. After only a year in California, what gold he managed to amass was gone, some by theft, the rest spent nursing himself back to health after contracting malaria. He returned to Hawaii penniless, earning his passage home by working aboard ship.[12]

Samuel and Pattie's engagement spanned nine years. The young Alexander was determined not to marry until he could support his bride and the family that he was sure would follow in rapid succession. Starting anew, he eagerly accepted an offer of a teaching position at the Lahainaluna Seminary. Teaching was not his first love, but he viewed the position as a way to ensure his fiancée a proper home and some degree of financial security. The offer, however, was contingent upon his spending several more years on the mainland completing his education. While Pattie waited patiently, Samuel sailed to New England. He attended Williams College for two years before accepting a residency in the teacher training program at the Normal School in Westfield, Massachusetts.[13]

Samuel returned to Hawaii eager to take up his new position. In keeping with his own interests, he began planting banana trees and raising sugarcane with his students as an adjunct to the school's more established curriculum. The scores of ripe bananas and rows of mature cane that carpeted the hillsides overlooking Lahaina did not go unnoticed. Two years after he had returned, Samuel was offered the position of operations manager on a sugar plantation in northcentral Maui where the enchanting and lush Iao Valley lay immediately to the west. With his father's blessing, he gave up teaching and moved with his bride to Waihee.

The energy and enterprise that William had foreseen in his third son quickly manifested themselves. Within four years, Samuel resigned his position at Waihee and resettled his family in the town of Haiku near the dry and dusty cane fields that he and his business partner, Henry Baldwin, had purchased. Samuel and Pattie were now the proud parents of three children—Juliette, born in 1865; Annie Montague, born in 1867; and Wallace McKinney, born in 1869. A third daughter, Martha Mabel, would follow in 1878, and a son, Clarence Chambers, in 1880.[14]

Samuel had met his business partner, Henry Baldwin, after the Alexander family moved from the Reverend Alexander's posting on Kauai to Maui. Dr. Dwight Baldwin, Henry's father, served as the community physician in

Lahaina, as well as one of its spiritual leaders. The two couples and their children quickly became close friends—so close, in fact, that Samuel's oldest brother, William, married Henry's older sister, Abigail, while Henry, in turn, married Samuel's younger sister Emily (see the Alexander family tree in Figure 1).

Samuel and Henry had more in common than missionary roots and close family ties. Both were bright and energetic, possessed of good natures and naturally green thumbs. As friends, they developed a bond of trust and confidence that neither time nor good fortune would diminish in any way. As individuals with complementary strengths, they were exceptional business partners. Samuel was the fountain of new and innovative ideas, Henry the partner who carried their projects through to completion. This fruitful collaboration was a model that Alexander would later emulate in her relationship with Joseph Grinnell, the man she selected to serve as director of her Museum of Vertebrate Zoology.

The fields that Alexander and Baldwin had purchased lay in a stretch of northcentral Maui between the towns of Paia and Makawao. The partners' ability to buy the land with relatively little capital had to do with its undesirable location. The dormant volcano Haleakala (house of the sun) rose majestically to the east and cast a long, dry shadow across their property; the paucity of rain limited the size and productivity of their fields.[15] The Victorian traveler Isabella Bird described the landscape in the vicinity of the family's new residence at Haiku as a "Sahara in miniature, a dreary expanse of sand and shifting sandhills, with a dismal growth in some places of thornless thistles and indigo, and a tremendous surf thunders on the margin. Trackless, glaring, choking, a guide is absolutely necessary to a stranger, for the footprints or wheel-marks of one moment are obliterated the next. . . . It is a hateful ride, yet anything so hideous and aggressively odious is a salutary experience in a land of so much beauty."[16]

To escape the dusty and barren landscape around Haiku, with its suffocating summer heat, Samuel built a second home at Olinda on the cool, damp slopes of Haleakala. The volcano towered more than 10,000 feet above sea level to dominate the landscape of eastern Maui like an awesome god. The children spent countless hours exploring the volcano's wooded slopes or turning over rocks and logs in search of land snails and ferns. Another of their favorite pastimes was watching the sun rise slowly and spectacularly over the "Big Island" of Hawaii. Often the children would pack camp gear, blankets, matches, and food and ride horses or mules to the top of the mountain before nightfall. On other occasions they would simply mount their horses at one or two in the morning, when the sky was black and no light

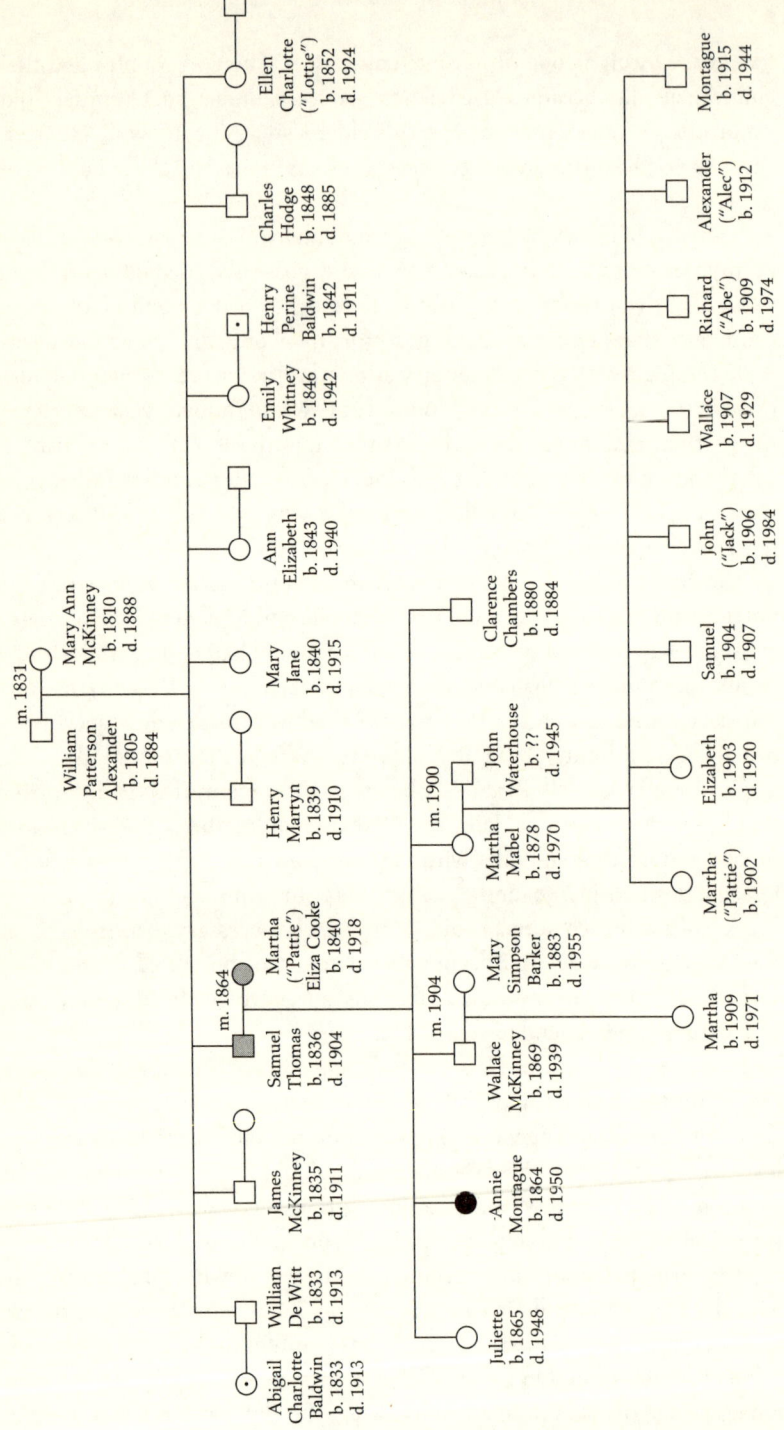

Figure 1. A simplified view of Alexander's family tree (it omits her many cousins and their offspring). Shaded areas highlight Alexander's lineage. Early marriages between the Alexanders and Baldwins are indicated by dots within a circle or square.

penetrated the dense woods along the steep mountain slopes, in order to arrive at the summit by daybreak. When they reached their destination there was no guarantee that they would actually see the sun rise. Instead, they might be greeted by dense clouds and fog, the spectacle reduced to little more than a gradual fading of darkness into the monotony of gray gloom.

Since the mountain itself was anything but dry, Samuel conceived of a remarkable plan to carry desperately needed water from its wet, windward side to his parched plantation below. By 1876, the property that he and Henry jointly owned had increased to such an extent that their only hope for survival, and for profit, was to devise a means to greatly enhance its irrigation capabilities. The rain that they depended upon for irrigation was now grossly insufficient for the size of their acreage.

The Hamakua ditch project became perhaps the most dramatic water project in Hawaii's history, and one that served as a model for later irrigation projects throughout the islands.[17] Though not an engineer, Samuel proposed to construct an aqueduct approximately twenty-five miles long, which would gather water from the dozens of mountain streams and rivulets on the wet, windward, eastern side of Haleakala and carry it through ridges and across ravines to the fertile, but thirsty, plantations lying in the volcano's rain shadow to the west. Samuel arranged a preliminary survey for the construction, worked out its financing, and negotiated a lease from the government for the land. In turn, Henry oversaw the digging of the miles of ditches and tunnels, the building of flumes, and the placement of pipes for the completed structure. Under his direction, the project was completed by deadline in an astounding two years.

The timing could not have been better. Construction of the aqueduct and the signing of the Reciprocity Treaty of 1876 between the United States and Hawaii marked the first steps toward profitability for Alexander and Baldwin. The ditch provided the water they so desperately needed to irrigate their crops. The treaty gave Hawaii a distinct advantage over other foreign producers by stipulating that the U.S. government agree to remit the duty on rice and sugar imported from the islands. Accordingly, the partners continued to purchase land in central Maui, increasing the size and importance of their plantation. They also bought out a third party's interest in a sugar mill. For the first time, they were now able to process some of their own cane.

In 1883, Samuel and Henry incorporated their informal partnership as a business. They chose to call it Paia Plantation.[18] That same year, for unspecified reasons of health, Samuel announced that he was moving his family to Oakland, California. The climate in Hawaii may have induced unpleasant recurrences of the malaria that he had contracted as a young man.

Samuel loved the climate in San Francisco, which he described as "splendid and invigorating." He wrote to his older brother William, "I am more than ever convinced that the climate of the Islands is debilitating;—the tendency is to sloth and vice." More to the point, the still vigorous Samuel was thoroughly convinced that California was "*the* country." Despite his somewhat harsh experiences in the state as a young man, he wrote to his brother James, "There is perhaps no country in the world better calculated to develop independence of character than California."[19]

The move to California made business sense as well. From Oakland, Samuel kept Henry well informed of developments in Congress with respect to sugar tariffs and the issue of annexation. When the Revolutionary Reform Party gained renewal of the Reciprocity Treaty in 1887, Hawaiians were convinced that their economic well-being was assured. Thus, when Congress suddenly revised its tariff policy a short time later and granted a bounty of two cents per pound to American domestic producers, Hawaiians were stunned. One stroke of a pen had shut Hawaiian growers out of American markets and put the issue of Hawaii's annexation to the mainland up for heated debate throughout the islands.[20]

Samuel's move to the mainland proved timely in other respects as well. Claus Spreckels, a wealthy San Francisco sugar refiner and an aggressive competitor, quickly established himself as a major sugar producer with the aim of monopolizing that commodity in the Hawaiian market.[21] When Alexander and Baldwin's purchasing agent in San Francisco teetered on the brink of bankruptcy during the depression of 1891–94, Spreckels's Hawaiian Commercial Company took over its affairs. This maneuver placed Alexander and Baldwin's debts to their agent in Spreckels's hands and left the partners in an unenviable and precarious position. Samuel worked feverishly to raise the money needed to pay off the debts, terminate Spreckels's firm as their agent, and establish a San Francisco purchasing agency of their own.

As a direct result of Samuel's efforts, Alexander & Baldwin opened its doors in San Francisco in 1894. Samuel's son, Wallace, and his nephew Joseph P. Cooke were picked to run the new purchasing agency. The two proved to be skillful businessmen. Marketing profits quickly increased and the agency began serving a growing number of smaller plantations on the islands. Less than three years after its founding, Joseph went back to Hawaii to open an A&B office in Honolulu, while Wallace remained in San Francisco to direct the company's affairs on the mainland.

With purchasing now under their control, the partners turned their attention to the problem of refining their sugar. Finding a buyer and securing the best possible price for their product were paramount concerns. Whether the partners executed a one-year contract or a multiyear contract with a refinery that would produce table sugar from the raw material they delivered, the purchase price for their sugar fluctuated unpredictably with the markets in New York and Manila. To make matters worse, there were only two sugar refineries on the West Coast, both in the Bay Area. Spreckels controlled one, the California Sugar Refining Company, and maintained a one-third interest in the other, the smaller American Refinery. To avoid Spreckels's virtual monopoly on the refining industry in California, the partners would have to ship their raw sugar to the lone refinery on the East Coast—two decades before the Panama Canal was completed, sending their product around the southern tip of South America—at a significant cost in time and profit.

In 1896, after devoting considerable thought to the problem, Samuel quietly organized a majority of the independent plantations in Hawaii and, without Spreckels's knowledge, secured a three-year contract with the American Refinery to handle the group's sugar. With that contract in place, the consortium of independent growers successfully maneuvered to acquire a controlling interest in the American Refinery and the following year named Samuel to its board of directors. Simultaneously, Alexander & Baldwin, in conjunction with several members of the consortium, purchased an old flour mill in the town of Crockett northeast of San Francisco and converted the mill into a refinery of their own. The mill became established as the California and Hawaiian Sugar Refining Company, predecessor of today's California and Hawaiian Sugar Company, better known as C&H Sugar—pure cane sugar from California and Hawaii.

Spreckels still maintained extensive land holdings on Maui and a partial interest in the Maui Railroad & Steamship Company. With these investments he was able to deny other growers access to the waterfront, leaving Alexander & Baldwin and their allies no way to get their sugar to port. The partners responded by buying out both the Kahului Railroad and the competing line and port facilities owned and controlled by Spreckels, thus securing complete control of their sugar from field through refinery.

When Samuel and Henry started their business, sugar was carried to the mainland on sailing ships that the partners chartered. With the passage of time, the partners came to rely increasingly on the new, larger and faster steamships, including those operated by the Swedish-born captain William Matson. Before long, Alexander & Baldwin became the agent for Matson Navigation and eventually acquired that business as well.

Alexander and Baldwin were now established as the largest sugar producers in Hawaii. In little more than six months, Hawaiian Commercial and Sugar Company (H.C. & S.) stock rose from $27 to $128 per share. Having outgrown their current organization, on June 30, 1900, they met to file papers of incorporation in Hawaii. Henry Baldwin was elected the corporation's first president and all the partnership's assets were transferred to the new company, Alexander & Baldwin, Limited.

While life in the Bay Area delighted Samuel, the move to Oakland never suited Annie. As a young girl growing up in Haiku, she regularly swam in the ocean and rode horseback through the dusty streets of town. She delighted in climbing to the roof of the upstairs verandah and entering her bedroom through an open window rather than using the stairs inside the house. Then, in the fall of 1881, at the age of thirteen, she enrolled at Punahou School in Honolulu, elated at the adventure involved in the arduous eighty-eight-mile ocean journey from Maui to Oahu. But her tenure and happiness there proved short-lived. After only a year, Samuel moved the family to California.

Perhaps the city of Oakland itself was to blame for Alexander's unhappiness or the role that she gradually felt compelled to play as an affluent, female member of society. Regardless, as a young woman she repeatedly flew from oppressive feelings that life in the city seemed to impose upon her. Increasingly, she exhibited her father's restless passion for adventure.

Oakland, however, did offer Alexander one incalculable benefit—its proximity to the newly established University of California in the neighboring town of Berkeley. Although Alexander could not quite come to terms with her life in the city, her introduction to the university in the fall of 1900 marked a turning point. The paleontology classes she began attending that semester did much more than simply stimulate her mind: they introduced her to fieldwork and a justifiable means of escape from her oppression.

2

Life in Oakland

In the 1880s Oakland was a fashionable place to live, a city reminiscent of the towns and villages on the East Coast from which its early settlers had come.[1] Its name reflected the groves of gigantic oak trees that lined its shore. Majestic redwoods still topped the gentle hills that rose to the east of town, and every spring wildflowers carpeted the fields surrounding the city as far as the eye could see. The city offered its residents paved roads, police and fire services, gasoline street lamps, regular hourly ferry service to San Francisco, and a steam railroad that connected to the ferry. Its convenient commuter connections and genteel atmosphere drew the families of prosperous San Francisco businessmen and many of that city's leading professionals. But perhaps more than any of Oakland's other attractions, its Mediterranean climate and educational facilities made it a bright, pleasant, wholesome family town—untouched by the excesses that were San Francisco's legacy of the gold rush days—"the Athens of the Pacific Coast," as it was known at the time.

Shortly after arriving in Oakland, Samuel set about building a house on the western side of town, on the northeastern corner of Sixteenth and Filbert streets.[2] Among the Alexanders' neighbors were the author Jack London, the architect Julia Morgan, philanthropists Jane and Peder Sather, a former city mayor, several university regents, and the family of Charles and Anita Kellogg, descendants of East Coast merchants and the cousins of Martin Kellogg, professor of ancient languages at the University of California and later the university's president.[3]

Details of Alexander's life in the years immediately following the move to Oakland are few. Although Samuel had come to California for reasons of health, he also wished to avail himself of the intellectual and social amenities that eluded an isolated plantation owner on Maui. In Oakland there were opera and symphony concerts to attend, musical instruction and dance lessons for the children, as well as afternoon teas and luncheons for his wife.

In the fall of 1887 Alexander traveled east to attend Lasell Seminary for Young Women in Auburndale, Massachusetts. Located nine miles west of Boston on the newly installed Worcester Railroad line, the school was named for its founder, Edward Lasell, a former chemistry teacher at Williams College. Lasell had come to believe that women as well as men could benefit from instruction after high school. His thinking was radical at a time when educating women beyond the elementary skills of reading, writing, and arithmetic was considered unnecessary, even wasteful, and certainly against human nature. Lasell embraced the notion that good academic training was in no way injurious to a woman's success in marriage or in life.[4] His seminary thus became the first four-year junior college for women in the United States and modeled its curriculum after the course of study at Williams College with which its founder was familiar. Samuel may have become aware of Lasell's relatively progressive program during his own brief attendance at Williams. Or Annie may simply have been following other missionary offspring to school on the mainland. When Alexander enrolled at Lasell in the fall of 1887, her close childhood friend Mary Beckwith was already a student there.

The mid-1880s ushered in a period of relative personal freedom for women at Lasell. Students were still expected to attend daily chapel and church on Sunday and participate in morning calisthenics and drill twice a week but, between 2:00 P.M. and 7:15 P.M. each day, they were free to go about on their own when not in class. This was a distinct change from previous years when every hour of the students' day was accounted for by special rules. This same freedom, however, was not yet accorded women's dress. Gloves were required and bloomers unacceptable during Alexander's tenure there. The mid-1880s also marked the introduction of military drills as a regular part of Lasell's program. The purpose behind these workouts was threefold—they were believed to be the best exercise for improving one's posture, they were deemed to be a good way to train young women to operate in an organized fashion, and they were a logical means to inspire patriotism.

Although Alexander never earned a degree at Lasell, during the two years she spent there she studied nineteenth-century history, French and Roman history, political economy, civil government, Shakespeare, English, German, French, composition, fair logic, moral logic, choir, voice, dress cutting, and photography.[5] In later years she was fond of quoting bits of poetry in French and German. But of all the courses in which Alexander enrolled at Lasell, it was photography that she pursued most faithfully in later life.

In the summer of 1889, after Annie had left Lasell, the Alexander and Baldwin families vacationed in Europe. When they returned to the States, Alexander remained in Paris to study French. Having exhibited some degree of artistic talent, she also enrolled in drawing classes at the Sorbonne. She spent the winter in Berlin studying German and the following year she returned to Paris, living with her Aunt Lottie, her father's younger sister Ellen Charlotte, on the fourth floor of an apartment building on Avenue Kléber. During this period, she began suffering from migraines and experiencing persistent eyestrain. The doctors she consulted warned of blindness if the strain continued.[6]

Reluctantly, Alexander relinquished any hope of becoming an artist. She destroyed all her paintings and drawings, feeling that her work lacked finish. She returned to Oakland and entered a period of searching. She attempted nursing school but found that the required reading put too much strain on her eyes. Fearing blindness, she set about memorizing poetry, committing to memory the whole first canto of Milton's *Paradise Lost* among other favorites. Emulating her father, she also began to travel.[7]

Samuel had long expressed his own father's yearning for adventure and deep-rooted passion for travel. The elder Alexander had been raised on the frontiers of Kentucky and, during his early missionary years, had sailed to the South Seas and lived among the cannibals of the Marquesas. Samuel felt similarly drawn to explore the wild, untouched places on earth. In a letter to his older brother William in 1866 he pronounced, "What is the use of settling down comfortably & leading a good virtuous & industrious life & then dying. No, I would rather start off in quest of the elixir of life, & roam ragged & hungry over barren mountain summits, than live the life of the most virtuous & useful men." Suiting his acts to his words, within a fifteen-year period Samuel visited England, Scotland, Ireland, Egypt, Palestine, and India; voyaged through the South Seas visiting the island groups between the Marquesas and Australia; sailed around South America; ascended one of the highest mountains in Bolivia; journeyed far up into the Andes in Peru; and touched Rio de Janeiro, Pernambuco, and the Antilles. Later in life he visited Java, China, Japan, Alaska, and Iceland.[8]

In the summer of 1893 father and daughter toured 1,500 miles on bicycles through England, France, and Switzerland with Annie's younger sister, Martha, and one of their cousins. Two years later Annie undertook a second bicycle excursion, this time with her cousin Will Cooke. In 1896 she sailed to Asia and through the South Pacific with her uncle James, Samuel's older brother, visiting Hong Kong, China, and Singapore before turning south to explore Java, Samoa, the Marquesas Islands, and, finally, New Zealand. Regrettably, no diaries or letters from these journeys survive.

In late May 1899 Alexander set out on horseback on a ten-week trip through northern California and southern Oregon with Martha Beckwith. Over the course of the summer, the women discovered themselves to be kindred spirits. "Had we not been the aspiring natures we were we never should have climbed the Lassen Buttes or ever reached Crater Lake," Annie wrote upon their return to Mary, her former classmate and Martha's older sister.[9]

Stowing their gear in packs and saddlebags, the women covered more than six hundred miles, camping as they went, sleeping on cushions of fir or pillows of tamarack boughs. The purpose of their journey was to collect plants, but Annie praised Martha's patience and exuberance in trying to teach her the names of all the birds that they encountered as well.

The pair stopped in Brandy City, Quincy, Greenville, Prattville, Fall River Mills, and Lookout, California, before wandering across the border into southern Oregon. From Klamath Falls they journeyed north to Crater Lake. In the high Sierra meadows, a few precocious blades of grass attempted to push their way toward a bright June sun amid scattered drifts of snow, often knee-deep. By their own reckoning, the most spectacular moment of the trip was their first glimpse of Mount Shasta silhouetted against the setting sun. The following morning, a three-hour hike up Lassen Buttes led them to the edge of a large meadow, beyond which dark hills forested with pines served as a stunning backdrop for the mountain. On their descent, they passed Tule Lake and from there continued on a distance through barren-looking country littered with fallen pinecones. When the weather turned inclement, evenings might be spent recuperating in sleepy little towns "where the men had nothing more to do than bask and blink in the sunshine and the women rocked in easy chairs and chewed gum." Accommodations in such towns were less than enticing, and Alexander displayed her playful humor when she wrote to Mary Beckwith, "To build little Swiss hotels in every town we had been in we considered to be on the whole the best missionary work that could be done."[10]

Almost immediately upon returning to Oakland, Martha boarded a train for the East Coast to resume teaching. With more than usual feeling, Annie mourned her going and counted the growing number of miles that separated the two. She penned increasingly fervent letters to her new friend, proposing long leisurely trips into the hills surrounding San Francisco Bay—they would hike, bring their lunch, glory in the sunshine, and take photographs together—if only Martha would return to California. In one letter Annie revealed her depth of longing for her new companion when she recounted, "I woke myself up last night calling out—Martha!—I was sitting up in bed looking around the room with the confused feeling that I had pushed you out of bed and you had slipped off to find another place to sleep. There was no Martha anywhere near to answer me."[11]

To Alexander's surprise and increasing dismay, after reaching the East Coast, Beckwith boarded a ship for Paris. Clearly distressed at the immense ocean that would be added to the continent already placed between them, Annie wrote passionately to her new friend. Understanding that "we only cast our lots together for the summer and the summer is over," she nonetheless now felt alone and bereft in Oakland.[12] Without Martha to guide her, Annie seemed at a loss for ways to develop the new interests that the ten-week trip to Crater Lake had added to her life.

The trip to Crater Lake marked the beginning of a lifelong friendship between the women. Yet Beckwith did not share Alexander's desire for a closer relationship. A letter from Alexander to Beckwith written in the fall of 1902 exudes her bitterness after confronting this reality and indicates that Beckwith precipitated the rift. As Alexander's letters make clear, she continued to care deeply for Beckwith. The correspondence went on unabated, albeit in a slightly less romantic tone. From Hawaii that winter she wrote most frequently about sugar—factors influencing the success of the current crop of cane or its price on the open market—and about mutual family and friends. Alexander's enduring fondness and commitment to Beckwith are perhaps most evident in letters written several years later in which she encourages her friend to become involved in a lasting relationship. She wished for Beckwith a union to ease the burdens and isolation that Annie felt were pervasive in Martha's life and impervious to the financial gifts that she herself continually proffered. She wrote, "Martha, dear friend, don't work too hard; find somebody to love you and do for you—you are off by yourself near that big University. I wish I might break in upon any hours of loneliness that might be yours. Think of me as loving you—Anne."[13]

Leaving the question of affection aside, the careers to which each of these women eventually devoted their lives were not suited to a partnership between them. Alexander later credited Beckwith with introducing her to geology and to an ensuing passion for fossils.[14] However, Alexander's pursuit of paleontology and zoology as vocations necessitated a partner who would accompany her on field expeditions and could share the tedium and sheer physical labor involved in the work. Not only did she face the issue of impropriety if she dared to travel either alone or as the only woman in the company of men but, once she began conducting fieldwork on her own rather than as a member of a party sent by the university, the possibility of equipment failure, sickness, or bodily injury made a compelling case for the presence of a collecting partner.

Beckwith's interest in fieldwork was of a different sort. After graduating with a bachelor's degree from Mt. Holyoke College in 1893, she began teaching, first at Elmira College, then at Mt. Holyoke, Smith, and Vassar colleges, before deciding to enter graduate school at Columbia University to pursue a degree in anthropology under Franz Boas. She published extensively on the folklore of Native Americans, Jamaicans, and indigenous Hawaiians, research that left no time for her to assist in the extended and arduous paleontological and zoological field expeditions that Alexander would eventually undertake for several months each year.[15]

Despite the physical distance between them, Annie continued to pour out her deepest feelings in letters to Martha. The increasing strain that reading now placed on her eyes led her to realize that she would never be capable of performing any close, detailed work. She wrote to Martha of her frustration with her optician and of her waxing and waning optimism:

> Am always preaching to myself that one must have resources within oneself for happiness but I've been distraught this week by the machinations of my Dr. B. (a stiff way to put it but it suits me). He tells me that I must have the muscles of my eyes cut. Is he honest! . . . He looks me straight in the eye as I look him. I never understood proportion very well but this is clear to me—doctor:patient::imposter:imposed upon. Proven by experience. Mama thinks I am cynical but it is a mistake. I've been brought face to face with my optimism so many times that I know I shall soon be embracing it again—perhaps before the end of this week.[16]

Writing to Martha became Annie's catharsis. She readily admitted that she had "not quite been able to adjust myself to things as they were" since their return from Oregon. Her letters that fall detail her increasing frustration with the conventional role that she was forced to play at home and led her to muse, "I think if I could tramp around on snow-shoes for a while I would get quite rid of my malaria [malaise?]—O you know I'm always thinking up some excuse to leave Oakland. It makes me feel ashamed of myself. I must fight it out here and attempt to do a little work of some kind. If I cannot here I should not expect to accomplish anything anywhere else."[17]

Many of the letters that Alexander wrote to Beckwith during this period express a similar yearning to be out-of-doors, to escape the confines of domestic and social activities that tired her eyes and drained her spirit. Alexander possessed the same restless spirit that had been the driving force behind the success of her father, yet comparable channels for her physical and mental energy did not exist. Perhaps equally frustrating was the realization that the problems with her eyes made acceptable professions for women, such as teaching, nursing, and the arts, unavailable to her.

After several pages, Annie's letter to Martha continued, "I decided to trust my doctor and have the muscles cut over two months ago—and my eyes have been better;—up to a certain point. All that remains is to pull what is left of the 'Old Man of the Sea' from my back and sling him away from me—I shall be as self-respecting as anyone but not until then."[18]

A number of Alexander's letters to Beckwith also reveal an insecurity about her own intellect, a belief that the intellectual pursuits she admired greatly in others were beyond her abilities, for example, "I'd hoe corn with you or milk cows and earn my wages if wages must be earned, but any labor of the head I shun." Even after Beckwith's prompting caused her to begin auditing classes at the university, her feelings remained unchanged: "Paleontology is getting too deep for me. I succumb before the complex structure of the echinodermata [starfish and their relatives]! This superficial way of doing things injures my self-respect but it is better than nothing isn't it and perhaps some day I can do laboratory work." Four years later, having already organized and led several highly successful paleontological field expeditions, Alexander would still feel compelled to confess, "I wish my work went further than to simply get the fossils to the University. I should like to follow the saurian [a fossil reptile] to the bitter end, chisel him out of the rock and write learned treatises on his venerable anatomy. But—."[19]

Why and how Alexander came to hold such attitudes about herself is the most vexing issue with respect to understanding this unusual woman. She was a creditable student, if not a brilliant one, and her lack of a diploma from Lasell was neither unusual nor indicative of poor scholastic performance. Many women attended the school who did not formally graduate. Whereas Alexander may have struggled to grasp the complexity of some subjects, she nonetheless had a comprehensive understanding of the major scientific issues of her day and held a detailed knowledge of many aspects of vertebrate biology and evolution.

Alexander did not view her shortcomings as related to her sex and did not harbor the illusion that a husband and family were the solution to her problem. She frequently expressed cynicism about the institution of marriage and the power wielded by men, as when she wrote wryly to Martha,

"I saw Mary McLean Olney this morning. She was Dean of Pomona College for a year, liked the work and would have enjoyed going on with it had not this other thing come up—that of falling in love and getting married."[20]

Apparently there was no pressure on the Alexander children to marry, and neither Juliette nor Annie did. Juliette was odd, "not quite right in the head," and her parents may not have wished to place this additional strain on their quiet and thoughtful daughter.[21] She lived with Samuel and Pattie until their deaths, after which time she lived in her own home but always in the company of a paid companion. For two weeks every fall, Alexander reportedly stayed with her older sister so that the companion could take a vacation.

In Annie's case, too, there was no financial incentive for marriage; the success of Alexander & Baldwin, Inc., ensured that each of the founders' offspring would be well cared for. And Alexander was not a weak individual who needed a husband to look after her personal welfare. Quite the opposite. As a young woman she had traveled abroad on her own and had already proved herself able to handle enormous physical challenges. If she was loath to abandon such adventures in favor of a husband, then apparently so were her parents to have her do so. Her purpose in life may have been unclear at the time, but she instinctively recognized that marriage would not quell her restlessness and pervasive malaise. Her continuing saga with the medical profession, at that time predominantly male, offers yet another clue to her attitudes about marriage. Cynically, if not a bit incredulously, she wrote to Martha:

> My pride has been sustained by quite a different diagnosis by a doctor in the City whom Miss Wilson has been wanting me to go to for two years.... The process was interesting so I will tell you. I was made to undress to the waist except for my undershirt and sat down in a great chair in the center of a room full of electric apparatus. The first step the doctor took was to look out my date of birth in a wizard volume and then he studied me while my eyes blinked with a new interest at this strange method that asked no questions. Finally he had it—arrested development at the base of the brain, congenital, and he was ready to swear by all possessed that he was right. So I am taking electric treatment of him [sic] three times a week and hope that the nerve cells in my brain under this stimulant will go to work and form new tissue as fast as possible. The doctor is a fatherly old gentleman but he has a fad with which he torments me every little while, namely—every woman should marry if it is only to live with a man three months—that they are not women until they have had the experience that comes with marriage. He calls this his discovery! He is satisfied from the lines in

my hand that I would make some one a very nice wife, but indicating the Mount of Venus—you will never attract a man until that is better developed—and he proposes to accomplish this for me! My love-nature is strong, proved by certain bumps on my head, but has not been able to express itself. Really, Martha, I would not prattle along in this way, but I think you need to be amused don't you?[22]

In December 1899 Martha left Paris for Halle, Germany. Annie continued to pen long newsy letters to her beloved friend. With the arrival of the New Year, she and Samuel traveled to the West Indies and Bermuda for several months. They rented rooms at a country boardinghouse along the edge of a small bay, eight miles outside of Hamilton, Bermuda. The ocean approached to within feet of their doorstep. Undeterred by the wet, windy weather, Alexander put on her waterproof cape and explored the surrounding countryside while other boarders stayed close to the fire. She walked along winding roads cut from coral rock and noted with pleasure inland ponds of clear salt water, patches of banana trees, a fruitless coconut palm, scattered mulberry bushes, and papaya plants, all of which reminded her of Hawaii and of the many hours she had spent as a child exploring its natural secrets. Yet the stone walls of white-washed coral against the pale blue of the ocean lent a picturesque feel to this landscape different from that other familiar island paradise.

Back in Oakland, once again Annie experienced that keen, sinking sensation of being trapped in an environment where she did not belong. Spring and summer passed slowly. Samuel and Pattie were planning to spend the fall of 1900 in New York City and the winter in India, leaving their daughter virtually alone in the big house on Sixteenth Street. Annie hoped to persuade Martha to travel to California to be with her. She wrote pleadingly, "I want you! Won't you come spend the fall and winter with me? What inducements can I offer? I would send you tickets several times over if that would bring you and go to Chicago myself to meet you."[23]

At Alexander's plaintive urging, Beckwith agreed. She herself seemed to be experiencing a hiatus. She had returned unexpectedly from Europe the previous spring and did not wish to resume teaching immediately, or at least did not feel compelled to do so. Annie was jubilant. Buoyed by Martha's presence and motivated by her encouragement, Annie began auditing paleontology lectures at the University of California in the neighboring town of Berkeley.

3

A Passion for Paleontology

Before the trip through northern California, Alexander had never verbalized an academic interest in natural science; her curriculum at Lasell was lacking in this arena. But her thrill at learning the names of the plants and animals that summer, and her obvious pleasure in being outdoors, may have prompted the choice. Study of earth's origins and the history of its flora and fauna was a logical prerequisite to full understanding of its more recent forms.

The course that Alexander chose to audit in the fall of 1900 was given by John C. Merriam, a faculty member in the Department of Geology who had acquired a reputation as an inspiring and captivating lecturer. Merriam had arrived in Berkeley in 1894 after completing his doctorate in paleontology at the University of Munich and by 1900 had added several advanced courses in paleontology to the department's curriculum, including Vertebrate Paleontology, the History of Vertebrate Life in North America, and the Geological History of Man. Merriam's presence in the geology department focused welcome attention on the discipline of vertebrate paleontology at Cal and he single-handedly sparked the development of its fossil collections.[1]

Merriam's lectures fascinated Alexander and she began to develop a passion for paleontology. At the end of the fall semester, Martha left for the East, and less than two months later Annie wrote to her delightedly, "What a fever the study of old earth that you thought should be a part of my education has set up in me! I am really alarmed. If it were a general interest in geology there might be something quite wholesome in it but it seems to centralize on fossils, fossils! And I am beleagu[er]ed."[2]

Armed with her "new and valued companions—the pick and collecting sack"—Alexander began to explore the geology and topography of the Bay

Area in neighboring Contra Costa and Santa Clara counties.³ Her first trips were to Los Gatos, at that time a small town at the foot of the gap that divided the Santa Cruz Mountains. From here she could look east across the broad, fertile Santa Clara Valley and, on a clear day, she could see Mount Hamilton twenty-five miles away. One Sunday she and Samuel rode with friends to a deserted resort in a valley of the Santa Cruz Mountains. There they took a long hike across the ridge, making their return by way of the gap. Ferns grew in mossy hollows along the road and quiet pools of water lay coated with a thin layer of ice. Samuel was entranced by the forest of redwoods through which they passed, and the two likened the thick evergreen to the lushness of the islands that were never far from their minds.

On another occasion Alexander accompanied several paleontologists from the university to the foot of Mount Diablo, the 1,173-foot peak that dominates the California Coast Range east of Berkeley. On the far side of the East Bay hills, the group passed through Redwood Canyon where dense tule fog hung draped like a thick mantle. They spent two nights on a ranch at the base of the mountain, climbing fourteen miles one day while breaking up rocks in search of Miocene shells. "To think that I should have let these Oakland hills shut out [this country] from my view all these twenty years! such a pretty country too yet such a tangle of rounded hills and ridges big and small lying in all directions that one must needs spend a week on old mountain to unravel it all. Surely this California is a crumpled piece of this round earth's skin as one ever did see!," Annie marveled in a letter to Martha.⁴

With the exception of these few trips, Alexander spent most of the winter and the early spring of 1901 in Hawaii. Her younger sister, Martha, had married John Waterhouse the previous year and the couple had made their home in Honolulu. Most of her aunts, uncles, and cousins still resided there as well. To Alexander, the islands would always remain her "only real home." She felt drawn to their beauty in an almost inexplicable manner, remarking on one occasion that "I was never in a place where nature seemed to have such a decided personality. She takes one out of oneself in spite of oneself."⁵

On this particular trip, the lure of paleontology accompanied her. Before her departure Alexander acquired and studied geologic maps of Oahu, directing her energy between family visits to unearthing invertebrate fossils from the Pliocene marine deposits at Pearl Harbor, specimens that she donated to the university upon her return.

But it was not just fossils that Alexander had begun turning over to the university in 1901. She had already begun funding some of Merriam's field expeditions. The moneys she sent him were used to pay the salaries of research assistants, purchase field equipment, and cover the cost of transporting hundreds or even thousands of pounds of fossils back to the university from the field, expenses not covered by the department's meager annual budget.

Although Alexander was pleased with the results that Merriam was achieving with her support, she wrote to the paleontologist in the spring of 1901 expressing her own desire to collect fossils closer to home and soliciting his advice. He responded immediately, inviting her to accompany him and his wife on a trip to Shasta County in northern California. But their departure date remained uncertain, and "besides it is better for me to be independent," Annie wrote to Martha in explaining why she had declined the invitation.[6] As an alternative, Merriam offered to guide Alexander in organizing a research expedition of her own to the Fossil Lake region of south-central Oregon, approximately eighty miles northeast of Crater Lake. This was a Quaternary site, quite recent from a geologic standpoint, and the fossils to be found there were mainly those of woolly mammoths, miniature horses, camels, and giant ground sloths that had roamed the North American continent within the last three million years (see Figure 2). A collection of similar material from the nearby John Day Formation had recently come to the university. Merriam felt that if he could procure additional specimens from this locale, scientists on the East Coast would finally be forced to travel to California to examine them, the quantity of material amassed having become too great for researchers to ignore any longer. He argued that such recognition would enhance the university's reputation (and, he undoubtedly hoped, his own) and would establish the Berkeley campus as an important center for paleontological research.[7] Merriam proposed that if Alexander were willing to finance the expedition and donate the specimens that she collected to the university, he would provide field assistants, logistic support, and some of the necessary equipment.

From the outset the 1901 Fossil Lake expedition was a valuable learning experience for Alexander. Herbert Furlong, the university's preparator of vertebrate fossil material, provided her with lists of field equipment, kitchen utensils, and provisions that would be needed. He then spent an entire day in San Francisco helping her select many of the necessary items. The most important of these was the wagon that would haul their specimens back from the field—they wanted one that could manage a load of at least two thousand pounds. They eventually decided on a farmer's wagon that could be covered with canvas should the need arise. So adorned, it was reminiscent of the Conestoga wagons that had carried the early settlers across the plains in their westward migrations.

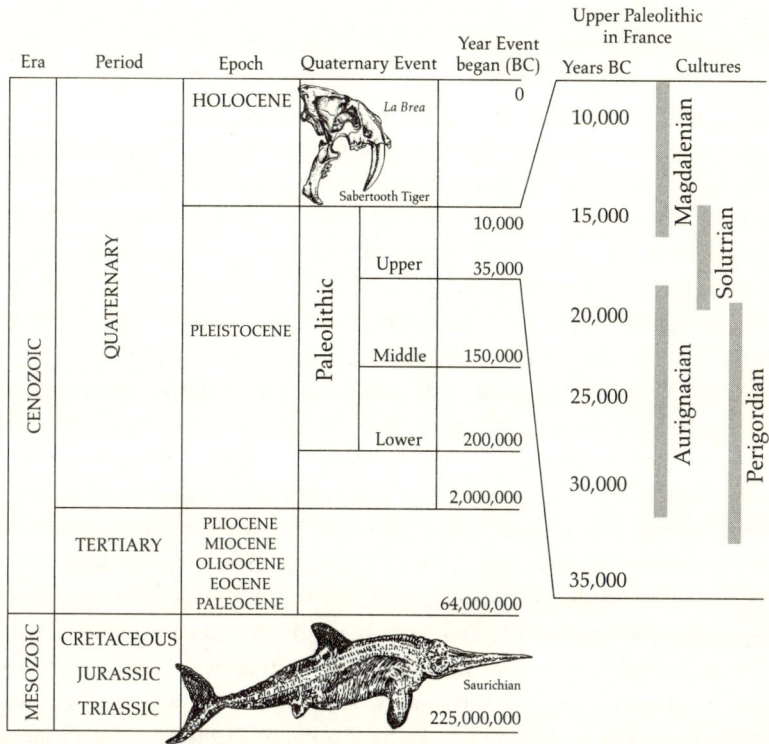

Figure 2. A chart highlighting the periods that relate to Alexander's paleontological fieldwork and the Paleolithic cultures that fostered her interest in Early Man.

Given the quantity of supplies and equipment needed to sustain a party of four to five persons in the field for several months, good logistic support was paramount. With space in the wagon at a premium, Furlong fashioned a wooden box to hold their various kitchen utensils compactly; in camp it doubled as a cutting table and work surface. Alexander purchased a small stove, and her brother, Wallace, loaned her a shotgun so that she could take geese, ducks, and rabbits to supplement the essential grains and canned food they would carry from home. She also ordered a tent, camp stools, a table, and two cots, the latter to be used by the women in the party if their campsite lacked pine needles or soft pine boughs, their preferred bedding.

Merriam believed firmly that it was inappropriate for a woman to travel alone in the company of men. What members of Alexander's family thought is not recorded, but they obviously did not prevent her from going, or from paying for the expedition. A second woman would be essential for the sake of propriety, but whom to invite?

Not surprisingly, Alexander's first choice was Beckwith. When it became clear that Martha would not travel west to join her, Annie persuaded an acquaintance, Mary Wilson, to accompany her. She wrote optimistically to Martha, "I introduced you to her once when we were in the Telegraph Ave. [street-] cars on our way to Berkeley. She teaches in the city but comes home every day—is a strong self-reliant person and I'm quite confident will help me to entertain the young men," a statement indicative of men's expectations of women at the time and of women's perceptions of their own role, even in unconventional circumstances such as these.[8]

As the expedition's sponsor, Alexander purchased all the necessary provisions well in advance of the group's departure. To minimize the tremendous cost of supporting a fully outfitted party in the field for approximately three months, nearly all the supplies and equipment were shipped ahead by slow freight, in this case to Yreka, California, a town just south of the Oregon border and the train station lying closest to their ultimate destination. Furlong left two weeks ahead of the others to lease horses and to hire a wagon driver who he hoped would look after the animals and function as cook.

In order to become more familiar with the fragments of bone and the different species of extinct mammals that she hoped to uncover, Alexander made two trips to the Academy of Sciences in San Francisco to examine its specimens from the Fossil Lake region. On the first visit Merriam accompanied her; on the second, Mary Wilson. She also inspected the limited amount of material from the site that was housed at the university. Would she be able to distinguish the toe bone of a horse from that of a sloth, she wondered? The tooth of a camel from that of a rodent? Or even a mammoth's wrist bone from its knee? Such routine identification of osteological material cannot be compared with its interpretation and synthesis in a broader evolutionary context, but Alexander's commitment to familiarize herself with these details contradicts her previously stated desire to shun "any labor of the head."

Alexander, Mary Wilson, and Will Greely, another of Merriam's assistants, left Oakland on May 29. Furlong was at the train station in Yreka to meet them, wagon, wagon driver, and horses in tow.[9] From the outset the going was slow, their progress plagued by somewhat predictable difficulties. Their first camp was in the vicinity of a flour mill where the muddy trail threatened to mire their wagon piled high with supplies. Then, as they lunched along the edge of a small lake, their two best riding horses, still saddled, fled. One of the animals had simply pulled up its sagebrush picket and run "like a thing possessed," the other following close on its heels. Neither returned and the group was now forced to ride with four persons atop the wagon until additional horses could be hired.

Continuing north, the party skirted Lower Klamath Lake, crossed the border into Oregon, and continued toward the town of Bly where they learned that their heavily laden wagon would be no match for the snow-covered mountains directly north. Consequently, they altered their course, going east to Lakeview and then north, at the cost of an additional hundred miles and several days collecting time. Annie was sanguine about their detour: surely the change in scenery would compensate her for the delay. Three hundred miles and twenty-four days after leaving Yreka the party finally reached Fossil Lake.

The site was an old lake bed, ten miles long by two miles wide. Over the centuries the surface of the lake had been blown free of loose sand, leaving fragments of fossil bone perfectly exposed on the surface or only partially buried in the densely compacted soil. The group established their first camp at a nearby spring where the air was cool and clear. Surrounded by sagebrush and sand dunes against a backdrop of low-lying mountains, Annie mused to Martha, "I wonder if this way of camping will spoil me for rougher camping? It's nice to have a cook and good things to eat, to have someone to look after the horses and packing the wagon but once in a while I have a wild desire to strap a roll of blankets on my back with two or three loaves of bread and alone make for some region where the firs grow thick and the streams flow from under banks of snow, leaving my party to shift for themselves."[10]

Alexander was not displeased with her companions, but it was "hard sometimes to keep up the pitch." Mary Wilson she found "to be a splendid woman, strong in her principles and very outspoken but always in good spirits." Herbert Furlong had a keen sense of humor. He and Mary maintained a lively interchange throughout the trip, initially somewhat barbed, but mellowing over time as they became better acquainted. "But she hasn't any poetry in her, Martha. . . . I miss you more than I can tell," Annie wrote.[11]

Alexander's musings reflect her own passionate, yet introverted, nature. She loved literature, theater, and music and her closest friends generally shared one or more of these interests with her. Her letters to Martha and Mary Beckwith frequently contained poetic excerpts, clever turns of phrase, or literary analogies. She and her brother often went to the opera in San Francisco and, in later years, she and her partner, Louise Kellogg, regularly attended motion picture shows. But Alexander seemed to avoid social affairs whenever possible and never mentioned attending parties. Even those functions that might be considered only nominally formal left her feeling ill at ease. On one occasion she wrote to Martha, "You already know my sentiment about teas. I always feel the masquerader among the masked."[12]

Alexander's party spent five days at Fossil Lake, beginning work along the western side of the site and moving around the bed in a clockwise direction. It was hard work, but at least they did not have to dig the fossils from solid rock. The first day alone, each member of the party recovered about twenty-five pounds of bones, a respectable haul considering that a number of field parties had previously collected in the region. Alexander became engrossed in the work and readily learned about all aspects of the operation.

After five days, the party decided to dismantle their camp and relocate to a nearby sandstone strata. It took two days to find the new fossil locale, and the campsite they chose proved to be too far from water, an essential ingredient not only for cooking and drinking but also for preparing the plaster bandages used in wrapping the specimens. Compounding these difficulties, four of their horses ran off. When the animals were finally located several days later, they were near death from dehydration. But Alexander was not deterred. She loved being in the field and at the conclusion of this chapter of the journey she reflected, "A trip could hardly have been more poorly planned but it has afforded us . . . more amusement than anything else we have done."[13]

Once the horses had recuperated, the party contemplated their return route. They decided to stop at Crater Lake, where Annie and Martha had camped just two years earlier. This time Alexander was profoundly struck by the beauty of the region, the intensity of the blue water and the visual impact made by the tremendous cliffs overshadowing the lake. Thinking of that first visit, she took long, leisurely walks in search of wildflowers, enjoying the solitude she found in the experience.

The party left Crater Lake for Yreka with great reluctance. Alexander's delight in a life lived outdoors, along with her distaste for life in Oakland, resonates throughout her letters. On the eve of her departure for Fossil Lake she had written to Martha, "Wish me good luck my sweet friend. My family are not enthusiastic and I'm lonely when shades of night fall and I think of the strangers with whom I am going to cast in my lot." In retrospect she need not have worried. Her trip "to the country of the saber-toothed tiger, those tigers whose sabers dripped with the blood of Oreodons and three-toed horses long before the mountains of Eastern Oregon were born" was, by all accounts, a success and she returned to Oakland elated. She had collected over one hundred bones, many of them knee joints of a variety of shapes, mostly *Hipparion* (the predecessor to modern horses), plus camel, elephant remains, rodents, and birds. And the group as a whole had collected "more than we expected after hearing how many parties had been in there on the same errand."[14]

Alexander's return from the field coincided with the beginning of the fall semester at Cal. She immediately began attending Merriam's General Paleontology lectures at 10 A.M. Tuesday and Thursday mornings and General Geology, taught by Andrew Lawson, at 10 A.M. Mondays, Wednesdays, and Fridays. As her passion to know and understand the history of life on earth increased, she wrote to Martha,

> I have not missed a lecture. I like it more and more, this study of our old, old world and the creatures to whom it belonged in the ages past, just as much as it does to us today. Perhaps the study is all the more interesting because it is incomplete, there is so much yet to find out—but I think it is wonderful, what all at once you might say, this tremendous desire to know has done for science.[15]

Any hesitation that Alexander might have felt about the unusual way in which she was conducting her life was countered by an increasing conviction that she had found her vocation. She wrote,

> Perhaps if I had a husband and family I should not be running after strange gods. It worries me a little Martha, folks do not seem to quite approve my new ambition. I do not want to be selfish yet it seems to me we have the right to a considerable extent of disposing of our lives as we think fit. As you yourself said—each one must live his own life and make effective whatever line of effort seems open before one.[16]

There is no indication that the members of Alexander's immediate family disapproved of her new pursuit. They did not prevent her from participating in field expeditions or attempt to control her finances. The fact that Alexander seems to have had a reasonably large sum of money to spend as she wished in 1900 is perhaps more remarkable than the question of whether or not her family showed interest in her work or in the object of her benefaction. One glimpse of their attitude may be gained in a letter that Alexander wrote to Beckwith from Shasta County in the summer of 1902. Elated over having found three unusual fossil specimens she remarked, "Dr. Merriam will send you his paper on the saurians gotten out last fall. You will note that I was responsible for that expedition [i.e., she financed it]. Mary W. [Wilson?] is the only one I have ever told. My family are not likely to know as I doubt they will ever take the trouble to read even the first page in the bulletin."[17]

Upon Alexander's return from Hawaii in late May 1902, Merriam proposed a second expedition, this one to gather specimens of the large marine reptile *Shastasaurus*, which he had finally concluded warranted recognition as a distinct subfamily within the fossil ichthyosaurs. During the Triassic, approximately 225 million years ago, land-dwelling reptiles had evolved their first aquatic relatives. The appearance of the fishlike ichthyosaurs was sudden and dramatic in the fossil record. Up to ten feet or more in length, these highly specialized marine reptiles had an elongated head that appeared to attach directly to their streamlined bodies. They lacked a distinct neck and all four of their limbs were modified into paddles. Posteriorly they sported a caudal fin, strikingly similar in shape to the caudal fin of many fishes living today, and a fleshy dorsal fin known from imprints left in rocks many millions of years old.

From mid-June to mid-July Alexander's field party scoured the strata in Shasta County, California, in search of these elusive creatures. Mary Wilson had not been free to participate in this expedition so Alexander asked Merriam to recommend a suitable alternative, a woman who knew something about the work and who would be interested in it. It is doubtful that many such women existed.

The person that Merriam selected was Katherine Jones, a single woman of about forty who had taken all the paleontology courses offered at Cal. Merriam had described Miss Jones to Alexander as being vigorous, fond of athletics, and enthusiastic. What he had failed to mention was that she had no camping or field experience and that she was equally inexperienced when it came to riding a horse, their sole means of travel in the field. In hindsight, her greatest contribution to the venture may have been the written account she left of the trip.[18]

Although Alexander served as the expedition's sponsor, the field party to Shasta County was led by Vance Osmont, an assistant professor of mineralogy at the university who had previously discovered ichthyosaur material in their targeted area. Professor James Perrin Smith of Stanford University, an ammonite (fossil mollusc) specialist; Eustace Furlong, Herbert's younger brother (and like him, a vertebrate preparator); and a Mr. Shallor (whose affiliation was never specified) rounded out the group. There was no cook or person whose job it was to attend to the horses and wagon on this trip. All party members became jointly responsible for camp chores. Consequently, after two weeks, Shallor left the group, declaring that fossil hunting was not what he expected it to be.

The party established their first field camp in a location known as Bear Cove. The limestone formation in which the group planned to work was about half a mile from camp, the fossil beds extending twenty-five miles in a north-south direction. The surface of these rocks was jagged and weathered, forming ridges in the topography of the surrounding landscape, and the work at this site was not like that which Alexander had experienced in the dry sands in Oregon the previous summer. The fossils here were entombed in their rocks and had to be pried at and hammered out with picks and chisels. The work was hard. It was also difficult to transport the specimens back to camp because of the added weight of the rock that encased each fossil. Consequently, the group relocated camp several times within the collecting area in order to be closer to the finds. Obtaining sufficient water also became a persistent problem. The days were pleasant at this time of year, but the afternoon sun often bore down mercilessly. There were also bears and rattlesnakes to contend with and, halfway through the trip, a fire of mysterious origin ignited, burning the ladies' tent and all its contents, including Alexander's camera and film. Summer is the dry season in California, the season when fire was, and is, a grave and reoccurring danger. The area in which the crew was working was not only dry but littered with pine needles as well. The collectors managed to contain the blaze but exhausted themselves in the process. And each day or hour spent attending to such activities was time lost to the equally demanding, but more important, task of finding and collecting fossils.

On this second expedition, Alexander proved to have an extraordinary knack for locating important material. Among other items, she uncovered three significant ichthyosaur skeletons—a complete paddle with digits, a partial skeleton, and an almost complete individual, lacking only a head and tail but otherwise beautifully preserved. No previous saurian specimens had ever been recovered with all the bones of the paddle intact. Pleased with her accomplishments, after exposing the paddle, Annie simply recounted to Martha, "They told me I had made the find of the trip." Her declaration gained substance when Merriam described a new species of ichthyosaur based on her find. He named it *Shastasaurus alexandrae*.[19]

Alexander enthusiastically pursued her studies the following fall. She set herself sixteen months to master the imposing volume *Osteology of the Mammalia*, after which time she proposed to spend another year learning something about "zoology and other parts of anatomy besides bones."[20] In keeping Martha apprised of her progress, Annie chided herself for having narrowed her interests to fossils and sugarcane. The former represented her intellectual interests, the latter her financial well-being.

Given her previous success, Alexander eagerly agreed to sponsor a second collecting expedition to Shasta County in the summer of 1903. On this occasion Edna Wemple, a paleontology graduate student, accompanied her. Other participants included Eustace Furlong, F. S. Ray, and W. B. Esterly.[21] As he had done the previous summer, Merriam joined the group for a brief period midway through the expedition. Alexander related her impressions of her companions in a letter to Beckwith, mixing her review with a smattering of philosophical reflection.

> About my party I find Miss Wemple more interesting than the men I'm afraid. Mr. Ray is just a boy. Esterly has a well trained mind but ventures few utterances. Furlong is good and patient and more interested than interesting—the more accomplished society man gets more credit for learning than the hard worked scholar. I wonder if there isn't more satisfaction in life in living to please and be entertaining than to be an authority on some one or two things![22]

The tension between the demands of scholarship and the expectations of society to which Alexander cynically refers shows how quickly she had come to recognize this conflict. Earlier in the year, she had tried to interest the university in funding paleontological research, as she grew convinced of its paramount importance to the interests of the university and to the scientific community in general. Benjamin Ide Wheeler, the university's president, had expressed his support for the program that Alexander wished to see inaugurated but, he added, with the university still in its infancy, funds for purposes other than instruction were sorely limited. However, as an indication of his own personal commitment to research and of his sympathy with her belief that the university should share in that expense, he offered to raise $1,000 for a research program in paleontology if Alexander would agree to donate an equal amount.[23]

Wheeler's unprecedented pledge marked the beginning of an unusual relationship that he and Alexander maintained until Wheeler's retirement in 1919. At that point, a changing social and political climate redefined both the university's administrative structure and the role of its administrators. These changes, coupled with the unique nature of Alexander's benefaction to the university (she was both patron and participant in her chosen area), would serve as the basis for ongoing friction between benefactor and recipient and would magnify the tensions of which she was already becoming aware.

Alexander's second paleontological expedition to Shasta County made its first camp at Black Oak Canyon. Almost immediately, she uncovered a fossil thought to be another specimen of the new genus of ichthyosaur, *Shastasaurus*, which Merriam had recently described. However, the paleontologist hesitated to make a definitive identification in the field. Pleased with her accomplishment, Alexander nonetheless suffered a twinge of self-doubt over the value of the specimen as a result of Merriam's scruple. Assuming it to be nothing more important than a *Shastasaurus*, she wrote to Martha from the shade of an oak overlooking the canyon, "You always experience keen disappointment that as long as you were finding something good it did not prove to be better."[24]

From the beginning, Alexander seemed to derive tremendous satisfaction if the material that she uncovered proved of interest or import to those whom she regarded as scientists. These were the sorts of data from which new hypotheses might emerge and on which papers were likely to be published. Accordingly, she strove to master identifying features that would allow her to distinguish one species from another and recognize important elements from common or useless fragments. However, her experience with this particular ichthyosaur in 1903 caused her to leave all final determinations to the scientists at the university and, perhaps, heightened her insecurities. The saurian specimen that she had unearthed proved far more significant than she had imagined at the time. The following year Merriam designated the fossil as the type specimen of another new genus of ichthyosaur, *Thalattosaurus alexandrae*.[25]

Whatever insecurities Alexander harbored, she did not for a moment wish others to misinterpret her feelings or her motivations with respect to her newly adopted vocation. She seemed acutely aware of the handicap under which women operated if they wished to define themselves outside the realm of domesticity. It was a man's world and, with few exceptions, men defined women's place in it. Women did not have the right to vote, own property, or work outside the home once they were married and, Alexander knew, those women who did not wish to marry faced the unenviable challenge of proving themselves to a patriarchy unaccustomed to dealing with women in any but the most traditional roles. The previous summer, Alexander had written to Beckwith, "I was very glad Dr. Merriam could spend two weeks with us. I wanted to know him better and I wanted him to know me better.

I had the disagreeable feeling that he ascribed a good part of my interest in the work to personal interest in himself. However that might have been I believe now he respects me thoroughly."[26]

As Alexander continued to pursue her passion, she remained acutely attuned to the frustrations of others of her sex who felt handicapped by the limitations that men placed on their freedom and opportunities for advancement in their chosen areas of interest. She replied to a letter from Martha, "I sympathize with you entirely. Isn't it most trying to have to sit calmly with folded hands and let our imagination do all the avenging of our wrongs! If we only had a little more power! There would be a good many headless bodies and battered frames lying around—serve them right!"[27]

Yet Alexander herself enjoyed a level of freedom that can only be characterized as unusual for a woman of her time—from instruction at Lasell to excursions around the world, and from lectures at the university to participation and sponsorship of paleontological expeditions. Her financial independence and unusual upbringing unquestionably fostered her autonomy and eventual success. Moreover, Samuel's willingness to defy convention, or simply ignore it, must have served as a remarkable example to his daughter. Even by today's standards, his unconventional attitudes about women and their place in society appear enlightened. One example was his invitation to Annie to accompany him on an African safari in 1904. The idea doubtless proceeded from the simple belief that she would prove the perfect companion on such an expedition.

4
Africa, 1904

It came as no surprise to Samuel's family and friends when he announced plans to embark on an African safari in the summer of 1904. Neither did the inclusion of Annie. If the invitation surprised her, she gave no hint of it. She readily assented, unable to forego the expedition's lure of adventure. Her purpose would be to collect wildlife, both on film and as trophies. "[T]he opportunity is one of a lifetime," she confided to Martha.[1]

Although the thought of reducing now scarce populations of game mammals for sport alone offends modern sensibilities, in 1904 the African landscape was relatively undisturbed and its wildlife flourishing. The part of British East Africa that Samuel had selected for their safari was considered at the time to be the greatest hunting ground on the entire African continent, if not in the world. He envisioned an expedition that would traverse a distance of almost 800 miles, beginning several hundred miles northwest of Mombasa near Nakuru and continuing west to the terminal point of the Uganda railroad at Port Florence, approximately 580 miles inland (see the route of the 1904 trip, in Map 1). Henry Stanley and David Livingstone had explored much of this region during the last half of the nineteenth century and their reports had piqued the interest and excitement of adventurers and armchair travelers alike.

Alexander wrote to Merriam about her change in plans for the coming field season. Wishing to be as helpful and solicitous as possible toward his new patron, the paleontologist took it upon himself to speak with President Wheeler about how the two might be of assistance to the expedition. They agreed to write letters of introduction, documents that might prove useful to Samuel should the party encounter any political entanglements or questions about the expedition's legitimacy while crossing the "dark continent."[2]

Unlike Merriam, Samuel did not feel obligated to include a second

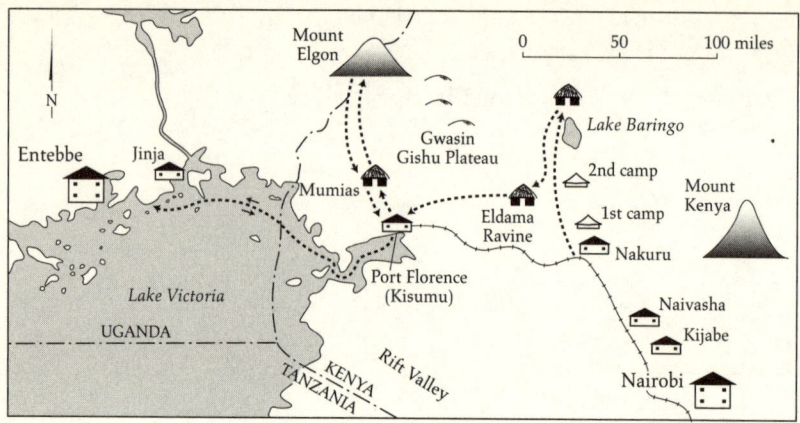

Map 1. The route followed by Alexander and her father during their 1904 safari across British East Africa, Uganda Protectorate (reconstructed from their letters).

woman in his party simply for the sake of propriety. Practically speaking, father and daughter were more concerned about the additional responsibility that such an individual would place on the expedition. "[N]o lady goes with me to Africa to burn up with fever or get lost in the tiger grass," Annie wrote to Martha before their departure. Instead, Samuel invited the Reverend Thomas Gulick to join them, a former Punahou classmate whom he considered a superb shot.[3]

The trio met in New York in early April 1904 and from there sailed to Europe. Rotterdam marked the real start of their journey, and on April 19 they set out from the Dutch port on an old German steamer, the *Kanzler*. More than a month later their ship dropped anchor at Mombasa, having made several brief stops at ports in the Mediterranean and along the Suez Canal.

Then part of British East Africa (now Kenya), Mombasa projected an air of unimagined familiarity to the excited travelers. Surrounded by lush vegetation, they recognized large groves of coconut trees and many of the fruits and flowers that they had known in Hawaii. Yet the swarms of mosquitoes and the ubiquitous cockroaches that they had assumed were an inseparable part of life in any tropical climate were absent.[4]

During the two days that they spent in Mombasa collecting their luggage and arranging their passage north, the three observed a number of hunting parties returning to their hotel. At the back of the building they came upon enormous bundles of skins, skulls, horns, antlers, and disarticulated carcasses, hard-won trophies now resting in silent homage to these successful hunters. Far from feeling put off by this carnage, Annie and Samuel's enthusiasm for their own impending adventure simply swelled.

The second evening after their arrival, the trio departed the city by train. By traveling at night through the low country, they hoped to minimize their chances of contracting malaria. Having suffered the disease as a young man, Samuel had no wish to repeat the experience and hoped that by morning their train would be at higher elevations where the dreaded mosquitoes would not be as prevalent. From there they could travel safely on foot to their designated hunting grounds.

As their train climbed toward the uplands, the air about them cooled and the scenery became one of gently rolling hills covered with scattered clumps of unfamiliar-looking trees and long stretches of green, grassy plains. Cultivated patches of bananas, papaws, and mangoes dotted the landscape. At station stops along the way the three walked up and down the platform enjoying the crisp air and views of the surrounding countryside—green mountains topped with white clouds and trees scattered among the tall grass that waved at their feet. Frequently, they bought papaya or the local pineapples and oranges from vendors along the tracks.

Their train passed through the Nyeri Desert (Taru desert), at this time of year still covered with trees and brush. Continuing northward through the Attic plains, their eyes met mile after mile of level grassy landscape. In these wide-open stretches came their first sightings of game—"at first a few scattering of deer, gazelle & antelope, then large bands, then bunches of ostriches, & then band after band of *hundreds* of zebras."[5] The animals seemed amazingly tame, coming so close to the sides of their railroad car that Annie felt she might be able to touch them merely by stretching her arms out the window. Protected within two miles of the railroad on either side, these were animals that had never known harm from hunters or their guns.

Any apprehensions that father and daughter might have experienced up to this point immediately evaporated and their eagerness to begin their own safari could hardly be contained. Seeing the animals that they had come so far to challenge, they now longed to stalk elands and elephants, confront rhinoceros and hippopotami, hear the roar of lions in the darkness and, if lucky, shoot one.

Located three hundred miles and twelve hours by train northwest of Mombasa, Nairobi was little more than a military town in 1904. Here Samuel set about hiring the sixty natives who would serve as porters on their long march across the continent. By law, each porter could not be required to carry more than sixty-five pounds on his head, but each was easily able to walk a distance of twenty miles a day carrying a box of that weight. In addition to the expedition's guns, ammunition, clothes, tents, bedding, cooking equipment, medical supplies, books, and foodstuffs, the porters also carried the three bulky cameras that Annie had brought with her from the States, along with all the supplies and equipment for developing her pictures in the field.

Travel photographs had become popular in the mid-nineteenth century, when the world seemed full of unexplored and exotic destinations. The advent of the steamship and the railroad had made many remote locales accessible for the first time and photography allowed explorers to document their travels faithfully and accurately. In 1901 Alexander had visited Yosemite and had heard the naturalist Ernest Thompson Seton present a lecture accompanied by wildlife photographs that he himself had taken. By then she was already taking her own floral snapshots and had set up a darkroom in the attic of her parents' home where she developed her prints. Inspired by Thompson Seton's technique of using observation to gain knowledge about the habits of wild animals, Annie had come to Africa eager to experiment with his approach and extend her photographic repertoire to capture animals as well as plants.

When Samuel had at last secured all the necessary guides and porters and had purchased the remainder of their supplies, the entire party, which now numbered almost seventy, set out by train for Nakuru, a site approximately six hours to the north where they planned to establish their first camp. For the first thirty miles or so after leaving Nairobi, only mile after mile of cultivated fields met their eyes. As they passed, native women waved and shouted to their departing sons and husbands. Then their train began to climb. After crossing the mountains at 7,800 feet and descending on the far side into the great Rift Valley, they entered a forested area heavily overgrown with vines. As they emerged once again into the bright African sunlight, they saw near the center of the valley a volcanic cone outlined prominently on the horizon, its shape reminiscent of the extinct Hawaiian volcano Diamond Head. Farther along, Annie could see Lake Naivasha from her window, and, near sundown, she watched as a crowd of baboons raced from the track as their train approached.

The party's first field camp was set on a broad, expansive plain in the vicinity of the lake. Although comfortable in all other respects, the site afforded little protection from the scorching sun. Annie, Samuel, and Thomas slept in a tent on folding beds, each complete with a thin mattress, the requisite mosquito netting, and plenty of blankets against the cold night air. A partition within the tent separated Annie's bed from the other two. Their porters slept eight to a tent and each man was given a single blanket. One man served as their cook and the trio ate their meals seated around a folding camp table.

The first morning they were up early, eager to embark on the real purpose of their journey. But hunting was not simply a matter of bagging trophies. As expedition leaders, the three were required to provide a daily ration of fresh meat for their porters and staff. Samuel had no hesitation about Annie's ability, either to walk their proposed 780-mile route over the next three months or to help secure the necessary meat. He wrote to Wallace from Nairobi, "An Austrian Count returned about two weeks ago from a shooting trip with *nine* lion skins. He came, however, in one case, very near being killed. If it had not been for his brave gun carrier, who brained the lion when it was on top of him, he would have been killed. I have the *same gun bearer*, & if he cannot handle the lion I know that Annie will sieze [sic] it by the tail and sling it 20 ft. away." Samuel's attitudes were certainly not characteristic of a man of the times, but they undoubtedly fostered the sense of independence and determination that allowed his daughter to pursue her unconventional life.[6]

Many of the group's first days near Nakuru were long and exhausting until the requisite number of animals had been shot. Often prey were elusive and the three might walk for hours before finally sighting a herd of zebra or hartebeest. Even then, their best efforts to shoot the animals before being detected by them often failed.

Compounding this difficulty, Thomas almost immediately began showing signs of fatigue and became unable to contribute significantly to their effort. He had not felt well during the trip from Mombasa to Nairobi and, as they continued by train to Nakuru, Annie prophetically wrote to her mother, "Papa is looking finely [sic] but Mr. Gulick looks like an old man. I'm afraid he's no match for a lion." On the evening after their first day's hunt, having walked twenty miles, Annie was not surprised when Thomas announced his intention to return home to the States. Among other less tangible ailments, his knee seemed to have given out completely. She wrote to her aunt, "I'm really very sorry to have him go but he is not cut out for roughing it and when a person isn't, you can't make him over."[7]

Throughout her life, Alexander made an almost unconscious distinction between those who were and those who were not "cut out for roughing it" (she did not categorize her family and close friends as belonging to one group or the other). She did not pass judgment on individuals who belonged to the second group. Rather the distinction had reference to her own strengths and proclivities and the people to whom she herself was consistently drawn. Because Alexander was not a social person, her close friends were few but cherished. Most were women she had known from childhood in Hawaii or from her days at Lasell. One exception was Armine von Tempsky, whom Alexander met on ship board en route to Honolulu in 1922. Armine was twenty-five years younger than Annie, but the two women had much in common. From her grandfather and father, Armine had inherited a literary talent and fearless horsemanship. She loved books and reading, but she also led hunting parties, helped with the work on her family's ranch, and competed in horse races. On subsequent visits to Hawaii, Annie often visited Maui specifically to spend time with Armine and her friends.[8]

Annie and Samuel continued to set out early each morning to hunt and quickly improved their rate of success. One day they brought back an impala from which the cook prepared steak, mincemeat, boiled meat, and roast. They shot ostrich and frequently saw gazelles and zebra. Applying Thompson Seton's approach, Alexander routinely observed and recorded the nature of her surroundings and the behavior of the animals she stalked while hunting. By June 4 the party had relocated their camp twenty-eight miles north of Lake Nakuru, to the banks of a stream beneath the spreading branches of an enormous shade tree. On either side grassy plains extended for miles and in the distance wooded mountains formed an idyllic backdrop. Mornings now were clear and fresh. In the afternoons clouds would gather and often, toward evening, the party experienced light showers. Annie was enthralled. Her face became burned, not so much from the sun as from the dryness of the wind at that altitude, about 5,500 feet. She likened the cloud effects to those she remembered from her childhood in Hawaii, and the climate in this part of Africa to that she had experienced in the high Sierra Nevada of California. Samuel wrote proudly to his wife, "Annie seems to flourish in this heat. She has a fine appetite, sleeps well & can stand any amount of tramping. Yesterday she shot a fine large stag (waterbuck) at 100 yards." Less than two weeks later he wrote, perhaps with a twinge of irony, "I have never seen her in better health. I am afraid this sort of life suits her too well."[9]

There is no indication that either Samuel or Pattie ever wished their daughter to be someone other than who she was. If ever such thoughts permeated their quiet moments, they certainly did not transmit them to Annie in any overt, or even subtle, manner. After all, they had sent their second daughter to Lasell Seminary, a school based on radical thinking with regard to women's education, and one in which military drill was a required part of the curriculum. Emulating his father, Samuel seemed to have imbued his children with the desire to lead honest lives and to make themselves of use in the world. William Alexander had recognized in his third son a vibrant spirit and Samuel, in turn, did nothing to quell that same spark in his own daughter.

By June 16 the party had reached Lake Baringo, having walked some two hundred miles. They had hunted Grant's gazelles, impalas, waterbuck, zebra, and oryx, but the lions, rhinoceros, and elephants that Samuel longed to encounter still eluded them.

Alexander had secured a fair proportion of the game that had been shot thus far, though she had equally often aimed one of her three large cameras. Not surprisingly, many of the photos she wished to take proved difficult to capture; as soon as she raised her camera to photograph several Maasai women in a Nairobi market, they fled. She did, however, secure poignant poses of several Maasai men. On another occasion she crawled along the bank of a stream and through the grass to within a hundred feet of a herd of fifty hartebeest. The animals did not stampede or drift away during the half hour she tried to approach them, but she was never able to get close enough to the group to take what she felt was a really good photograph. Often she would stay up into the early hours of the morning developing her photos in order to see whether or not she had captured her prey. An article in the *San Francisco Chronicle* published a year after her return noted that the nearly two hundred photographs that Alexander took on the expedition constituted the most complete set of pictures yet recorded from that region of Africa.[10]

But the party's stay at Lake Baringo was not entirely idyllic. Thomas had not in fact left after their first day's hike. After being carried a mile or so from camp by several porters, he began to feel foolish about his knee and embarrassed by his lack of physical stamina. Then and there, he jumped off his stretcher. Finding that his knee now gave him no trouble, he promptly walked back to camp and cheerfully greeted his companions when they returned that evening from their hunt.

His joy was short-lived. While Samuel and Annie thrived, Thomas grew increasingly fatigued and began suffering severe gastritis. Gravely concerned about his condition, Samuel again arranged for porters to transport him by litter back to Nakuru. From there he would be able to secure passage by train to Mombasa and to return home by ship.

En route to Nairobi from Nakuru, Thomas stopped in Kijabe to visit the American missionary family stationed there. From Kijabe he sent word to Samuel that he was recovering nicely and that he now hoped to meet up with them at Entebbe toward the end of their trip. Feeling reassured and relieved, father and daughter continued to hunt. They were at the extreme northern end of Lake Baringo, approximately 110 miles from the railroad at Nakuru, when a runner brought word that Thomas Gulick had died.

Stunned and distressed, father and daughter anxiously awaited a second message containing the details surrounding Thomas's unexpected death. No details were forthcoming. As they were disappointed and confused, their first impulse was to return home. With the gradual acceptance that nothing would be gained by this action, Samuel wrote to Gulick's family and to his own. He and Annie then continued their journey.

From the Baringo district of Uganda, at approximately 3,600 feet, Annie and Samuel commenced their long walk southwest toward the Eldama Ravine. During the next five days, the trail climbed to 7,300 feet, an elevation that afforded them magnificent views of the surrounding countryside. To the west lay vast cedar forests and the Uasin Gishu District (Gwasin Gishu Plateau), a fertile and picturesque part of the country teeming with wildlife. The few tribes who inhabited the region hunted with poisoned spears and arrows. Lying at 6,000 to 7,000 feet, the plateau was covered with rich alluvial soil, scattered groves of trees, and tall grass. Now the days were cool, with heavy rains and thunderstorms each afternoon.

The pair had intended to walk across the ravine directly to the plateau but the trails penetrating the forest at that point were poor, the region cold—almost to freezing—and the rain virtually continual. Under such conditions it would have taken them six days of hiking to reach their next hunting ground. They rejected this plan, choosing instead to walk twenty miles south and then directly west to Port Florence (now Kisumu), the small town marking the terminus of the Uganda railroad on the northeastern shore of Lake Victoria. From Port Florence they turned north, toward Mumias, and spent a month hunting in the vicinity of Mount Elgon.

On the plateau they now heard lions roar night and day and they made sure that their campfire burned brightly at all hours. Annie and Samuel no longer hunted together as a general rule. Each morning father and daughter would take off in opposite directions, each accompanied by a gun bearer, hoping to maximize their chances of bringing down game. Samuel encountered lions three times on these solo expeditions, but at each opportunity his shots missed their mark. Sorely disappointed, he still had nothing but pride and admiration for his daughter's prowess and her growing list of accomplishments. He wrote to Pattie, "If Annie had had my chance she would have got her lion sure, as she has developed into a very fine shot. She can put a ball through an animal's neck at 100 yards most every time. In fact, on this trip, she has brought from 200 to 400 lbs of meat into camp most every day.... Should Annie meet a lion by herself, I don't think that she would hesitate to open fire on it. She is made of good stuff." Alexander herself wrote more modestly to Mary Beckwith, "You may grieve to learn that your friend has become a woman of blood, enamored of firearms and the chase! I know now why I have dreamed so often of being in battle. But my experience has been too small for me to do much boasting."[11]

The plateau also offered father and daughter their first glimpses of graceful and awe-inspiring herds of giraffe, "with their heads 20 feet in the air, walking in Indian file like so many ships of the forest." Although Alexander had several opportunities to fire at the elegant and lofty creatures, she could not bring herself to take aim. She could only sit motionless and watch as they seemed to float across the plain within a hundred yards of her. She wrote to Mary Beckwith, "My gun bearer was terribly excited and kept saying—Shootie, shootie!—but I said—no, I don't want to shoot, I want to look at them. O, if I had my camera, it was five hundred yards away with one of the porters. What a picture of pictures I could have gotten."[12]

It was not until the day they descended from the plateau into the lower plains that they finally encountered a rhinoceros, an experience Alexander later referred to as their "sorrowful adventure," sorrowful because not only did they lose their rhino but they also became separated from their party for two and a half days, during which time they were without food and had only a small quantity of water. Finally, they stumbled weak and exhausted into a native village. A short while later there was shouting, and running. Out of the bush came their cook and two porters, breathless and excited, having trailed the wanderers for almost two days. A fire was started, water was immediately put on for tea, and, after a meal of cooked rice, pears, meat, and hard bread, neither father nor daughter seemed worse for the experience. Upon their arrival in Port Florence a few days later, Annie confessed to Mary Beckwith,

> My shoes were about reduced to sandals by the time we reached here. They were out at the toes when we began our first expedition and the amount of care I've bestowed on them to make them hold out and the sticking plasters to my poor blisters! But this trip closes a great chapter. We didn't kill our lion tho we wounded him and we didn't see elephants but we saw several large herds of giraffe, and a wild people who went about dressed in skins and hunted with bows and arrows and spears. And we saw every morning the sun rise over miles of tall grass laden with dew ... and set behind Mt. Elgon whose gradual slope reminded

us very much of Haleakala. . . . But as we did not go simply for the hunting but to see the country and the people every day had its interesting features.[13]

From Port Florence, Annie and Samuel decided to board a steamer bound for Entebbe on the northwestern shore of Lake Victoria. At Jinja, they viewed the Ripon Falls as well as the "ugly huge profile and bulging eyes" of a hippopotamus.[14] The pair had originally contemplated a week's trip by steamer around the perimeter of the lake but this alternative proved less attractive when the time actually came to embark. Entebbe appeared more like their idea of tropical Africa than anything they had yet encountered. They found the scenery in the surrounding countryside pleasing but the city itself seemed possessed of a terrible gloom, undoubtedly brought on by an epidemic of sleeping sickness that had already taken the lives of 250,000 natives around the lake. Islands in the lake that had been well populated were now without inhabitants. Villages and their adjacent plantations had been abandoned to their animals.

Arriving at Port Florence after their brief excursion, Annie and Samuel boarded the Uganda railroad for the return trip to Mombasa. En route they decided to make one additional stop at Kijabe. They wished to visit Thomas's grave and thank the Hurlburts, the missionary family that had cared for their friend during his final illness. The cause of Thomas's death was never specified, but after meeting with Mr. Hurlburt, Alexander wrote simply, "There seemed to be a sudden collapsing of all his powers and he apparently did not realize that his end was near."[15]

The village of Kijabe lay partway out of the Rift Valley at almost 8,000 feet. In 1904 the area was still relatively undeveloped and the woods surrounding the Hurlburt home were filled with wildlife. Bands of colobus monkeys played in the cedar trees and leopards came down toward the house almost every evening. On the road to and from the railroad station, a distance of approximately three miles, it was not uncommon to confront wild buffalo or rhinoceros.

At the Naivasha station one of Hurlburt's assistants, Mr. Stauffacher, boarded their train and, for the remainder of their journey to Kijabe, the three talked not of mission work but of hunting. Stauffacher insisted that the pair pitch their tent in the Hurlburts' yard and remain in the area for several days to try their luck. Samuel did not need convincing. Lacking his father's missionary zeal, he had been dreading this obligatory visit and readily accepted the diversion from psalm singing that otherwise seemed inevitable.

The following morning Samuel immediately shot two colobus monkeys but Annie confessed to her mother, "I must say, though I've grown somewhat hardened, I shrank from killing a monkey—it seemed more like murder." However, Stauffacher's real enticement to the pair had been the prospect of an elephant hunt. Of all of Africa's large and exotic beasts, this was the only creature that had eluded them. That afternoon they set out, and, after only a quarter of an hour, the three found themselves confronting fresh elephant tracks along their path. In a matter of moments eight or ten of the superb beasts charged directly toward them, announced by the sound of crashing hooves and snapping branches. With no place to run, the trio froze behind a tree, stunned as the herd suddenly parted in their midst, some animals veering in one direction, some in another. The amazed hunters were unscathed and, seizing the moment, they took aim. Amid a flurry of bullets, a midsized bull with tusks a bit over 4 feet in length slumped to its knees. Several days later, Alexander wrote elatedly to her mother, "Here we are back in Mombasa without having been attacked either by lions or the fever during three months stay in British East Africa. . . . We have made every day count. I doubt if there are two more delighted people in this country than Papa and I."[16]

Samuel's original itinerary did not call for their journey to end in British East Africa but rather to travel south from Mombasa by steamer and to visit Victoria Falls. Although he and Annie now held less enthusiasm for this significant addition to their journey, at her behest he agreed to go, fearing that if they did not they might always regret it.

The two secured passage on the German steamer the *Konig*, sailing to Beira. From there they traveled by rail to Bulawayo and then to Matapao, a small town on the Zambesi River where they visited the grave of Cecil Rhodes. Rhodes had died two years earlier, a mere three years before the completion of the famed Zambesi Railway Bridge. Through his foresight and backing, the bridge would span the canyon below Victoria Falls, promoting commerce and uniting the areas that became Northern and Southern Rhodesia (now Zambia and Zimbabwe).

On their return from the grave site, Samuel seemed depressed for the first time since they had set out, and he remarked to Annie that he "felt a foreboding of disaster."[17] Nevertheless, that evening they departed for the falls. As they walked toward their first view of this spectacular cascade the following morning, Samuel again spoke of his premonition, this time informing Annie where she might find his money, letters of credit, and the steamer tickets for their voyage home from Cape Town on September 28.

It appears from Alexander's photographs that she and her father viewed the falls from a variety of vantage points over the next two days. On the morning of September 9 they crossed to the northern shore of the Zambesi River by cable in order to view the great chasm into which the water was falling. Work on the famous 900-foot bridge across the river had barely begun. Finding a trail leading into the Palm Grove Ravine, the two descended to view the falls at their exit point, the narrow chasm from which the water emerges after falling over the precipice. Just as the pair reached the point that afforded them their best view into the chasm, they became aware of small rocks falling from the heights directly above them. Before their descent Annie had noted that workers excavating for the foundation of the bridge had been dumping debris into the ravine but had given them little thought. Now, seeing the falling rocks, she and Samuel immediately ran thirty yards to where they felt they were out of danger.

Annie proceeded to set up her camera. Samuel stood leaning against a rock six feet in front of her. Hearing a sharp noise, Annie looked up. A three-foot boulder was hurtling down close to her father. Seemingly in an instant, it collided with another boulder, veered violently, and struck Samuel on his left side, crushing his left foot and leaving him writhing in pain.

Horrified, Annie fled for help. Three men who had witnessed the accident descended the trail to help. The task of scaling the precipice—lifting and carrying Samuel with great care up to a small bamboo hut on the summit—took an hour and a half, and an additional half hour passed there before the doctor who had been summoned arrived. During that time Samuel bled profusely. There was a severe bruise on his left hip, his left leg had been fractured just above the ankle, and his foot had been crushed. The doctor applied a tourniquet and directed that the patient be carried to his own home six miles away. The decision proved unfortunate. The region was one of deep sand and burnt brush and the caravan proceeded slowly. By the time they arrived at the doctor's residence, Samuel was almost unconscious. He was given chloroform and his left leg amputated immediately. Alexander remained with her father throughout the ordeal but the following morning, September 10, at 2:30 A.M., he died. The funeral was conducted the same day and Samuel was buried in the cemetery at the old city of Livingstone, four miles from the falls.[18]

The family learned of Samuel's death by cable on September 12, and a letter from Annie followed shortly thereafter. It began, "My very very dear Mamma, Julie, Wallace & Martha & John, It seems as though I could not bring myself to write of what has happened—it is all so dreadful & yet I know that you will want to know it all, & if anything should prevent my reaching home who would there be to tell you."[19]

Alone, using the tickets her father had purchased, Alexander traveled to Cape Town and from there embarked on the British Mail Line to London, a sixteen-day voyage. When her ship arrived in port, her brother, Wallace, and his wife, Mary, were on shore to meet her. She later wrote to Martha and Mary Beckwith, "I could scarcely believe my eyes. It was so, so good of them to come to me. I have felt so terribly alone. It is a month today since my darling father died and was buried. It doesn't seem as if I could ever make it seem true. He was so much alive. I knew him so well—every curve in his face and every tone in his voice. It was very, very dreadful to believe."[20]

We can only guess at the tremendous sense of loss and disorientation that Alexander felt in the wake of her father's sudden death. In a society where women were expected to marry, raise a family, and remain subservient to men in virtually all aspects of their lives, she had experienced an atmosphere of nurturing and social freedom that differed markedly from the norm. Bolstered by a financial independence available to few women of her age and time—but feeling terribly alone—she contemplated the future in a new light. Immediately, she turned to the only vocation that had given her pleasure and a sense of accomplishment in her adult life. Many years later she would write about this period, "I felt I had to do something to divert my mind and absorb my interest and the idea of making collections of west coast fauna as a nucleus for study gradually took shape in my mind."[21]

5

Meeting C. Hart Merriam

Alexander quickly realized that simply collecting fossils would not "absorb her interest." By 1904 the focus of her study had broadened to include Recent animals[1] as well as fossil forms, and during the next year she began to contemplate how she might make a more significant contribution to research in paleontology. In 1905 she also met two biologists, William Bryan of the Bernice Pauahi Bishop Museum in Honolulu and C. Hart Merriam of the Biological Survey in Washington, D.C. Both men influenced the development of an idea that had already begun to germinate in her head, but Merriam's encouragement and support ultimately inspired Alexander to carry out her plan to found a museum of vertebrate zoology on the Berkeley campus.

Although Alexander's passion for paleontology remained undiminished, before the Africa expedition she had written enthusiastically to Beckwith about a new interest. The detail in her letter indicates the enthusiasm and intensity with which she pursued this new occupation:

> What do you suppose my latest fad is? Collecting skulls of wild animals! I have about 45 now representing 40 different species, mostly carnivores. Really you don't know how fascinating it is. . . . I'm expecting a box over tomorrow from the city and shall tell Charlotte to have it secretly conveyed to the attic. I think my room will have to be remodeled to accommodate these new acquisitions. My study of osteology led me first to get some dog skulls. My experiences connected with this first quest were too revolting to relate—suffice it to say that I made two visits to the Pound, bought a new clothes boiler at Mamma's request and the result was scientific—I have one complete skull, another in which a longitudinal median section has been made, a third in which the various bones have been separated from each other. The study of

these gives me work for a month yet with no immediate necessity of other skulls for comparison but my exploring propensity took me [to] a taxidermist in the city one day and as luck would have it, he had always taken a special interest in collecting skulls and had put them aside as he came across them in his work so that the majority of what I have came from his collection though I have visited several other taxidermists. My most beautiful set is that of cats. I have the wild cat, Lynx, Felis pardalis (I don't know common name) Panther (from Cal.) and Jaguar. I have five kinds of bears largest from Alaska—16 inches long, a raccoon, badger, marten, mink, wolverine, civet-cat, coati-mondi, fox, coyote, 2 kinds of sea-lion, one Mastiff and a skunk. These are my carnivores. Of the large group of ungulates I have only one—a peccary. Of the Cetacea—a very young dolphin. Of the Insectivora—the mole. Of the Chiroptera—a small fruit bat. Three marsupials—one very imperfect Echidna—two rodents (squirrels), seven monkeys.[2]

Skeleton preparation was a cathartic exercise, and putting her hands to the material that she and her father had collected obviated the need to dwell on events of the recent past. Annie wrote to Martha,

I've been hard at work over the trophies from Africa. Such a boiling and a stewing that has been going on! I used a strong solution of ammonia, salt petre and soap with the result that I had a handful of teeth to locate. The next question is what to do with them all! A good share I shall send to the University but I doubt if my closet will hold what is left over and just as present I don't want them anywhere near me. I've had too intimate an acquaintance with them with the scrubbing brush.[3]

Alexander's immediate solution was to have a room built in the garden of the family home to house the African trophies as well as her other collections, among them a number of Indian baskets.[4] As the year drew to a close, she again traveled to Hawaii. The days she had once spent visiting with her sister and brother-in-law were now shared with an increasing number of nieces and nephews. Martha and John Waterhouse's first child, also named Martha (but called Pattie like her grandmother), had been born in 1902 and a second daughter, Elizabeth, had followed in 1903. Their first son, Samuel, had been born in November 1904, a month before Annie's arrival in Honolulu. Within a decade Martha would give birth to five more sons, John ("Jack"), Wallace, Richard ("Abe"), Alexander ("Alec"), and Montague.

Alexander felt close to her sister's family. She recognized that there was "something sane about family life," especially when children were present.

"Children help to keep an even balance of things," she wrote reflectively and with a heavy heart following her nephew Samuel's untimely death little more than a month before his third birthday.[5] Nonetheless, she recognized that children were not the answer to her own unhappiness.

Alexander loved Hawaii and the time spent with her nieces and nephews, and the realization that Hawaiian society held no place for unmarried women pained her. Slowly her unease and uncertainties about the future gave way to an acknowledgment that "[e]very drop of blood in me . . . is for an active out-of-door life since study is out of the question."[6] Whether this was simply an expression of her intellectual insecurities or referred to worsening problems with her eyes is unclear.

What Alexander described as an "active out-of-door life," the fieldwork that she undertook with increasing regularity, was also an avenue of escape from the conventional expectations defined by her wealth and social class, one that provided an opportunity to appropriate an additional cultural identity—that of field biologist-cum-amateur naturalist.[7] The ease with which she moved in that direction is clear. Within two months of her return to Oakland from Hawaii, Annie was off to the field.

Alexander's 1905 paleontological expedition confined its work to a single outcrop of limestone dating from the upper portion of the Middle Triassic in American Canyon in the Humboldt Range in northwestern Nevada, an area less than two square miles in extent. Her crew included Edna Wemple; the vertebrate preparator Eustace Furlong; two field assistants, Malcolm Goddard and Bill Boyton; and James Perrin Smith, the faculty member from Stanford University who had participated in the 1902 expedition. Merriam joined the party briefly midway through the summer.[8]

For several weeks after their arrival in early May the weather was unpredictable. At one point, an icy wind blew through the canyon with occasional sleet and frequent thunder showers during the day. Temperatures at night dropped to below zero. An abandoned miner's cabin provided welcome refuge and the group cooked their meals indoors over an open fire. Cooking outside on the camp stove left them vulnerable to the penetrating wind and persistent cold.

The site they worked had been dubbed Saurian Hill for the incredible number of ichthyosaur specimens unearthed there. But the extraction of complete skeletons from their limestone crypts often required strength in wielding a pick and shovel that Alexander was loath to admit exceeded her physical capabilities. Not infrequently, the men resorted to putting in small blasts of gunpowder to loosen a desired slab from the surrounding matrix. As the blocks of rock were slowly separated from the hillside, she and

Wemple took on the task of wrapping them in plaster jackets and carefully marking the parcels. When the greater part of one such specimen was secured intact as a 500-lb block, it required no small effort by the entire party to move the fossil from its rocky embankment back to camp. During the course of this single two-month excursion, her group excavated twenty-five ichthyosaur specimens as well as other valuable bones and fossil fragments.

After the 1901 trip to Oregon, Alexander had reflected, "It's lucky I've promised my summer collections to the University. The fever for amassing these strange treasures might make me a collector of the most greedy type, unmoved by 'threats of Hell or hopes of Paradise.'" Yet four years later, her perception had altered. She confessed to Martha, "We were almost too successful on our expedition this time. It seems to have taken something away from the elation of discovery to have found as many saurians as we did. It seems in this world as if the surest way to make a person dissatisfied were to gratify his ambitions too completely."[9]

When the members of the 1905 expedition got back to Oakland, Alexander wrote an account to accompany the photographs that she had taken in the field. In addition to providing scientific details that continue to be of interest and value to paleontologists and historians alike, the account mixes wit with bits of personal philosophy and reflection. For example, in describing the difficulty of extricating some of the fossils from their rocky encasements she wrote, "Pure grit, pluck, dogged perseverance—whatever you may call it—is what accomplished wonders in our prosaic day. With the Greek heroes it was brute force and occasional help from the Gods which wasn't fair and I think they do not half deserve their fame."[10] In contrast, the closing paragraph of her narrative reflects her joy at being in the out-of-doors and the magic of sleeping beneath a blanket of stars:

> People naturally count it among their blessings to have a roof over their heads at night, but how oppressive that roof seems to you, and the four walls of your room after a month or two in the open! Half the universe shone down upon us those clear nights in Nevada; not a tree to break the wonderful arch of the Milky Way reaching from horizon to horizon. The same constellations seen night after night as we lay on our backs on the ground made their impress on our minds that a casual view of them from a bedroom window or city street could never make.[11]

Within just a few years Alexander would drop her humorous and at times metaphorical style of recording field observations. After the establishment of the Museum of Vertebrate Zoology in 1908, her field notes took on the terse, impersonal, and descriptive tone prescribed by academic convention.

In truth, the fossil expedition to Humboldt County had not been Alexander's first choice for fieldwork in the summer of 1905. She had discussed with Merriam the possibility of a hunting expedition in Alaska. But each time she contemplated the ocean voyage, the rugged and unfamiliar wilderness, and the difficulty of the quest for game in that region, the proposition depressed her; her father's death was still fresh in her mind. Early in January 1905 Merriam once again took the initiative, hoping to please his unusual benefactor. He sent a letter to Alexander in Hawaii, informing her that he had written on her behalf to Edward W. Nelson of the Biological Survey in Washington, D.C. Merriam also noted that he had mentioned her name to his cousin, C. Hart Merriam, then chief of the Biological Survey, as both men were familiar with Alaska and with the conditions to be faced by fieldworkers in that remote northern wilderness.

Growing up in the shadows of the Adirondacks in upstate New York, surrounded by acres of woods and streams, C. Hart Merriam quickly developed a passion for nature and wildlife, birds in particular. His parents encouraged these nascent interests and, while still in his twenties, Merriam became a founding member of the American Ornithologists Union (AOU). An interest in mammals also evolved and his private collection of specimens grew to unprecedented proportions.[12]

In the early 1880s there were few professional biologists to study the native American plants and animals. However, rapid growth and development of the country were already beginning to take their toll on its wilderness areas. Merriam's personal awakening to the crisis coincided with a similar realization at the national level. Although he had attended medical school and was a practicing physician, he became deeply involved with the AOU's plan for a national study of bird migration at this time. In 1885, the year in which Congress authorized the establishment of a section of ornithology within the Division of Entomology (then a branch of the Department of Agriculture), the thirty-year-old Merriam was invited to assume the newly created position of ornithologist and take charge of the work. Disenchanted with the practice of medicine, he accepted the post.

Within a year, the section that Merriam headed had been elevated to the Division of Ornithology, separate from Entomology and, in 1888, it was expanded to become the Division of Economic Ornithology and Mammalogy. Under his management the work carried out by the division received increasing recognition and its mission continued to expand. In 1905 its title changed to the Bureau of Biological Survey and he was serving as head. It was in this context that C. Hart Merriam and Annie M. Alexander became formally acquainted.

After her return from Humboldt County in mid-July, Alexander found herself "idling at home again now with vague plans as to what I'm to do next." She wrote to Martha, "Dr. C. Hart Merriam of the Smithsonian is camping just at present with his family on the Russian River and I must see him on his return to talk about Alaska."[13] The paleontologist had mentioned Alexander to his cousin on several occasions, undoubtedly in conjunction with the financial support and fossil material that she had been providing to him and to the university for several years. Hence, by the time the two finally met over dinner at the home of Judge Theodore Hittell in San Francisco in October 1905, each was well known to the other by reputation. Despite Alexander's initial trepidation about the dinner, the evening proved a great success. In a letter to Martha, she recounted the events leading up to it this way:

> I must tell you an interesting experience I had the other evening. To go back a little—about two years ago I went on several long walks with the "across country club" of S.F. Miss Alice Eastwood, botanist in the Academy of Sciences and a Miss Hittell were the leaders. Miss Hittell had heard about me through C. Hart Merriam of Washington who always makes their home his headquarters when he comes out to California. I had hardly been introduced to her before she began telling the forty things I had done in a most exaggerated way. I was disgusted and took pains after that to avoid her. I had not heard of her since then until the other day when I received a most eccentric letter from her inviting me to take dinner with them to meet Dr. Merriam, and to spend the night. It was impossible for me to go that evening so I declined but Miss Hittell was not discouraged and later repeated the invitation. I could hardly refuse again so I went.[14]

Alexander's initial impression of Hittell's daughter remained unchanged at the conclusion of the evening. However, the judge's home was one of distinction and he and his family regularly played host to an exclusive circle of Bay Area writers and painters, as well as many of the leading naturalists of the day, including John Muir. Hittell was a well-known and well-respected attorney in San Francisco and a prodigious author of law books. But his first love was writing general literature and books about history. In 1905, at the age of seventy-five, he had just began work on a "History of the Hawaiian

Islands," a labor that would take him seven years to complete. The judge had never set foot in Hawaii and he told Alexander that evening that he now felt too old to embark on the journey. Instead, he had satisfied himself by collecting practically all the information needed for the book from extant literature about the islands. Annie found herself both amused and intrigued that a man felt he could describe in detail something that he had neither seen nor experienced.[15]

For Alexander, the high point of the evening began only after the remainder of the dinner guests had departed and she and Merriam were free to retire to the study and exchange stories, hers about hunting big game in Africa, his about previous biological expeditions to Alaska and his interest in the bears of that region. Much to her delight, the survey chief provided exactly the kind of detailed information that she needed about conditions for fieldwork in the northern territory. Consequently, she followed up the evening's events with an invitation to Merriam to visit her home so that he might personally examine the collections that she herself had amassed over the past half-dozen years.

On February 22, 1906, Alexander wrote two letters from her sister Martha's home in Hawaii. To C. Hart Merriam in Washington, D.C., she wrote seven pages, outlining her plans for collecting specimens in Alaska that summer and soliciting additional advice about fieldwork in the region. In reality, the letter asserted a great deal more. It revealed her plan to develop a museum of natural history on the West Coast. In asking the Bureau of Biological Survey to grant her a collecting permit, she justified her request by stating:

> My object in making this collection is to form the nucleus for a collection representing the fauna of the Western coast. I would like to keep it in my own hands for at least three years, if it could be possibly arranged. I would meanwhile loan it to the Academy of Sciences for the benefit of the Public and employ a skillful taxidermist to mount and look after the specimens. If, however, the U.S. Department of Agriculture could not issue my agent a special permit or commission unless I donated the collection to some museum, might I not incorporate with a few others under the name, say of West Coast Museum?
>
> Or could I do this:—collect for the University of California on condition that the specimens be loaned to me until such time as a suitable museum be built? The reasons for my unwillingness to turn the collection over either to the Academy of Sciences or the University of California are these:
>
> The curator of the Academy of Sciences has no interest in mammals. The University of California has no natural history museum. The

skins would be laid away and the skeletons consigned to some basement. It seems to me that some arrangement satisfactory to me might be made under the circumstances. The work is something that especially interests me and I would like to make it a life work. I need only a little cooperation.[16]

The sentiments expressed in Alexander's letter to Beckwith the same day were terse and far more to the point—"This collection which I intend for a private Museum collection I shall keep in my own hands as long as I can—neither John C. nor C. Hart will prevail against me."[17]

Whether or not Alexander would have made such a bold proposal to C. Hart Merriam had the two not previously spoken at length is questionable. What is indisputable is her total commitment to a plan that had been gestating for more than a year. During her visit to Hawaii in 1905, Alexander had met William A. Bryan, a researcher studying the collection of Philippine birds at the Bishop Museum in Honolulu. The two fell to talking about the future of the museum and the practicality of undertaking a comprehensive survey of the island groups in the South Pacific. It was Bryan's wish to make the Bishop Museum a center for research on all scientific subjects relating to Polynesia. The material amassed during the extended expeditions that Bryan envisioned—biological, geologic, and anthropological—would become the focal point for the museum's research collections. Bryan himself was extremely eager to collect in the New Hebrides (Vanuatu). Because of their relative geographic isolation and the limited size of the islands, he felt that they held tremendous promise for resolving several interesting evolutionary questions. Bryan further proposed that the workers chosen to conduct these extensive surveys should be students rather than faculty. He reasoned that by allowing students to apply the materials that they collected to their graduate research, a tremendous incentive would accrue for them to do their best work and, at the same time, make a lifetime commitment to basic research.[18]

Alexander was intrigued by Bryan's vision for the Bishop Museum and the role that he foresaw for the collections that would result from such an effort. She later adopted several of Bryan's innovations for use in her own museums. Chief among them was the belief that hiring younger researchers, those whose careers lay ahead of them and who were willing to become deeply involved in the work, was preferable to hiring individuals who had already attained creditable academic standing.

By the turn of the twentieth century the fauna of the West Coast, and of California in particular, was rapidly disappearing, succumbing to the ever-increasing development. In 1850 the population of California was approximately 90,000. By 1880 it approached 1 million, owing to the discovery of gold and to intensive recruitment efforts from real estate promoters and railroad developers. Between 1880 and 1890 it had increased to more than 1.2 million. By 1910 almost 2.5 million people would claim California as their home. Alexander was clearly cognizant of this dramatic rise. She realized that if a complete and accurate record of the fauna was to be made for posterity, work would have to begin at once. Even before 1910 several large mammals such as the grizzly bear, the California white-tail deer, and the California valley antelope could no longer be found in the state. Several indigenous Native American tribes had already vanished as well.

From the paleontology lectures that she had attended, Alexander came to understand that it was necessary to compare fossil specimens with what seemed to be their closest living relatives in order to deduce the evolution of different taxonomic groups. In addition, she recognized that it was possible to extrapolate knowledge about the distributions and habits of extant vertebrate species and apply that information to their fossil relatives. In 1906, however, all the major American collections of living vertebrates were on the East Coast or in the Midwest and she quickly realized the disadvantage faced by university researchers who lacked sufficient comparative material for study. Before writing up any treatise on new fossil taxa, John Merriam was forced to visit the American Museum of Natural History in New York City, the Philadelphia Academy of Natural Sciences, the Biological Survey in Washington, D.C., and the Field Museum of Natural History in Chicago to examine specimens. Alexander now realized that the presence of a suitable collection of such material on the West Coast would greatly enhance his research capabilities and those of the other members of his department. Certainly the creation of such an important collection would prove more personally satisfying than simply gathering large numbers of fossils in the field.

C. Hart Merriam responded almost immediately to Alexander's proposal. He concurred that there was a "pressing need" for a natural history museum on the West Coast and evinced his support for her proposition and his belief in her ability and willingness to create such an institution by stating, "the time has come for people of means and influence to join hands and work for its accomplishment."[19]

As Alexander had noted in her letter to C. Hart Merriam, there already existed a natural history museum in the Bay Area. However, the vertebrate collections of the Academy of Sciences in San Francisco to which she had referred emphasized amphibians, reptiles, and aquatic birds, but not terrestrial birds or mammals. In addition, the curators in charge of those collections rejected specimen donations, including an earlier offer by Alexander of her growing collection of Pacific Coast vertebrates. The curators also refused to loan specimens for study to external researchers, a stance that placed the academy at odds with the majority of research museums, where such practice was common. Consequently, as their exchange of letters suggested, neither Alexander nor Merriam felt that the academy, satisfactorily met their criteria for a much-needed research collection of vertebrate specimens on the West Coast. In any case, within little more than a month of Merriam's reply to Alexander, the point would be moot. The San Francisco earthquake in April 1906 leveled most of the academy, and the ensuing blaze consumed the majority of its collections. In the interim between the Academy of Sciences' founding in 1853 and its destruction in 1906, several research facilities representing a variety of disciplines had sprung up on the West Coast or, like the projected vertebrate zoology museum in Berkeley, were about to make their debut. The presence of these institutions, coupled with longstanding policies that had previously fostered the academy's insularity and had hampered its cooperation with other organizations, later handicapped that institution in its efforts to rebuild.[20]

Throughout his career, C. Hart Merriam maintained a close professional affiliation with Alexander and with the Museum of Vertebrate Zoology that she created. When consulted in 1911 regarding the academy's future, barely three years after the MVZ's founding, Merriam cited the role that the university museum across the bay had already established for itself as a research institution and encouraged the academy to channel its energies in other directions.[21] This advice, coupled with Alexander's decision to dispense with exhibitions in the MVZ that same year, provided the necessary opportunity and inspired the California Academy of Sciences to develop itself as a great public exhibition museum.

Reminiscing in 1932 on the occasion of Merriam's receipt of the Roosevelt Memorial Medal for Excellence for his work in natural history, Alexander credited the survey chief with giving her the idea for a research museum on the Pacific Coast and encouraging her to carry the project through to completion.[22] In retrospect, Merriam's influence seems to have been just one of several factors that came to bear on Alexander's final decision. However, before she would write to President Wheeler with her formal proposal to found such a museum, Alexander would spend two summers in Alaska. Her objective—to collect bears for C. Hart Merriam.

6
Alaska, 1906

While Alexander labored to extract fossils from the limestone beds of northern Nevada in the summer of 1905, her father's sister Mary and several friends spent the summer in Alaska, traveling as far north as Dawson in the heart of the Yukon Territory. True, she had not felt up to the journey earlier in the year but, finding herself back in Oakland once again, Annie wrote to Martha, "I should have gone with them! but my heart is set on Kodiak Island and the Alaskan Peninsula where I can combine bear hunting with fossil hunting."[1]

Rethinking this proposition after meeting with C. Hart Merriam in early October, Alexander confided to Beckwith, "Bear hunting in Alaska seems to be a rather dangerous undertaking especially in the willow regions near the coast. I don't know that I am particularly attracted towards it but I should like to drift up into that country some day with no particular preparation and take whatever happens to come along, be it to hunt or to prospect or to explore the vast solitudes. Once cut loose from everybody who knows me, might I not become a stronger and braver woman!"[2]

Alexander continued such ruminations throughout the winter of 1905–6. She wrote to Beckwith from Honolulu with a renewed sense of commitment toward the venture and increasing clarity about her presently untenable situation,

> Here I am visiting and at home I am visiting. I want to be mistress even if it be only of a small tent. . . . I haven't anything in particular to demand my time and attention hence it is at the disposal of anyone who wants it. I'm glad to be of use but I do declare I believe it is weakening to the will to be always conforming to other people's wishes—so I am going off all by myself where I won't be influenced by anyone I know to see what I can do. It will be in the line of exploration, the

only thing I am fitted for.... This nomadic life may seem a poor sort of dream to you but if by means of it I can get back my grip on myself it will mean my salvation.[3]

From that point on, Alexander began making practical preparations for the trip.

Alexander's preoccupation with collecting bears on Kodiak Island seems to reflect Merriam's persuasion and research interests, whereas her request to the Biological Survey for a collecting permit that included large game mammals seems to parallel her experiences in Africa and her own idea of adventure. Not yet fully committed to hunting bears or ready to abandon the plan, Alexander also researched what was known about Alaska's geology with reference to a limestone bed reportedly full of ammonite fossils and possibly ichthyosaurs. But with the exact locality of the site undocumented she vacillated. In the end, she based her decision to go to Alaska in 1906 on the value of the specimens she planned to collect for Merriam's ongoing research on the taxonomy of bears.[4] Her specimen collections might make the greatest contributions to science, she reasoned, if they came from regions largely ignored by earlier expeditions like that of Edward H. Harriman, chairman of the board of Union Pacific Railroad, in 1899.

When the steamship *George W. Elder* set out from Seattle on May 30, 1899, it carried 126 individuals: Harriman's family, a select group of friends, servants, scientists, artists, photographers, stenographers, medical and religious personnel, hunters, packers, camp hands, officers, and crew members, plus a research library of five hundred volumes.[5] Harriman had originally commissioned the vessel to take his family on a summer vacation but, as a highly successful businessman accustomed to grand and expansive ventures, decided to enlarge its mission. He planned to build himself a public image as a philanthropist. Accordingly, in late March 1899 he traveled to Washington, D.C., to confer with C. Hart Merriam about using the ship as a research vessel. After several days of intense meetings, enthusiasm grew for the novel and impulsive scheme. Harriman delegated full responsibility for the research aspects of the expedition to Merriam, and the survey chief immediately set about securing the coterie of elite scientists who would accompany him that summer.

Over the course of its nine-thousand-mile journey, the *George W. Elder* stopped often, usually for a short time at each landing. There was little opportunity for the crew to hunt big game or for the scientists to explore the interior of Alaska's vast northern territory. At Glacier Bay, Yakutat Bay, Prince William Sound, Kodiak Island, the Alaska peninsula, and the Shumagin Islands the ship dropped anchor for longer periods of time, and larger collecting parties were put ashore to carry out more intense sampling. When the vessel returned to Seattle at the end of two months, its scientific exploration of the Alaskan coastline stood unmatched. Its participants had made extensive and important collections of birds, small mammals, marine animals, insects, seaweed, and plants from the coastal regions, as well as smaller collections of fossil shells and plants. The voyage resulted in the description of more than thirteen new genera and almost six hundred new species of plants and animals.

Other, more modest, expeditions soon followed. Between 1901 and 1903 the American Museum of Natural History in New York sent three expeditions to Alaska and northern British Columbia to secure specimens for its own collections.[6] Not to be compared with the Harriman expedition in size or scope, each of these smaller field parties nonetheless managed to accumulate fine series of large game mammals and several thousand small mammals and birds—a great increase in the knowledge of Alaska's flora and fauna. Yet when Alexander contemplated her trip in 1906, many areas of this northern territory remained unexplored and little was known about the presence of big game or about the flora and fauna to be found in its more interior regions.

In much the same way that Alexander's first paleontological expedition proved to be a tremendous learning experience, her first expedition to Alaska was equally instructive, albeit more painful. The exact sequence of events is difficult to reconstruct as no field notes from the trip, and only minimal correspondence, survive. Apparently on Merriam's recommendation, Alexander contracted with a hunter, L. L. Bales, to collect for her over a period of eight months, beginning March 1, 1906. Her plan was to join Bales at Sitka somewhere between the middle of April and the first of May and work with him along the Alaskan coast hunting birds and mammals.

From the outset, Alexander regarded the arrangement with a fair degree of skepticism. She perceptively described Bales to Beckwith as "a shrewd little Yankee frontiersman" based on a meeting with him early in the year. Her initial impression of the man hinted at the trouble that she would face: "His knowledge of wild animals and their habits is astonishing but he won't believe that a glacier moves! He is a man who has attained a certain self reliance from living alone a great deal in a wild inhospitable region, he has learned some of the woodcraft of the animal but at the same time he has lost with humanity. I think the time may come when I shall seize him by the collar and pitch him overboard or come down to his level in unworthy dispute. Well, he is a strange character," she wrote to Martha of her gamble.[7] In applying for a federal permit to collect the larger game mammals she desired such as moose, caribou, mountain goat, and mountain sheep, Alexander requested a *"personal* permit," one in her name only rather than in his.

Barely two weeks after writing to Merriam requesting the permit, Alexander humbly wrote to him again, this time stating that he need not bother securing it for her. She had cabled Bales to discontinue the expedition and placed the matter in the hands of her attorney. Alexander wrote to Merriam that she felt almost unbearably humiliated at the manner in which Bales had conducted himself but never explained what caused her to take legal action against him. Had he collected game illegally in her name? She also bitterly regretted her haste in employing Bales on the strength of a single recommendation.

In her next letter to Merriam she returned the collecting permit she had obviously received from him and stated that she had discharged Bales "because he is an impossible man to have any dealings with" but again gave no details.[8] However, fully appreciating the opportunity for collecting to which the permit entitled her, she went on to ask Merriam if the Department of Agriculture (the Division of Economic Ornithology and Mammalogy's parent organization) would extend to her the same privileges at a later date should she find a competent agent to carry out her original plan.

The events of the next two months are lost, and details of the expedition are sketchy, but in May 1906 Alexander, Edna Wemple, and Alvin Seale, a government-appointed scientist to whom a restricted federal permit to take big game had been issued, arrived in Juneau, Alaska. Juneau was a small town in 1906, its fortunes having faded with the passing of the gold rush boom. Its population numbered less than two thousand. Alexander was anxious to visit the Malaspina Glacier, so the group wasted little time in town. They bought provisions and immediately set off again by boat, their ultimate destination—the Kenai Peninsula.[9] As it was still early in the season, the weather slowed their progress and forced them to camp for several weeks in a meadow at Disenchantment Bay. There they waited for the ice to break up sufficiently so that a westward crossing could be made between floes. On the morning when it appeared that the ice had cleared sufficiently to attempt a departure, their boat remained firmly lodged amidst the frozen bergs. A joint effort required all three of them to enter the frigid water to lift the craft onto and over the ice that blocked their way.

What Alexander had anticipated would be the negative aspects of this particular venture, that is, swarms of ravenous mosquitoes and having to walk around perpetually in rubber boots, proved to be not as disagreeable as she had feared. What she gained in return for these moderate discomforts was a backdrop of spectacular scenery—pristine forests, rugged peaks, magnificent glaciers—coupled with the inestimable pleasure of watching bears as they browsed along the not-so-distant hillsides. However, the animals remained out of range of her gun.

At Disenchantment Bay a local hunter, Bill Dawson, joined the three collectors and accompanied them as far as Skilak Lake on the Kenai Peninsula. Traveling by boat until they reached the peninsula, carrying their heavy gear and equipment in backpacks after that point, the party camped above snow line and subsisted primarily on bread, small tins of Carnation milk, a generous supply of wild berries, and the fish and game that they caught or shot. Wemple recalled that "the hemlocks were so snow-laden that their bowed-down branches made beneath each tree a walled-in room."[10]

Having established a camp near Skilak Lake, they spent their days collecting mammals and birds while they explored the general area. By this time it was August. The group had selected this particular site in the hope of securing several mountain sheep from the surrounding slopes, statuesque animals that would form the basis for a natural history diorama in the atrium of the museum that Alexander now envisioned with increasing clarity. It happened to be Wemple's birthday when Alexander located a desirable group of the animals within rifle range. As Edna recalled, in Annie's gracious and generous manner, and in honor of the celebratory occasion, she insisted that Edna be allowed the first shot.[11]

In September the trio moved south to Seldovia and then across to Kodiak Island where they continued to hunt. It was late October when they returned to Seldovia for the last time and, from there, set out for home.

Although a primary objective of the expedition had been to collect bears, the only ursine specimens that the group acquired were those that Alexander had been able to purchase from local hunters. In addition to the three sheep that they had shot, the party trapped and hunted marmots, ermines, squirrels, sea otters, moose, porcupines, minks, woodpeckers, and jays, all of which would become incorporated into the Museum of Vertebrate Zoology collections. Without Goretex or Polarguard to protect them from cold in the frequent rain, without lightweight gear, plastic, fiberglass, preprocessed food, or the comfort and luxury of Harriman's steamship, Alexander and her party had spent four months traversing the Alaskan wilderness. She had learned much from the experience and planned to return the following year to collect scientific specimens in another portion of the territory that still remained largely unexplored.

7
Meeting Joseph Grinnell

As 1906 drew to a close, Alexander traveled east rather than west for the New Year, visiting Mary Beckwith in New England and attending the theater in New York. Then in mid-January she took the train to Washington, D.C., to confer with C. Hart Merriam about her plans for a second expedition to Alaska the following summer. This trip's scientific scope would be more extensive than her previous venture and she again asked the survey chief for suggestions as to collectors who might accompany her. She also needed to find a female companion. Annie had been pleased with Edna's company but much to her dismay the paleontology student planned to spend the summer of 1907 abroad. Wemple married following her return to the States and, although the women remained close friends for life, the 1906 expedition was the last one they made together.

This time Merriam recommended a mammalogist, Frank Stephens, to accompany Alexander. Stephens had begun his career as a naturalist in the 1870s collecting scientific specimens of the relatively unknown birds and mammals of the West for museums in the East. By the 1880s he was collecting specimens for the U.S. Biological Survey. The 1906 publication of his volume entitled *California Mammals* had recently won him recognition and praise from biologists nationwide and it was this acclaim that presumably brought his name to mind when Alexander asked Merriam for a recommendation.[1]

Stephens and his wife, Kate, a malacologist, were living in San Diego County when Alexander paid them a visit on her way back to the Bay Area in late January 1907. Circumspect after her experience with Bales, Alexander wanted to talk with the couple directly. She sensed during their meeting that the Stephenses would be valuable additions to her party and extended them an invitation to accompany her. Annie also realized that Kate

Stephens's membership in the group would obviate any further search for a female companion.

Stephens told Alexander that he wished to concentrate on collecting mammals and recommended Joseph Dixon, a biology student at Throop Polytechnic Institute in nearby Pasadena, as someone who could collect birds on the expedition.[2] Alexander agreed to interview Dixon before returning to Oakland. Arriving on campus, she inquired of the biology instructor, Joseph Grinnell, how she might locate the young man.

To those outside the disciplines of evolutionary and conservation biology, the name Joseph Grinnell will not be familiar. To others, he was a researcher without peer who fostered an academic dynasty that remains vibrant even today. His seminal publications on systematics, speciation, geographic variation, ecology, conservation, and collections management of vertebrate taxa are still frequently cited. Often his hypotheses were prescient. Biologists familiar with his legacy will find it inconceivable that in 1907 this driven field biologist, consummate museum curator, meticulous editor, and incomparable authority on the birds and mammals of the West Coast was nothing more than a biology instructor at Throop, in a sense languishing there, before Alexander selected him to become the first director of the Museum of Vertebrate Zoology.[3]

Grinnell was born in 1877 in the Indian Territory, about forty miles from what is now Fort Sill, Oklahoma. His father served as the agency's physician before moving the family to Pasadena, California, in 1885. By the time that Joseph graduated from high school he had been acknowledged as "having captured, preserved, labeled and classified more specimens of our native birds than any other person."[4] By Grinnell's own admission, his happiest times were those he spent in the field.

Grinnell received his bachelor's degree from Throop in 1897, the year following his first trip to Alaska. His time in the territory had been brief, much too brief for his liking and, in April 1898 he eagerly boarded ship a second time, on this occasion hoping to spend a more extended period in the frontier. But it was not gold that lured Grinnell to Alaska. The young biologist spent every spare moment there collecting birds and recording observations about their natural history.[5]

At the end of almost two years in Alaska, Grinnell reluctantly found himself back in California. He completed a master's degree at Stanford University and began work toward his doctorate. He quickly became friends with Walter Fisher and Wilfred Osgood, men who had collected bird specimens for C. Hart Merriam, the man "whose writings were Joseph's gospel."[6] But in 1903, midway through his studies at Stanford, Grinnell suffered an attack of typhoid fever. Recovering at home, he was offered the position of biology instructor at Throop. He accepted the post, assuming he would return to Stanford within a year to complete his degree. He did not. In January 1907, when Alexander visited the campus, Grinnell was still teaching there.

Learning of Alexander's plans for Alaska, Grinnell vividly recalled his own forays into the northern territory and expressed a great deal of pleasure at the opportunities and experience that the expedition would afford Dixon. Sending his brother-in-law in search of the young man, Grinnell then invited Alexander home. There, meticulously arranged in the back parlor like a miniature museum, was his personal collection of specimens, numbering in the thousands. Grinnell proudly showed Alexander the Alaska material and the two spoke at length about their respective journeys.

After Alexander and Dixon met, and tentative arrangements for the summer had been made, Alexander returned to her hotel. As Grinnell pondered the events of the afternoon, he was somehow struck by the familiarity of Alexander's name. Finally, linking it with that of John C. Merriam, he dove into a pile of reprints until he recovered one in which the paleontologist expressed gratitude to Alexander for her help in collecting specimens and for her financial support, both of which had made his own work possible. Merriam's acknowledgment was sufficient to convince Grinnell of Alexander's genuine interest in research and of her financial ability to follow through with her stated objectives. Satisfied as to her commitment, he immediately sat down and penned a long letter to Alexander on the intricacies of conducting fieldwork in the arctic wilderness. In particular, he enumerated many things that she might do to make her expedition maximally useful to science.

The thoughtful and thorough nature of Grinnell's letter was not lost on Alexander. Impressed with his initiative and with the attention to detail reflected in his prose, she replied by inviting him to visit her home the following fall to view the collections that she and her party would make.[7]

Lying wedged between the Pacific Ocean and the Canadian coast, southeastern Alaska is frequently considered the most scenic part of the state. It contains 10 million acres of forest, 10,000 miles of shoreline, 50 to 70 glaciers, and 25,000 brown bears. A chain of 1,000 islands forms the Alexander Archipelago, named for Czar Alexander of Russia, and an intricate network of waterways and fjords known as the Inside Passage separates the islands from the mainland. In 1907 the mountainous and densely forested interiors of these islands remained virtually unexplored by those of European descent.[8]

Just west of Juneau, Admiralty, Baranof, and Chichagof islands compose the primary land masses in the northern portion of the Alexander Archipelago. Because bears were reported to be plentiful on the three large northern islands, C. Hart Merriam had recommended this area to Alexander for her second expedition to Alaska (see the 1907 expedition's route in Map 2). Grinnell had visited several points along this panhandle a decade earlier and he enthusiastically supported Merriam's recommendation. He also sent Alexander a list of the birds that he had collected in the region. She responded that she would be happy to collect additional specimens of both birds and small mammals if he would commit to producing a publishable manuscript on the material. She clearly noted, however, that her primary objective was bears. Grinnell agreed.

Alexander left for Seattle in mid-March 1907 to examine a "power schooner," a gasoline-powered launch that had been recommended to her for navigating among the islands of the Inside Passage.[9] Her thought in providing a boat for the expedition was that the party would not have to depend on hiring local craft that might not be suitable or readily available. With their own boat, several members of the group could easily explore new collecting localities along the coast of a given island or on smaller nearby islands. As she had done the previous year, Alexander also provided all the requisite camping gear, field equipment, and supplies that her group would need, including guns and ammunition. She paid each collector a monthly salary and she covered the expenses of the launch, its captain, and the various fishing boats that were eventually hired to transport the party and all its equipment from island to island and camp to camp during their five-month stay. Once in Alaska, she also purchased desirable specimens from among the variety of animals, including bears, that were brought into camp by local hunters and native Alaskans.

Dixon, the Stephenses, and Chase Littlejohn, a collector from Redwood City, California, who had been added to the party at the last minute, joined Alexander in Seattle, and on April 10 the group departed.[10] Weaving their way north along the Inside Passage, the six Californians stopped first at Juneau. After purchasing last-minute supplies in the local shops, Alexander hired two small launches to carry the entire group to the first field site—Windfall Harbor on the eastern shore of Admiralty Island, approximately one hundred miles southwest of Juneau.

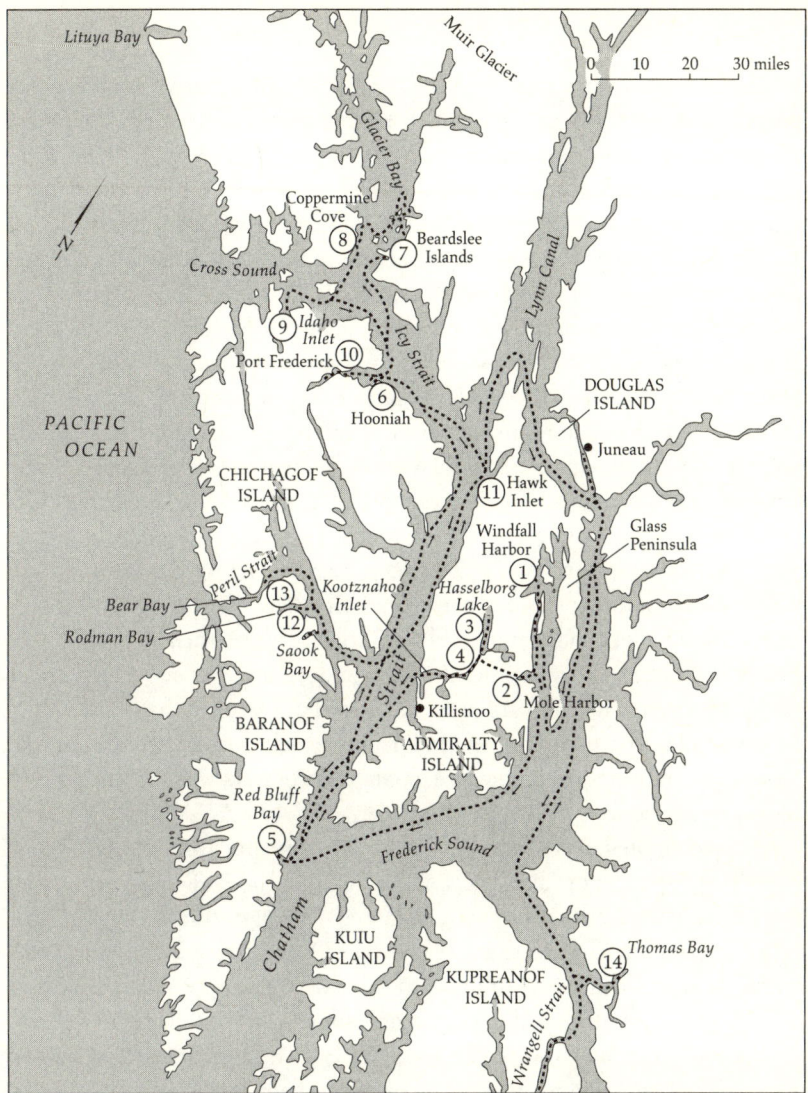

Map 2. The route of the 1907 Alexander expedition to the Sitkan District of southeastern Alaska. Helm Bay, Stephens's last collecting stop, is not shown on this map (based on the original USGS topographic map that Alexander used and annotated on that trip). Circled numbers indicate the order of the expedition's collecting localities.

As the boats approached the spit of land that would serve as their temporary home for the next few weeks, the group was struck both by the severity of the surrounding landscape and by its astounding beauty. To begin with, there was no level ground in obvious view on which to pitch their tents. Given this limitation, the party finally made camp on the western side of the harbor, midway between its upper and lower bend, their tents tucked close against a hillside of spruce and fir so dark in color that the trees appeared to be almost black. Behind them, jagged mountains rose abruptly from the beach, while directly across the harbor the rugged snow-capped peaks of Glass Peninsula gave Alexander the feeling that their camp was set entirely apart from the outside world. While eagles soared overhead, the tap, tap, tap of woodpeckers clashed with the drumming of grouse in the surrounding woods. Snow, in places ten feet deep, extended from the cliffs at the base of the mountains to the watermark at high tide. Even in places where the snow was only 2 or 3 feet deep, its crust was barely thick enough to support an adult for more than a few steps. Wearing hip-length rubber boots, the travelers had to put extraordinary effort into each step before the brittle crust gave way, plunging them thigh deep into wet snow. Alexander had not brought snowshoes on the expedition but reasoned that these would have been of little use given the steep and densely forested terrain on the island. The only strip of land where it was clear to walk was the thin stretch of beach that became exposed at low tide.

Camp consisted of six tents—three for sleeping, one for cooking, one for putting up specimens, and one for storing provisions. Ducks, geese, and grouse were easy dinner targets, and clams were plentiful for anyone ambitious enough to dig for them at low tide. Occasionally a native Alaskan would bring a halibut or salmon into camp. If walking proved a problem, obtaining sufficient food along the beach directly in front of them did not.

Scientific specimens were equally plentiful. The harbor itself contained a number of tiny islands where the party immediately set traps for mice and mink. Small mammal trapping was still relatively new to Alexander and she was equally unfamiliar with the birds that Dixon, Stephens, and Littlejohn were bringing back to camp. Her challenge now was to learn how to put up a good bird skin. It was not just the basic mechanics of preparing a study specimen that she strove to master, but the ability to add the finishing touches, that is, shaping the skins and making the feathers lie exactly right, details that would make the finished product as much a work of art as of science.

One day in mid-May Annie sat in her tent wrapped in blankets, a hot-water bottle comfortably tucked beneath her legs. She was writing to Martha and awaiting the return of the captain with their launch. It was their first real stormy day. A strong southwestern gale was stirring up water in the harbor and causing large swells to crash along the beach. Visibility was drastically reduced and she could no longer see the smaller islands in the harbor that had been a constant presence during the past three weeks. In one respect she welcomed the rain. Although the snow on the beach had gradually begun to melt, the forests and trees remained draped in a heavy blanket of glistening white. Thus far, travel to the interior of the island had been impossible. With the advent of rain, she hoped that the situation would change. Because many of the birds and mammals confined themselves to the edge of the forest and along the beach, Alexander had not been disappointed with their success to date. But she also realized that to hunt bears, and make an accurate faunal survey of the island, the group would have to move inland. She was increasingly anxious to begin. She had already made one unsuccessful attempt at bear hunting in a canyon to the north of camp and was ready to try again. Undeterred by her initial lack of success, she concluded the tale in her letter to Martha by reflecting, "There is a curious fascination for me in attempting something where the chances for success are but the very slightest. It is the mark of the dreamer."[11]

Given the close quarters of their camp and the relatively strenuous conditions under which the group had had to live for the past month, Annie was not displeased with her companions. With her discerning eye she wrote to Martha, "Littlejohn is the life of the camp, a close observer in his work and full of energy and enthusiasm. Dixon comes to life when he has an eagle's nest to rob and some one hundred and fifty feet to climb in order to do it. . . . Stephens is a hard worker but he has spent his best strength long before I ever knew him. . . . His wife is a funny little English woman with foreign notions and foreign ways of expressing them that makes her not altogether uninteresting."[12]

Alexander described one additional member of the party. Like others who had wandered into their camp at Windfall Harbor, Allen Hasselborg had come with a bear and a beaver to sell.[13] He had originally made his way to Alaska in search of gold. For a while he had held an assortment of odd jobs—fishing, sailing, woodworking, mining. Most recently, he had taken up hunting. Something about this particular man impressed Alexander, though she could not as yet clearly articulate it, and he had more or less remained with the group since the day he first appeared in camp.

A month after the arrival on Admiralty Island, Hasselborg arranged for a fishing launch to relocate the party to Mole Harbor, approximately twenty miles south of their present camp. This was a warmer locality, with more accessible low ground, and the hunter felt that it would provide the group more convenient access to the interior. He now acted as their guide and Alexander felt inclined to offer him a salary for the duration of the summer. He was strong, could paddle a canoe "to perfection," and could steal silently along on land as well as in the water. He was also "the only white man around here who has dared make interior trips alone," Annie confided to Martha.[14] A lot of backpacking remained to be done and Alexander recognized that Stephens and Littlejohn were already at an age where such work was becoming increasingly difficult for them. Hasselborg had also proved to be wonderfully observant. Dixon remarked that if one told him the name of a bird just once, he would recognize the species the next time he saw it and remember its name.

As the weather warmed, Stephens and Dixon were the first to head inland to hunt. Birds were not as plentiful at Mole Harbor as they had been to the north, but bears were what the group now hoped and expected to find. Much to everyone's surprise, approximately three miles west from the head of Mole Harbor, they found beaver. The animals were living on a series of three lakes forming an L-shaped chain about fifteen miles in length and bordered by a dense forest of spruce and hemlock. Alder and a little willow grew along stretches of shoreline where the banks of the lakes were not too steep. A thick layer of underbrush was pervasive and left relatively little in the way of beach along their periphery. Previously unknown to cartographers and rarely visited by native Alaskans, the lakes had never been explored by Europeans other than Hasselborg. Stephens immediately suggested that the largest lake be named in the hunter's honor. The party then promptly named the two smaller lakes "Alexander" and "Beaver."[15] These two were practically one lake, connected by a long, shallow channel, with the far end of Beaver Lake flowing into the lower end of Hasselborg Lake via a thirty-foot waterfall and a short, rapid stream.

The discovery of beaver on Admiralty Island marked the first time that these rodents had been recorded on any of the islands in southeastern Alaska, and the party now became as interested in collecting beaver as they were in collecting bears.[16] Evidence of beaver activity was everywhere—from freshly cut hemlock trees to runways between dams and small patches of willow. The group immediately caught three animals and took several photographs of others sitting on the banks of the lake gnawing sticks. They also saw evidence that an increasing number of bears were emerging from their dens—the hungry carnivores had torn open at least one beaver den, presumably in search of food.

Using the bird list that Grinnell had provided as a guide, Alexander determined that the group had already collected a large percentage of the species listed, and at least four that were not. Adding to their collection, Dixon and Littlejohn had also secured a pair of sharp-shinned hawks and Dixon had captured three ptarmigan. Alexander wrote to Beckwith that those birds cost him "the hardest climb I guess he ever had—but he told me as he went off with Hasselborg that Grinnell would say he was no true pupil of his if he knew ptarmigan were to be had and made no effort to secure them."[17]

Hasselborg River, as the party now called it, emerged from the northern end of Hasselborg Lake, draining a broad area and carrying a large volume of water. Because of its swift descent to sea level over the course of just a few miles, its headwaters were filled with rapids. Stephens suspected that the river flowed into the long Kootznahoo Inlet on the western side of Admiralty Island. However, at the time, no one knew for sure where the outlet for the river lay. This unknown variable, with its promise of discovery and adventure, was sufficient to lure Alexander to paddle the river from east to west, accompanied by Dixon and Hasselborg. The remainder of their party would go by launch around the southern tip of Admiralty Island and would meet the trio three days hence at the native village, Killisnoo. (No provision seems to have been made should their assumption about the river's outlet prove incorrect.)

After loading their canoe, the three set out from Mole Harbor just before noon. It took an hour and a half to traverse the rough terrain between their camp and Alexander Lake as they had to portage their gear "through thick forest, over logs, up hills, through underbrush and thorny 'devil clubs.'"[18] They followed a well-worn beaver trail for part of the way. The beaver continued their industry unabated as the explorers paddled the lengths, first of Alexander and then of Beaver Lake. A portage of three hundred yards was necessary to circumvent the waterfall that connected Beaver Lake to the southern tip of Hasselborg Lake. As the afternoon progressed, they easily paddled across Hasselborg Lake and out into the main river channel. The sun shone brightly and the worst of their effort seemed behind them. Where the water was smooth and their paddling almost silent, they were able to catch glimpses of deer along the sandy banks of the river.

About a mile downstream they decided to camp for the evening. Alexander awoke the next morning as raindrops splattered across her cheek. Continuing throughout the day, the rain stirred up the river and made paddling difficult. Despite the swiftness of the current, Hasselborg was able to negotiate some of the rapids that they encountered with extraordinary skill. Others he chose to portage. Late in the day the group stopped for supper, but that evening they made an additional portage of almost a mile and a half before finally setting up camp.

On the morning of the third day the surface of the water was smooth and paddling relatively easy once again. Only a logjam that Hasselborg successfully circumvented caught them by surprise—until they reached the fifty- to sixty-foot waterfalls marking the western terminus of the river. Here they made a short portage. They continued paddling for several more miles before realizing that they were now in salt water. They had, in fact, reached the western side of the island and had entered a narrow arm of Kootznahoo Inlet. For several hours they explored various arms of the inlet and paddled around the islands at the head of Mitchell Bay. Late that night they made camp just above the point at which the bay narrowed.

Rising at 3 A.M., the three hastily loaded their gear. They had six miles of inlet narrows in Chatham Strait to cross before the ebbing tide made their passage impossible. Even with the tide in their favor, paddling the currents in the strait proved challenging and treacherous. By the time they reached Killisnoo at the mouth of the inlet three hours later, the tide was falling fast. Despite the tremendous physical effort expended, or perhaps because of it, at the conclusion of their journey Annie wrote to Martha, "[N]ot before or since on this trip has three days been more happily spent."[19]

Feeling thoroughly satisfied with their adventure, the trio eventually met up with the rest of their party at Red Bluff Bay on the eastern coast of Baranof Island, southwest of Killisnoo across Chatham Strait. Here the shoreline of the bay rose ominously and precipitously to the mountains above, once again leaving no stretch of beach on which the collectors could pitch their tents. At the head of the bay a small mud flat had formed, the result of streams rushing down from the overhanging mountains, and a few acres of grassland bordered the tide line. It was on this small plot of land that the group set up camp. The snow in this region arrived early and obviously remained on the ground until late in the season. This was the coldest and gloomiest site that they had encountered thus far. In the days following a solitary hike around the bay, Annie experienced recurring nightmares in which she was trapped atop the steep cliffs hovering above their camp with no possible means of escape. Others, too, found the canyon's walls and overhanging glaciers oppressive, almost claustrophobic and, as the collecting in the area proved poor, the group left the site after only a week.

After more than two months in the field, Alexander once again pondered the nature of her companions. Overall, she was pleased with the way that they had all gotten along under the strenuous circumstances of the work and their austere living conditions. For a party of seven there had been relatively little friction among individuals in the group. She wrote to Martha:

> [Hasselborg] is a strange young man. Nothing seems too difficult for him and he seems never weary. Yet why has he not made something more out of himself than he has? With his determination and alertness there must be some weakness. He has learned the names of all the birds from us now and is constantly on the lookout for new forms. If he does discover one, he worries the men until they go after it. He never idols [sic]—I never knew a man leading the vicarious life he has been leading, so insistent that every moment of time be spent to good purpose. He keeps us all up to the mark. Dixon is slow but thorough and he has remarkably good judgment for so young a man. Littlejohn is emotional, sometimes up, sometimes down, but generally in the best of spirits, overflowing with vitality. He is certainly an intense lover of bird and animal life.[20]

While always an admirer of physical prowess and stamina, particularly under strenuous circumstances, Alexander's comments about Hasselborg suggest an ongoing preoccupation with her own intellectual insecurities and highlight the fundamental difference between the two individuals. Alexander failed to comprehend why anyone with a sharp eye and a quick mind would have chosen to live in a remote wilderness rather than exploit such talents within the framework of society. Living a life of almost complete solitude on Admiralty Island, Hasselborg garnered no money or fame from his superior abilities. In contrast, Alexander aspired to do more than simply collect specimens. She felt impelled—as her father had been, and his father before him—to make a substantive contribution, in her case to science. This value underlay her sense of well-being and helped to offset her insecurities.

Traveling north by launch through Chatham Strait, then west into Icy Strait, the expedition made its next camp about five miles northwest of the Indian village of Hooniah near Port Frederick on Chichagof Island. From there they planned to explore the small islands at the entrance to Glacier Bay. Because icebergs were so abundant just a few miles inside the bay, their launch was forced to make its way into Bartlett Cove instead, a comparatively ice-free anchorage on its eastern shore. Dixon and Hasselborg crossed the bay to work at Coppermine Cove to the west, while the others camped and collected specimens on two of the Beardslee Islands.

In some respects this was the most productive site the group visited. Because Glacier Bay is part of the Alaska mainland, the group was able to add several new species of mammals to their collection, taxa that presumably did not exist on the adjacent islands. This was also the most northern locality visited by the expedition. With the approach of summer the usable daylight hours increased. The sun rose now before 3 A.M., and at 10 P.M. there was still enough light for Stephens and Dixon to continue preparing the specimens that they had collected. Weather in the bay remained wet and rainy but on July 4 the temperature reached a balmy 60° F. To celebrate, the group made ice cream by mixing condensed milk with fragments broken off a small iceberg floating near camp.

But their camp at Glacier Bay was surrounded by a dark, intricate forest of deep peat moss that yielded beneath every step. Alexander found tramping in heavy rubber boots in this vegetation "a drag," the brush and briars unrelenting.[21] It was a bewildering place and only Hasselborg willingly ventured into its depths—and even he did not dare to do so without a compass. Alexander preferred to collect in more open country and she decided that Hasselborg, Dixon, and Littlejohn should accompany her to Idaho Inlet on Chichagof Island. She had seen a trail there that climbed 1,200 feet and wanted to explore it. The Stephenses would remain at Glacier Bay and continue to collect in the open area along the beach.

Before Alexander could implement her plan, however, the mail launch arrived carrying a cable urging her to return home at once—her mother had been involved in a serious accident and Sam, her sister Martha's little boy, had died. She left immediately. From Juneau she caught the mail boat to Seattle, and from there she took the train back to Oakland.

Following her sudden departure, the remaining members of the expedition crossed to the western shore of Glacier Bay and continued trapping. Crossing back to Chichagof Island, the group then spent four days at Idaho Inlet before relocating to Port Frederick. From here they headed south through Chatham Strait, then west into Peril Strait along the northern coast of Baranof Island. They camped and trapped for several days at Rodman Bay in a nearly deserted mining camp before making their last camp at Bear Bay on Peril Strait. It was now early September and the weather was becoming increasingly stormy, the rainfall more intense. Dixon, Littlejohn, and Kate Stephens returned to Juneau where they caught the mail boat to Seattle, while Frank Stephens returned to Glacier Bay and continued trapping for several more weeks. He collected additional specimens at Thomas Bay and Helm Bay before turning south and heading home.

It took more than a month for the members of the Alexander expedition to reach the Bay Area from Juneau, given the logistical difficulties of transporting their gear and all the specimens that they had amassed. But even before the 476 mammal specimens, 532 bird skins, and 33 sets of eggs that had been collected could be critically examined, Stephens felt sure that the group had obtained several new species of both birds and mammals.

Despite such success, Alexander expressed her disappointment that the expedition had not been able to cover more ground; there were at least two or three additional localities that she had hoped to visit. But after returning from her sister's home in Honolulu, she wrote to Grinnell as promised, inviting him to Oakland (at her expense) to view the specimens that she and her party had collected during the summer. Alexander also intended to use the opportunity to discuss with Grinnell her plans for founding a museum of vertebrate zoology. Having carefully weighed the concerns and considerations that both John C. and C. Hart Merriam had previously expressed, she wrote to him,

> I should like to see a collection developed (more especially of the California fauna) and would be glad to give what support I could if I could find the right man to take hold; someone interested not only in bringing a collection together but with the larger object in view, namely gathering data in connection with the work that would have direct bearing upon the important biological issues of the day. Work systematically and intelligently carried on is the work that counts.[22]

Responding positively to her invitation, and excited by the prospects that her letter portended, Grinnell visited Oakland over the Thanksgiving holiday. By the time he left, Alexander felt that she had found a field biologist who shared her vision, as well as an individual with the experience necessary to serve as her museum director. With increasing conviction that her museum would prosper under his direction, she and Grinnell began to correspond regularly and often.

8

Founding a Museum of Vertebrate Zoology

Teaching at Throop offered Grinnell the opportunity to make brief collecting trips in and around the Los Angeles area or more extensive excursions into the California desert, but his wish was to devote himself exclusively to research. His desire to see a museum of vertebrate zoology established on the West Coast stemmed from his need for a large, comparative specimen collection for his work, studies that focused on elucidating an evolutionary theory in which speciation was viewed as a function of an organism's geographic locality, influenced not only by latitude and ambient temperature at a given locale but, more complexly, by a multiplicity of biotic variables that included humidity and vegetational distributions.

Grinnell also insisted on studying animals in relation to their environment and he recognized that museum specimens, properly documented, not only provide insight into the biology of individual organisms but, in a much broader context, reveal biological interactions that are key to understanding how the natural world evolves. As he asserted in 1910, "It is quite probable that the facts of [a species's geographic] distribution, life history, and economic status may finally prove to be of more far-reaching value [. . .], than whatever information is obtainable exclusively from the specimens themselves."[1]

Grinnell's desire to establish a "center of [scientific] authority" on the West Coast and his belief that Alexander shared in this purpose were pivotal to the development of policies and methodologies that would shape the new museum's future. In 1907, both Grinnell and Alexander recognized that the vertebrate fauna in the United States was vanishing. Evidence of human impact was particularly acute in the new territories of the western United States. In their early correspondence, Grinnell remarked to Alexander, "Fieldwork, too, can be carried on now perhaps with better results than a few years later when so much more of the country is cultivated or deforested." Equally important, both individuals recognized the value of permanently documenting these dramatic changes for posterity. However, Grinnell also understood that any documentation must be sufficiently detailed and organized so that, "200 years from now," people would know the conditions as he and Alexander observed them. Careful field observations were to be recorded, both on specimen tags and in field notebooks. With astounding foresight, in 1908 he noted, "Our field-records will be perhaps the most valuable of all our results."[2]

The regular correspondence that Grinnell and Alexander initiated in the fall of 1907 continued throughout the winter and well into the following spring. Not surprisingly, Grinnell favored establishing the new museum at nearby Stanford University, his alma mater, but he acknowledged to Alexander, "we can *do good work anywhere*. I shall bend my energies to accomplish *results*, wherever we locate."[3] He listed the advantages that he foresaw the Palo Alto campus providing, for example, "vast areas" of unused space in the Zoology Building that he felt would be ideal for storing research collections and for laboratories, excellent lighting and heating, custodial services, and what he considered to be the best library of vertebrate zoology on the West Coast. Grinnell also believed that Stanford might provide all the zinc cases necessary to house the specimen collections that would form the basis of the new museum. The Zoology Building itself was of stone and what little damage it had incurred during the 1906 earthquake had already been repaired. Last, he argued that all Stanford's income now went into developing its buildings and infrastructure. Therefore, within a relatively short time, all construction would be complete and the surplus funds currently allocated to construction would be made available to the departments. During that interim period, Grinnell believed that the new museum would be able to justify its worth and, at the appointed hour, be granted a sizable appropriation from those funds.

While conceding that Grinnell's points about locating the new museum at Stanford were "good," Alexander almost immediately rejected the idea of affiliating with that institution. She did not "feel drawn" toward Stanford, she was not familiar with the researchers there, and she had already established relations with the University of California at Berkeley. Were the museum located on the Palo Alto campus she sensed that she would "feel quite 'out of it.'" As for the quality of the library at Cal, if it were as poor as Grinnell implied, then perhaps she had a responsibility to "stir them up on that score."[4] And, she argued, the University of California had thus far neglected the study of extant vertebrates.

Both Alexander and Grinnell emphatically rejected any notion of affiliating with the Academy of Sciences in San Francisco, given its restrictive policies for specimen acquisitions and loans. He wrote to her, "I should far rather see a purely *private* institution thruout, than that!" She replied, "As regards affiliation with the Academy of Sciences you need have no fears whatsoever."[5]

In reality, Grinnell's arguments in favor of locating the new museum at Stanford were moot, as the date of his letter to Alexander followed by one day that of her own to Wheeler. Dated October 28, 1907, her letter began:

> Dear President Wheeler:—
>
> Should the University of California within the next six months erect a galvanized iron building furnished with electric light, heat and janitor's services and turn it over to my entire control as a Museum of Natural History for the next seven years, I will guarantee the expenditure of $7000.—yearly during that time for field and research work relating exclusively to mammals, birds, and reptiles of the West Coast, with the understanding that the University of California would be in no way responsible for the management of the funds for carrying out the work, or selection of collectors.[6]

The funds that Alexander was offering to the university would be used to pay salaries of the museum staff; make and prepare specimen collections from throughout the state; purchase supplies, casework, and equipment for the housing and long-term preservation of specimens; and cover the costs associated with producing publications that would result from research based on the collections. Not mentioned in her letter to Wheeler was the fact that she had already purchased a dozen specimen cases, at $50 per case, which she was storing in her Oakland home pending construction of a museum building. Her personal collection of 3,424 specimens, plus 1,300 specimens that would be acquired from the Department of Zoology on campus, would form the nucleus of the new museum.[7]

While Alexander's offer stipulated that she would retain complete control over the moneys provided and the museum's program of research, her letter went on to emphasize that the specimens obtained during the course of that research would become the exclusive property of the university. However, harking back to a criticism of the Academy of Sciences, she stated explicitly that all "materials would be and remain accessible to any qualified student within or without the university wishing to carry on special lines of investigation."[8] In other words, she was inaugurating the policy, generally accepted in the East, of sending catalogued specimens on loan for purposes of research, even though the specimens loaned would remain the legal property of the university.

While wishing Wheeler to feel comfortable with her selection of Grinnell as director, Alexander looked upon his appointment as an essential component of her proposition and she was unwilling to negotiate this point. Although he was still five years away from completing his Ph.D. and had not yet made a name for himself within the discipline of evolutionary biology, Alexander perceptively discerned in him the qualities that she felt were prerequisites for a successful administrator.[9] If Wheeler was unwilling to accept her entire proposal as stated, Alexander was prepared to withdraw her offer.

John Merriam staunchly supported Alexander's efforts to found a museum on the Berkeley campus. To support her letter to President Wheeler, Merriam added an outline that stressed the value of such a collection to the university and the manner in which the specimens gathered would enhance both scientific research on campus and the university's standing within the academic and local communities alike.[10] From this outline, Merriam's early influence on Alexander's thinking about the role of research museums within the discipline of natural history and their contribution to understanding evolutionary theory becomes apparent.

In his statement Merriam argued several points. Alexander's collections would be used in instruction and research, thereby making the university a recognized center for the study of mammals, birds, and reptiles. Her specimens would provide extensive knowledge of mammals and birds as part of the nature study curriculum in local secondary schools.[11] This popular program was aimed at familiarizing students with the local flora and fauna. The specimens would have economic importance by leading to an increased understanding of the relationship of birds and mammals to the agricultural interests in the state and finally, he reminded the president, they would provide education to the general public as to the meaning and importance of wildlife.

Merriam elaborated on the contributions of the collections to research, arguing that they would be of great value to studies of morphological variation, geographic distribution, and the evolution of species, as well as to systematic studies of the fauna in the western United States. He noted further that the collections would also serve as the basis for all paleontological, geologic, and anthropological studies in which it was necessary to make comparisons with recent taxa. Last, the publications that would result from such studies would appeal to a segment of the scientific community that had not previously been interested in the work of the university. He also remarked on the availability of the collections to external researchers and their potential for making the university a recognized center for the study of birds and mammals in America. To a university still in its infancy and lacking a national reputation, these were compelling arguments.

Although final decision in the matter rested with Wheeler, Alexander proceeded to contact several of the regents and present her proposal to them in person. Such lobbying is reminiscent of one of her father's successful business strategies and characterized her patronage style. After her 1903 exchange with Wheeler (in which she had requested that the university commit funds for research in paleontology and was told that it would not), she assumed that the administration never fully comprehended the value of basic research in either paleontology or vertebrate zoology. Thus, when she did not receive an immediate or positive response to a given proposal, she often lobbied directly those in positions of power for its support.

When no reply from the president was forthcoming after more than a month, Alexander wrote to Wheeler a second time, on this occasion requesting a formal interview and an acknowledgment of her offer. Alexander possessed a tremendous sense of what was necessary for the success of any venture she undertook. Recognizing the stringent nature of her proposal, as 1907 drew to a close she wrote to Wheeler elaborating her position:

> I am not so disinterested a giver as the University might like to have me. The plan is wholly my own, matured after many months of thought and some experience in the field. Were I to turn the funds over to the University to control, they having all the say and getting all the credit for the carrying on of the work, I should feel quite out of it. Some responsibility in the work is necessary to my own well-being and a legitimate incentive to see the enterprise develop.[12]

Alexander returned from a holiday visit to the East Coast in late January to find that Wheeler had appropriated $7,000 for the new building though, she wrote to Grinnell, "the President would like to have it come under $6,000."[13] Construction was to begin as soon as specifications were complete but, Alexander was quick to point out, the regents had not yet formally accepted her proposal.

Feeling that she had reached a critical juncture in her negotiations with the university, Alexander asked her attorney to draw up an agreement specifying the terms of her offer. She understood that, from Wheeler's perspective, the difficulty in her original proposal stemmed from her position that the museum director would not be a university employee but would remain under her direct control. She clearly perceived the conflict in authority that the regents envisioned if she were to retain complete control of the funds and the direction of the museum's research program while all specimens collected became the property of the university.

The letter in which Alexander conveyed these legalities to Wheeler further specified that, in the event of death, her estate would pay the yearly amount that she had guaranteed for support of the museum throughout the duration of the seven-year period of her initial agreement. Likewise, it stipulated that the university would, in good faith, be committed to continuing the work of the museum in the spirit in which she had conceived it during that seven-year period. Finally, it stated that the proposed name for the new building and its program would be the California Museum of Vertebrate Zoology.[14]

In early February 1908 Alexander spoke with Wheeler and learned that the regents had formally accepted her "gift," acknowledging it with an appropriation of $7,000 for construction of the new building. However, they had not consented to all of her stipulations. Insisting that her offer, as presented, was nonnegotiable, she informed Grinnell, "I foresee trouble unless we do have it. I may be asking too much but it is absolutely necessary in that nest of schemers. Once [we] have a clear business understanding we are all right!"[15] The prudence evidenced by this statement would prove well taken in time. The business acumen and candor that it reflects seem equally astonishing for a woman at that time.

By midmonth Alexander had received a formal response to her proposal. The regents balked at turning over complete control of the museum to her. Wheeler conveyed this sentiment by stating that the regents had found her stipulations in violation of the Organic Act under which the university had been created. After consulting briefly with her attorney, Alexander responded almost immediately by withdrawing her offer.[16]

Wheeler countered the following day, regretting deeply the light in which Alexander had interpreted the regents' response to her offer. He concluded, "Nothing vital in your plan was changed. There is no intention or expectation that any limitations will be set upon your management of the work you propose. . . . I guarantee that there will be no interference whatsoever by the Regents with the progress of the work under your management."[17] Wheeler then attempted to downplay the incident, remarking that the wording of the initial acceptance letter had been advised by the university's general counsel, based upon what was thought necessary to remain in compliance with the Organic Act. He did not feel that the wording should be interpreted verbatim.

Alexander again consulted with her attorney before responding to this latest volley. Her lawyer was of the opinion that the regents would come to terms with her on the agreement by ignoring law rather than acting contrary to it. Thus, after careful consideration, Alexander accepted Wheeler at his word. In her reply to the president, she reiterated what she felt were the "vital conditions" of her offer and acknowledged his concessions to legalities. On March 23, 1908, she telegrammed Grinnell in Pasadena, BEGIN WORK NEW MUSEUM TODAY YOUR APPOINTMENT AS DIRECTOR WILL BEGIN FROM APRIL FIRST.[18]

There were several aspects of Alexander's proposal that continued to be fleshed out with the university administration in the following months and years. One, of particular import, involved the procedure by which museum staff were to be hired and consensus as to their chain of command. Compromise was reached on this first issue after Alexander and Wheeler agreed that all museum appointments would be formalized through the office of the president—after having been approved and confirmed by her. On the second issue, Alexander was willing to entertain the possibility that Grinnell report directly to the university's president, but she remained adamant that he not be subordinate to the chairman of the zoology department. She cautioned Grinnell that, until his position in relation to the university and to the work was solidified, she would not acquiesce on any additional points. Although it proved to be many years before a regularized relationship developed between museum employees and the university, Alexander's foresight again proved auspicious.

As a preliminary arrangement, Alexander guaranteed Grinnell a salary of $200 per month for one year, with an additional $300 per month during that period for carrying out research. As director, Grinnell was to be given full responsibility for disbursing the funds, but, simultaneously, he was accountable to Alexander for every penny expended. She also gave him complete charge of the work during that time, "but should any matter of moment arise I should like to be consulted."[19] This arrangement would characterize their relationship for over thirty years.

In early 1908 Grinnell was still without a replacement at Throop. Alexander supplied him with additional funds to hire a student assistant to help with his classwork so that he would be able to devote more time to writing up the results of the 1907 Alaska expedition. That manuscript eventually became a model for many of the research publications emanating from the museum and reflected the curatorial practices that Grinnell developed and adopted in conjunction with his philosophy about the roles of fieldwork and specimen collecting. In time, these subjects became inextricably linked.

The system of curatorial and field practices that Grinnell honed and implemented in consultation with Alexander aimed to facilitate the investigation of natural history as an objective scientific discipline. Adopted by his students and disseminated through their academic "progeny," these practices and philosophy became deeply embedded in the discipline of vertebrate natural history and are generally referred to under the rubric of "Grinnellian methodology."[20] In short, Grinnell believed that museum specimens act as vouchers for facts ascertained and, at the same time, become permanent objects for later study. He further argued that the simple act of collecting specimens also conveys to the observer unique and intimate knowledge about each organism. Series of specimens document the range of morphological variation within a taxon and provide clear evidence for the evolutionary processes acting to produce intra- and interpopulational variation. Extensive and detailed field observations on the ecology and morphology of each individual also document species and subspecies distributions, habitat parameters, and the range of morphological characteristics. Finally, the systematic organization of these data within a museum, coupled with specimen examination, allows for broad-based, conceptual analyses of the geographic variation and distribution of vertebrates. Such analyses, in turn, lead to the construction of testable hypotheses about speciation and vertebrate evolution.

Throughout the fall of 1907 Grinnell and Alexander corresponded almost daily about this topic and the multitude of affairs surrounding the birth of the new museum. Together they discussed and designed its every detail, from the cases in which specimens would be housed for safekeeping, to the catalog cards on which specimen data would be recorded, to the ink that should be used in writing on the cards. For lack of a museum building in which to immediately begin work, Grinnell remained in Pasadena. Both correspondents faithfully dated each of their letters and, at Grinnell's request, Alexander saved all correspondence for subsequent deposit in the museum's files. Grinnell stamped each of Alexander's letters upon receipt and annotated the stamp with the date of his reply to her. In so doing, he provided an invaluable transcript of the growth and development of the museum as well as insight into his own growth and development as an evolutionary biologist.

Because the material collected by the 1907 Alexander Expedition to Alaska was to become the museum's first accession, a system for cataloguing specimens was Grinnell's first priority. Throughout the fall of 1907 he read the available literature about museum catalogs and weighed the merits of a "pure card system" against those of a "loose-leaf system of recording," writing to Alexander his impressions of each and soliciting her opinions.

At Alexander's suggestion and expense, Grinnell toured the major natural history museum collections on the East Coast and in the Midwest during the Christmas holiday to talk with scientists and observe the methodologies that they employed in caring for and cataloguing specimens. During Alexander's own visit to Washington, D.C., the previous winter, C. Hart Merriam had given her a tour of the Smithsonian collections. She chronicled her impressions in a letter to Beckwith noting,

> Dr. [C. Hart] Merriam took me through the Smithsonian and we had a look at the bear skulls in the morning. After lunch, [Vernon] Bailey, his brother-in-law, gave me a general idea of the cabinet collection from mice up. They are so crowded for room that they have the attic of the national Museum stacked up to the ceiling. The bear skulls are put in a little room by themselves, shelves to the ceiling, requiring a step ladder to get at anything. All Friday I spent taking measurements and yesterday photographing. The floor was dirty and the skulls covered with dust and I wished I could lock everyone out—all the funny little scientists with their long faces. Merriam is the only one who has any life to him—[21]

Before writing up a description of the Alexander material for publication, Grinnell took advantage of his visit to the eastern museums to examine specimens from Alaska in their collections. He felt strongly that if the MVZ intended to establish itself as a "center of authority," descriptions of the specimens collected for the museum should be written up for publication by MVZ staff members. This was counter to the custom of the day whereby all such reports on vertebrates were written by researchers at the Biological Survey. "I should want our institution to turn out results of recognized value and authority," he wrote to Alexander. She concurred. To that end, a year and a half after her return, a scientific report on the results of the 1907 Alexander Expedition to Alaska was published by the University of California.[22] The paper emerged as a multiauthored work. Stephens and Dixon produced the locality descriptions, Edmund Heller described the mammals, and Grinnell wrote the introduction and described the birds. The accumulated specimens, field notes, and photographs on which the report was based became the first numbered accession in the new museum.

In Pasadena, Grinnell also set to work reviewing building plans for the museum. The blueprints showed a two-story structure, 50 × 100 feet, to be located in a portion of campus known as Faculty Glade. Facing east, the entrance opened into a 30-foot corridor with two rooms opening off it on either side. To the southeast was a storeroom that gave access to the "skin room," a walk-in cold room designed for the storage of large animal furs such as the bear hides that Alexander had acquired in Alaska. Opposite the storeroom, on the northern side of the corridor, a "prep room" was planned, a working space 20 × 20 feet that would be used for construction of museum exhibits. The prep room led to the room that would house the museum's collection of fluid-preserved specimens. The "alcohol room," as it was affectionately dubbed, would be the primary storage facility for reptiles and amphibians that were preserved in formalin but permanently stored in 70 percent ethanol. Such specimens would be arranged in jars along rows of braced shelving, secured against toppling in the event of an earthquake. Birds and mammals might be stored in ethanol as well but this method of preparation was less common for those vertebrates. The remaining square footage in the northeastern quadrant of the building was devoted to a room for storing curatorial supplies. This space would be accessible both from the main corridor and from the alcohol room.

At the end of the corridor was a large exhibition area, 50 × 70 feet. A staircase to the museum's second floor rose immediately to the left. A wooden gallery, 12 feet wide, ringed the western end of the second floor overlooking the exhibit space. A slanted roof rose 22 feet above the floor, and skylights had been drawn into the plans on each half of the ceiling to allow for the passage of natural light into the atrium.

The majority of the research collections were to be housed on the eastern end of the second floor above the workrooms. Mammal and bird specimens that had been stuffed with cotton—"study skins"—were to be stored in insect-proof cases made of zinc plied over wooden frames. A library/"study laboratory" equipped with benches, chairs, and a light table opened off this collections-storage space and would be used for specimen examination and the writing up of reports.

In accepting the terms of Alexander's proposal, the regents had agreed to start construction of the requisite "galvanized iron building" within four months of their jointly signed agreement, dated February 17, 1908. Yet Alexander returned from her third expedition to Alaska in the fall of 1908 to find that ground had not yet been broken for the new building. Surprised and disappointed, she wrote immediately to Wheeler requesting a satisfactory explanation. He responded that, in her absence, the regents had realized that the cost for the new building would be twice their original appropriation, or $14,000. Not having had additional funds to apply to the project, Wheeler awaited her return to discuss the predicament.

Despite her frustration, Alexander reevaluated the museum's budget as set forth in her proposal. After assuring herself that she would not be hampering the work of the new museum before it had begun, she offered to decrease her stated annual contribution by $1,000 per year and to apply the $7,000 remainder to the university's appropriation.[23] The regents gratefully approved this financial rearrangement and construction went forward.

Also included in Alexander's original proposition to Wheeler was the statement that a portion of her funds would be directed toward the creation and installation of several dioramas for exhibition purposes. These would be available for viewing by the general public in the main atrium of the museum during specified hours each week. Alexander's thought in creating public exhibits was "to show the people ... that within the limits of our West Coast there is a great deal worth while in the animal and bird line to attract their interest."[24] Once this task was accomplished, she felt that the need for a natural history museum would become apparent. Public appreciation would lead, in turn, to an understanding of a need for conservation, habitat preservation, and a reversal—or at least a slowing, she thought—of the species declines that she and Grinnell were witnessing. She may also have harbored the hope that public appreciation would lead to public support for her museum.

In the fall of 1907 Alexander had contracted with John Rowley at Stanford University to begin work on three large habitat groupings of California mammals, one of the mountain sheep that had been collected during her 1906 Alaska expedition, one of California sea lions, and one of Steller sea lions. Although Rowley had perfected the techniques of his day and was considered the finest preparator on the West Coast at the time, he was not an innovator or a naturalist. His groupings lacked the feeling and subtleties of integration with their surroundings that the famous taxidermist Carl Akeley imparted to his work. Akeley had been hired as the chief taxidermist for the Field Museum in Chicago when it opened its doors in 1896. Practically unknown at the time, his talent as an artist, coupled with his abilities as a naturalist, inventor, sculptor, and craftsman, quickly attracted worldwide attention and elevated taxidermy to a level of fine art. Alexander recognized Akeley's skill and the superior quality of his work, but when she made discreet inquiries as to his availability in the fall of 1907, the prices that Akeley commanded vastly exceeded what even she was willing to expend on exhibitions.[25]

Although Alexander recognized the weaknesses in Rowley's technical ability, the taxidermist's lack of innovation was only one reason that she began to grow disillusioned with supporting exhibition work almost before she had begun. Rowley was now also employed by the Academy of Sciences in San Francisco and Alexander feared that a duplication of groupings might engender bad feeling.[26] It was not in her nature to compete for the sake of competition. From its inception in 1908, the museum restricted its investigations to terrestrial vertebrates (mammals, birds, reptiles) because Grinnell's major professor at Stanford, the evolutionary biologist David Starr Jordan, had already amassed a large fish collection and had developed a strong program in ichthyology there. By the same token, if the academy's plans to rebuild focused on the creation of public exhibitions, Alexander may have felt no need to do likewise.

Alexander's abandonment of an exhibition program at the MVZ made practical sense in other respects as well. She now realized that the thousands of dollars that she was expending to create dioramas could instead support additional field expeditions or purchase private specimen collections, functions that would enhance the MVZ's primary mission of research. And there was the issue of space. Because the operating funds she provided allowed the museum to keep at least one party in the field almost year-round, it would soon need to use the square footage allocated to exhibits as storage space for its rapidly expanding collections. Thus in 1911 she wrote to Grinnell, "I confess to having lost interest in exhibition work. The sheep and the two sea lion groups at the Museum show what can be accomplished in that line. The invitation is open to anyone to step forward and shoulder that end of the work. As the scope of the Museum widens it will become necessary to encourage outside support while I pledge myself to the support of the main issues of the work."[27] Grinnell himself raised no objections.

The three original groupings that Alexander commissioned from Rowley remained in the MVZ until 1930 when the museum moved into the west wing of the newly constructed Life Sciences Building on the Berkeley campus.[28] By that time the museum had established its reputation as a research institution and no square footage in its new home was allocated for exhibitions. Between 1930 and 1989 when the museum occupied the address 2593 Life Sciences Building, the opaque glass of its unadorned entrance bore the words that became known to generations of biologists from around the world: MUSEUM OF VERTEBRATE ZOOLOGY, NO PUBLIC EXHIBITS.

9

An Unusual Collaboration

From the outset, an unusual synergy characterized the relationship between Alexander and Grinnell. Their shared goals and complementary strengths allowed each to concede areas of expertise each to the other.[1] In some arenas the two collaborated as equals, for example, fieldwork. In other spheres, growth and development of the museum's program was left to the individual better qualified to carry out that portion of their plan: fiscal management of the museum versus research. In this sense, the relationship between Alexander and Grinnell closely paralleled that between Samuel Alexander and Henry Baldwin.

Alexander's desire to create a museum of vertebrate zoology dedicated to the pursuit of "pure science" coincided with Grinnell's ambition to establish a systematic research center on the West Coast of equal importance with the major natural history museums of the day: the Smithsonian, the Philadelphia Academy of Natural Sciences, the Rothchild Collection at Tring, and the British Museum of Natural History in London. Grinnell's rejection of an offer to write a popular book on California birds in 1910 typified his thinking on the matter. Harking back to the conflict between science and society to which Alexander alluded, he explained, "The completed work would be more spectacular than scientific, more popular and amusing than generally useful and used. . . . What time and talent I have would better be spent in more serious research, especially in view of the wealth of opportunities at hand. . . . I got a pretty definite notion the other evening that you have about the same ideas as above."[2] Research would be the MVZ's primary mission and creditable scientific publications the primary objective of its staff.

From the museum's inception, Alexander's management of its finances and her deft handling of the political issues attending its peculiar position within the university were crucial to the museum's success. Perhaps she learned about business and the stock market through lively discussions with her father (an unusual pastime for a young woman at the time). Or, being an astute observer, perhaps she merely watched and listened as he transacted business. What is clear is that Alexander not only knew and understood business but was interested in it as well. She followed the market closely and directed her investments personally. At times she traded on margin. She maintained a broker in San Francisco and, on occasion, discussed investment options with her brother, Wallace. In 1899, following their return from Crater Lake, she wrote to Beckwith, "H.C. [Hawaiian Commercial] stock advances and retreats—unstable as water. It has been over 100 but is back in the nineties somewhere again. If it should rise to what you paid for it sell out and buy again when it drops—good advice."[3]

Thirty years later she continued to ponder the market with more than casual interest, noting,

> I have to confess that a great deal of my spare time is spent poring over *Barron's Financial Weekly*, instead of *reading* uplifting literature. There is something that captivates one's imagination in the big propositions that are being undertaken these days and in the struggle for mastery, as in the case of Standard Oil of Indiana, Stewart against Rockefeller—an era of expansion—everyone is caught in the whirl of it.[4]

However, when consulted by the regents as to the administration of her endowment funds a decade after they had been given to the university, her response was to inform those gentlemen that she no longer took responsibility for management of the moneys—she merely expected the university to earn at least the same percentage on them that she, personally, would be able to achieve.[5]

But Alexander was unable to remain on the sidelines completely. In 1947 she wrote to MVZ Director Alden Miller, "I must say here confidentially, that I hope the Regents do not contemplate disposing of any of the H.C. & S. [Hawaiian Commercial & Sugar] Company stock of my original endowment of M.V.Z. in the near future. It has always irked them that the price at which the stock was turned over to them at that time was at an unusually high figure. A merger is in prospect which will materially strengthen the company."[6]

The extent of Alexander's benefaction to the university in any given year was tied to the market, sugar in particular, and, like her father, she followed closely any tariff issues that were pending in Congress. The proposal of new sugar tariffs, or the renewal of old ones, often caused her to postpone additional outlays for research and fieldwork that she might otherwise have made. Although the monthly stipend that she provided to the museum was intended to cover field expeditions, as a general rule she supplemented this amount annually. In 1909, for example, over and above her regular appropriation she contributed eight months' salary and field expenses for Frank Stephens in southern California, six months' salary and expenses for Harry Swarth and Allen Hasselborg (who were working in Alaska), two and a half months' salary for MVZ staff member Walter Taylor, and salary and expenses for the exhibit preparator, John Rowley. However, in years when the market fell, or when sugar did poorly, she was forced to forego such supplementation. In 1920 she wrote to Grinnell, "With regard to Oregon—I should like nothing better than to finance an expedition into that country for the Museum but am sailing pretty close to the wind this year and think it is wise not to attempt too much."[7]

Alexander's ability to sense what was necessary for the success of any venture she undertook, and act on that sense, informed her basic approach to the university's bureaucracy. Grinnell deferred the handling of all high-level administrative negotiations to Alexander, declaring from the outset that he was "more interested in *natural history,* than in politics." He fully credited her with exacting from the regents their unqualified commitment to her original proposal and he voiced his admiration for her accomplishments in that endeavor, writing, "I am very much elated over the way things are working out; and the success of it all so far is due to your persistent and cautious engineering. . . . Let me express again my appreciation of your persistence and good judgment in bringing thru the arrangements with the University so successfully."[8]

In view of his eventual stature as one of the leading evolutionary biologists of his day, the degree to which Grinnell deferred to this petite, quiet woman for more than three decades is extraordinary. When acknowledging his preference for locating the new museum at Stanford, Grinnell closed his letter to Alexander by writing, "I shall always be very frank to you in expressing my ideas, as I know that is what you want. But of course I am amenable to your personal wishes in every respect, *always.*" In closing a letter to her on the eve of his assuming the directorship of the museum, he wrote, "I have tried to be careful and assume no authority or responsibility beyond your instructions."[9]

Grinnell's willingness to take direction from Alexander and be accountable to her in almost all respects is an astonishing and key factor in the success of their venture. Before his official appointment as director he wrote, "You can trust me, Miss Alexander, to maintain the initial policy of the Museum consistently in spite of any possible 'suggestions' to me from the University people. At least, I will do nothing without your entire knowledge and approval, and I shall tell you everything of the sort I may experience." On another occasion he wrote, "I know that you are not given to snap judgment or impulsive expression one way or the other. So it gives me an exquisite feeling of elation at having in a degree fulfilled your expectations. My earnest hope is that I conduct myself in the closest accord with your plans and wishes."[10] Such deferential expressions were counter to the manner in which Grinnell treated all other women (with the possible exception of his wife) and many of his colleagues. Yet these excerpts are not isolated examples. Similar expressions of admiration and deference can be found throughout his thirty-year correspondence with Alexander.

From the museum's inception, Grinnell accepted Alexander as an active partner in the work of building the institution. Her skill and competence as a field biologist did more than greatly enlarge its collections; they enhanced the museum's reputation and highlighted its presence. The success of the 1907 and 1908 Alexander expeditions to Alaska led C. Hart Merriam to report in *Science* magazine that "Miss Alexander has amassed the largest and most important collection [of bears] in existence after that of the United States Biological Survey."[11]

Grinnell's acceptance of Alexander as an equal partner in fieldwork extended to his assumption that she would wish to publish reports on the specimens that she collected. When noting that no provision for private offices had been made in the building plans he was reviewing, he inquired whether she wished to have a separate study for herself. His inquiry does not seem to have been merely solicitous but prompted instead by the assumption, natural for an academic, that Alexander would be pursuing research and that, because of her standing as the museum's patron, she would prefer to have a private office in which to work. She did not.

It would be several years before Grinnell accepted Alexander's reticence to publish as a sincere expression of her wishes. After her return from Alaska in 1907, he suggested that she become acquainted with bear systematics and prepare a monograph utilizing the specimens her party had collected. Neither individual believed that C. Hart Merriam would complete his work on the subject any time soon. She immediately deflected his suggestion. Although the following spring he would allude to the topic once more, little if any mention was ever made of it again.

The intellectual insecurities that Alexander had harbored since youth and disclosed in her earliest letters to Beckwith persisted throughout her life, although no basis for them can ever be firmly established. Fourteen years after Grinnell proposed that she execute a manuscript on bear systematics, she responded to a letter from him on an unrelated topic in a manner that focused on the larger issue of her self-perception:

> The only trouble, if there is any, is with myself.... I haven't mental vigor enough to pursue any study alone and would have to put myself under someone's guidance and get away from all present connections until I had gotten a footing in something—and it would not be science. I am writing this so that you may understand me better.[12]

Even before he had officially assumed the title of director, Grinnell publicized the establishment of the MVZ within the scientific community by placing an announcement in the ornithological journal, *The Condor*, for which he was the editor. Shortly after its appearance, Alexander wrote to him,

> O, about that little notice of our work in the "Condor." It was nice except for one mistake I regret very much—I have never attended any classes at the University except as a *visitor*. I don't like to appear as sailing under false colors. It was natural for you to suppose from my interest that I had more of a connection with the University—and if it ever comes in opportunely to contradict the statement without seeming to make a point of it, I wish you would do it.[13]

Alexander perceived herself as an amateur naturalist and, given her honest and forthright nature, it would have been inconceivable for her to represent herself in any other manner. Accordingly, she consistently made a distinction between herself and those whom she characterized as trained scientists. This may have been one reason why she never enrolled as a student at the university. From the perspective of an amateur, enrollment may have seemed presumptuous or even inappropriate.

Why Alexander would not have made science her chosen profession is more difficult to fathom. While she felt passionate about science, she clearly seemed to favor physical activities over those that were cerebral. Being intellectually insecure, she may simply not have wished to subject herself to the necessary examinations required of graduate students or the criticism of colleagues. She may also have recognized that the constraints imposed upon women in science at the time would have been intolerable for her to bear.[14] Her patronage and amateur status gave her access to a world she loved while allowing her to control the nature and extent of her participation in it.

Alexander's insecurities notwithstanding, she was far more knowledgeable about science and about her chosen field of patronage than many, if not most, benefactors. Certainly, she did not lack understanding about the important biological questions that were being addressed by the museum staff. She read broadly, including scientific journals such as *Science*. She was a life member and fellow of the American Museum of Natural History (AMNH), a life associate of the American Ornithologists' Union, a member of the Cooper Ornithological Society, the Society for Vertebrate Paleontology, and both a charter member and the first female life member of the American Society of Mammalogists.[15] Grinnell sent her copies of all papers published by the museum staff and students. She read and acknowledged each in the copious correspondence they maintained. Grinnell's respect for Alexander's intellect may be glimpsed in the following excerpt from a personal memorandum that he recorded after a meeting with her at the museum. It is representative of the many such memoranda that he wrote over the years:

> Throughout her visit, Miss Alexander showed lively interest in the Museum's current activities; also a keen appreciation of those features in our results which have to bear on evolutionary problems. . . . She also expressed concern in regard to the great amount of Museum work involved in working up the various collections and seemed to fully appreciate the great amount of time that would likely be consumed, to get worthy results.[16]

Alexander's name appears as author on only two papers. Both are short notes and relate to the controversial issues of hunting and pest control in this country. One is merely the excerpt of a letter from her nephew to his parents concerning the summer migration of passenger pigeons on the East Coast. This was submitted for publication because, as stated in the introduction, Alexander felt that it might add to present knowledge about the biology of that species. The second article, advocating control rather than extermination of ground squirrels, probably was written by Grinnell despite bearing Alexander's name.[17] For reasons not specified, as a scientist and journal editor he felt that the article would have greater impact if her name, rather than his, were associated with it.

Whereas Grinnell deferred completely to Alexander's financial and management expertise, Alexander's contention that "[t]he Director is the *life* of an institution" exemplified her understanding that she was dependent upon Grinnell to make something of the collections scientifically, to give the MVZ intellectual prominence on par with the eastern museums.[18] His mental acuity, his skill as a journal editor, and his ability to achieve multiple objectives from a single initiative, coupled with his organizational and leadership capabilities, contributed equally but in different realms to the museum's development and success. In this sense their relationship somewhat resembled a traditional marriage in which one partner (Alexander) supplied the livelihood while the other (Grinnell) ruled the domicile.

Alexander always maintained the expectation that Grinnell would devote all his time and energies fully to the museum enterprise, that he would remain an active researcher while carrying out his administrative duties, and that he would continue to publish on museum specimens. In turn, Grinnell trusted that Alexander would maintain continuous financial support of the museum's staff and its collections, as well as the research program that he was building. He also expected that she would be fair and reasonable in the degree to which she intended to make him subject to her counsel and to her approval on all matters concerning the welfare of the museum.

But this unusual collaboration between patron and benefactor was not achieved without sacrifice. The price each paid was a loss of autonomy. In all Alexander's other ventures she was, in essence, directly responsible for the success or failure of their outcomes, for example, her investments, her many field expeditions, and the farm that she would establish in the Sacramento Delta. This was not true of her museums, either the MVZ or the Museum of Paleontology that she established later.

Similarly, Grinnell was required to consult with Alexander on "any matter of the moment" that arose at the museum throughout the duration of his tenure as director. All expenditures, selection of staff, plans for fieldwork, or external activities with which he wished to become involved—teaching, journal editing, or advocacy for the National Park system, for example— were subject to her prior approval until the day he died.[19] Despite the long-term research gains that accrued to him as a result of the MVZ's founding, it is difficult to imagine how this giant in his field, a man revered by his colleagues and students, comfortably acquiesced to such control for the duration of his professional life.

Throughout the thirty-two years that Grinnell served as the MVZ's director, on only one occasion did he and Alexander have a serious misunderstanding, serious enough for Grinnell to draft a letter of resignation. Two days later he drafted a second letter, at the top of which he scrawled the word "Sent," underlined twice.[20] Its tone was conciliatory, fully expressive of his desire to rectify the wrong and focus on the museum's primary mission. The conflict arose from a misunderstanding about the source of funding for salary increases for museum staff. Alexander had stipulated that the university should cover such raises as of December 21, 1921. However, then-UC President Barrows, Comptroller Sproul, and Grinnell were under the impression that the university became responsible for such increases only as of July 1, 1923. By the time Alexander confronted Grinnell with the resulting accounting discrepancies in the fall of 1923, she was convinced that both he and the university administration had conspired to disregard her stipulations.

Alexander's close attention to detail, so clearly evident in this instance, is a reminder that she remained acutely aware and in control of all aspects of the museum's program, that is, its budget, the hiring of museum personnel, and the lines of research pursued by its staff. Those who were not privy to the unique relationship maintained by Alexander and Grinnell might have viewed her usually tacit approval of each of the proposals that he put forth as simply perfunctory, the gratuitous but unthinking concession demanded by a generous patron of a grateful recipient. Nothing could have been further from the truth. His deference to her in all matters relating to the conduct of museum business and her usual approval of his proposals were a well-choreographed point and counterpoint between persons of like mind. Although Grinnell had voiced personal feelings to Alexander on previous occasions, upon the resolution of this incident his words and feelings seemed to take on greater urgency as he reiterated his commitment to their enterprise:

> What I value, and what I *want*, is your confidence in my general integrity; for I have a conscience. . . . If there be any suspicion in your mind that I am not a real friend to you in every sense of the word—that I am merely predacious or parasitic—put me to any sort of appropriate test. . . . This is *your* enterprise. You have sacrificed a very great deal to bring it to its present estate. The Museum's assets in the way of scientific materials are enormously valuable. We have only begun to tap their potentialities. As regards my sincerity, again, there is nothing I would have done otherwise with my own energies and substance than you have with yours. Indeed, I have poured what of these I have myself into it. I have faith in the worthiness of your undertaking, have had for approaching 16 years now. . . . Personally, I want you to know that I am not unmindful of my debt to you for the opportunities I have had here to work in exactly the field I have always preferred to any other. In other words your getting me here meant the realization of my dreams. I am keenly grateful.[21]

The obvious respect that Grinnell and Alexander maintained for each other and their commitment to a shared vision were key to overcoming the difficulties that inevitably arose as the museum developed and its reputation grew. These elements are visible in Alexander's arrangements for another season of fieldwork in Alaska during the spring of 1908. She outlined her plan in a letter to Grinnell, soliciting his opinion on her choice of the islands in Prince William Sound as a collecting locale and on their potential scientific merit. For example, she felt that any bears collected from that region would be extremely valuable, but what of the birds and small mammals? Did he think that California would be a more productive environment in which to concentrate her efforts? By the same token she cautioned that he might not be able to influence her decision. She rather liked the idea of making Alaska her area of specialty and, never having collected mammals or birds in California, did not feel as comfortable with the idea or as drawn to the region.[22]

Far from trying to persuade her to give up Alaska, Grinnell gave the expedition his hearty endorsement and asked Alexander to collect as many birds as possible. The Prince William Sound area remained a largely unexplored wilderness. Little had been published on its flora or fauna and nothing, to his knowledge, was known about the biology of the archipelago. He envisioned that the museum might become an "authority" on the vertebrates of Alaska through Alexander's efforts.

Pleased with his reply but unaware of his disdain for women as field biologists, Alexander went on to solicit Grinnell's recommendation of a female companion who might accompany her. However, in the interval between her request and his unexpected reply, she met a woman who would not only accompany her on this trip but also become her partner for life.

10
Louise and Prince William Sound

Alexander's discovery of a companion who could share both the physical labor and the joy of fieldwork assured her success as a field biologist. A collecting companion with physical stamina *and* aesthetic sensibilities was Alexander's ideal; the companions thus far had satisfied her need for propriety but only Edna Wemple had come close to filling that role. As planning for the 1908 expedition to Alaska proceeded, it was by sheer coincidence that Annie met Louise Kellogg.

Alexander's choice of Kellogg to accompany her to Alaska seems as spontaneous and fortuitous as her decision had been to appoint Grinnell as museum director. And it differed markedly from the objective and deliberate manner in which she and Grinnell had set about hiring staff for the new museum just a few months earlier. Recalling her conversation with William Bryan of the Bishop Museum, Alexander and Grinnell agreed to hire young men, "those with their accomplishments ahead of them."[1] Grinnell was familiar with a number of suitable candidates from his years of teaching and conducting fieldwork in southern California and he recommended that the museum hire Charles Richardson and Walter Taylor, biology students in their early twenties. He also suggested that successful offers might be made to Edmund Heller and Harry Swarth. Both men were on the staff of the Field Museum of Natural History but neither was happy in Chicago.

As Grinnell expected, both Heller and Swarth were pleased to have the opportunity to return to California and to work for the new museum. Grinnell believed that field parties should always consist of two individuals. Not only was this more cost-effective, but he also felt that an area could be worked more thoroughly and with greater efficiency that way. Moreover, with a staff of five, he would be able to maintain one party in the field year-round, or two parties simultaneously, and still keep one or two individuals at the museum to work and write up reports.

Alexander made clear that she would cover all expenses for members of her own field party—over and above the museum's monthly stipend. Grinnell was delighted. He himself had hoped to spend the summer collecting in the San Jacinto Mountains, and her additional support for fieldwork in Alaska would enable him to keep Taylor and Richardson in the field during the summer, as well as himself.

Grinnell felt that Dixon would be eager to accompany Alexander again and she stated her willingness to have him. As an additional collector, Grinnell suggested Swarth. But Dixon quickly made clear that he did not like Swarth and believed that his presence would disrupt the expedition's "perfect confidence and good feeling."[2] Alexander agreed; harmony among individuals in the group was paramount. She was willing to forego Swarth, whom she did not know, and to travel with Dixon alone but, alternatively, she suggested taking Heller if Grinnell felt that he could spare the biologist for several months. Heller had been a classmate of Grinnell's at Stanford and had been hired as the museum's chief mammalogist. He was presently writing up descriptions of the mammals collected in southeastern Alaska the previous summer and, in Grinnell's mind, the matter of his going hinged largely on his ability to complete the manuscript before Alexander planned to set out.

Frank Stephens was not anxious to return to Alaska for a second adventure of that sort. And Alexander did not wish to engage anyone who did not feel up to the challenge. Instead, she offered to pay him on a contractual basis to trap mammals in southern California during the summer. She discussed with Grinnell what a reasonable cost per specimen might be and again agreed to provide the necessary funds apart from the museum's regular budget.

Without Stephens, however, there would be no Mrs. Stephens and, without Kate, Alexander again faced the problem of finding a female companion. A trip alone to Alaska with Dixon and, possibly, Heller was too unconventional. Not that heading an expedition to hunt bears and to trap birds and mammals in the wilds of the northern territory was conventional for a woman in 1908. But only the issue of traveling alone in the company of men seemed to worry her. At one point she reasoned that Hasselborg would join them in the field, but perhaps even a party of that size would cause gossip that might disrupt the effectiveness of their primary mission. And what if she were forced to travel alone to Alaska with Dixon?

While weighing her options and carefully evaluating each of their consequences, Alexander wrote to Grinnell inquiring whether or not he might be able to recommend a woman for the expedition. The volley of letters in which they discussed this topic is delightful (as the excerpts here attest). Grinnell's obvious respect for, and deference to, Alexander's ability as a field biologist clearly did not apply to women in general. Throughout his tenure as the MVZ's director, women were not permitted to participate in museum-sponsored field trips or expeditions. They were not even accepted into the museum's graduate program until the 1930s. Perhaps more surprising, Alexander never challenged these policies. But in the spring of 1908 Alexander's relationship with Grinnell was still evolving, and their interchange on the topic of women as naturalists and competent field biologists must have proved enlightening for both. It proceeded as follows: in mid-April Alexander wrote to Grinnell:

> I've decided to take a lady with me on the Alaska trip if I can find the right one. I don't want my collectors to suffer any unpleasantness from talk on my account. Do you know anyone among the members of the Cooper [Ornithological] Club or elsewhere who would meet the requirements? It would be so much better to have one who was interested in the work as well as being a good camper, for the trip is not in the nature of an "outing" at all, this would have to be clearly understood. I would pay all the expenses of anyone who went with me. Please help me if you can! Mrs. Bailey might know of someone, if you can't think of anyone.[3]

In less than a week, Alexander had her reply:

> Now as to your decision to take a lady with you—I am *sorry!* I have been thinking over every one of my acquaintances, with the result that I can think of no one qualified to go with you. You know, as well as I, that even the best intentions fail, when a little hardship comes along, and then—homesickness, unhappiness, discontent—unless the person in question has a definite purpose in view, as *you* have. I will not take the responsibility of suggesting anyone, though I will keep the matter in mind, and report, if the unexpected happens. I can think of no one now whom *I* would risk taking on such a trip. Of Cooper Club ladies, all are "parlor naturalists"! Or, at most, "opera-glass observers" who get as far as the Yosemite by stage occasionally! It would be better for *you* to ask Mrs. Bailey whether or not she knows anyone than for me to. I am sorry not to be of use to you in this matter; but I feel very shy of incurring such a responsibility.[4]

Grinnell's response elicited a sharp reply. Alexander fired back almost immediately:

Am rather relieved you could not recommend a lady for our trip though regret your evident contempt of women as naturalists. If I elect to go to such a place as Prince William Sound it is up to me to make the expedition as far from being a failure as possible, so looking at it from a business standpoint I am perfectly willing to be the only lady in the party. We shall let the discussion end here shall we not!⁵

Yet just a week later Alexander wrote again to Grinnell, "I have a little friend going with me after all who wants to learn how to put up specimens and to take notes. She will be with us two months. She is a dandy girl but I never dreamed her folks would let her go when I half jokingly asked her to join us—so that's one on me!"⁶

From that point on the matter was dropped, or almost so. In a reply that seemed to plumb the true depth of Grinnell's feeling on the issue of women as serious and capable field biologists, he could not refrain from remarking, "I am glad to hear about your arrangements for the '1908 Alexander Expedition,' and that things are shaping themselves so satisfactorily. . . . I do hope your discovery [a companion for the trip] proves tractable and industrious. One good test might be to have her string tags [specimen labels] for five hours straight!"⁷

Alexander's trip to Alaska with Louise Kellogg in 1908 marked the beginning of a forty-two-year relationship between the women that ended only with Alexander's death in 1950. With a thirteen-year age difference between them (Alexander was the older), it is unlikely that the women were well acquainted before Kellogg's decision to accompany Alexander to Alaska. But on April 25, Louise paid a call to the Alexander home, noting the visit in her line-a-day diary. Two days later she penned, "Decided to go to Alaska with Annie A." On May 15 her bags were taken to the Alexander residence and from there transported to the ship that would carry the expedition party to Seattle. Three days later, as their ship hoisted anchor, the women stood on deck in the cold and rain and waved to the relatives who had come to see them off.

Louise Kellogg was born on August 27, 1879, the youngest of five children, four girls and a boy. Her parents, Charles W. Kellogg and Anita Flint, owned a home in Oakland not far from the Alexander's. The house had been built by her grandfather, James P. Flint, who had come west during the gold rush and had established a line of clipper ships that carried goods between Boston and the Bay Area. Louise's father was an officer of Tubbs Cordage in San

Francisco and founder of the Cordelia Shooting Club, the first duck-hunting club in California.[8] It is from him that she learned how to hunt and fish. She excelled at both but fishing became her passion.

Kellogg graduated from the University of California in 1901 with a degree in classics. Her cousin Martin Kellogg was a faculty member in the classics department and may have encouraged her attendance. Her enrollment was also an indication of her parents' affluence and their social standing within the community.

Soon after graduation, Louise began teaching school. Single, and not engaged to be married, she occupied her leisure time with church activities, dance lessons, and musical instruction. She was adept at playing the violin and the ukulele. She also enjoyed working with her hands. That skill translated well in the coming age of the automobile when she would be the one to fix the crankcase, change a tire, or replace a broken fan belt in the field. And when Alexander purchased a farm in the Sacramento Delta, Kellogg fixed the machinery, painted the interior of their farmhouse, and planted most of the crops.

It is not clear when Kellogg's interests in hunting and fishing turned more broadly and seriously to the disciplines of biology and paleontology. Perhaps it was not until she met Alexander. Yet a distant relative of her father's, Dr. Albert Kellogg, was a botanist and founding member of the Academy of Sciences in San Francisco while her mother's brother, Edward Flint, is reported to have helped found the Oakland Public Museum through contributions toward the purchase of a large ornithological collection.[9]

The expedition to Alaska in the summer of 1908 was Louise's first exposure to fieldwork. Why Charles and Anita Kellogg assented to their daughter's participation in the venture is unclear. They were progressive enough to have allowed their youngest to earn a bachelor's degree at a time when relatively few women attended college. As she approached her thirtieth birthday, they may simply have allowed her to make her own decisions. The Alexanders were reputable members of the community. If Annie Alexander was leading a scientific expedition to Alaska in which their daughter was invited to participate, they may have reasoned that adverse publicity would be unlikely. Alexander herself was not eccentric; only her life was.

After the 1908 expedition, Kellogg's presence in Alexander's life and her contributions to the museum's collections were never questioned by family, by Grinnell, or by any of the museum's staff. For many years the fact remained that it was far more appropriate for a woman to be accompanied by a companion than for her to travel alone. Other dimensions of their relationship were simply never discussed. For his part, Grinnell based his impressions of Kellogg on her performance in the field and her contributions to the museum's collections. To him, little else mattered.

Alexander's original expedition itinerary called for her party to leave Seattle by steamer in late May, pick up Hasselborg in Yakutat, a town directly north from where the panhandle of Alaska meets the mainland, and then continue on and have the steamer drop them at Nuchek, a small village on the western coast of Hinchinbrook Island in Prince William Sound. Here, after sufficient time spent collecting, Alexander planned to rent a boat and relocate the party to Montague Island, the largest and southernmost island in the sound and the one most promising for bears. They would then move to Knight Island, Chenega Island, Green Island, Latouche Island, and so on until they had thoroughly sampled the entire region. They would complete their survey of the area by collecting on the Kenai Peninsula that jutted south from the mainland, with plans to return to California in mid- to late September.[10]

Alexander had intended to rent a twenty-ton sailboat in Alaska to carry the group between the islands in Prince William Sound, but by April she was having doubts about the availability of a suitable craft. Alternatively, she considered bringing the lumber necessary to build one. In the end, she opted to have a small, 30-foot, 10-ton sailing skiff built in San Francisco and bring it north with her. With the exception of the boat, Alexander intended that expedition expenses would be slight since she planned to "travel light and do away with everything but the most necessary comforts."[11]

Heller had proved agreeable to Dixon as an addition to the party and he arrived in Oakland from Pasadena the evening of May 17. The following morning their ship hoisted anchor in the bay and quickly passed beneath the Golden Gate Bridge. Out on the open water the sea was rough. Alexander was not prone to seasickness, perhaps as a result of having grown up on the islands, but Kellogg was. On their first day out, she ate lunch but no dinner.

As their ship sailed along the Inside Passage, the weather improved. The group took advantage of the opportunity to sit on deck, enjoying the scenery. While talking about the possibility of collecting bears, they strung specimen tags. Alexander impressed Heller during this relatively quiet period as being "well contented and jolly and terrifically interested in zoology." He remarked further to Grinnell that "Miss Kellogg [. . .], is apparently the right sort, without any yellow streak or quitting propensities."[12]

At the northern end of the passage, their ship dropped anchor at Yakutat. It was here that Alexander had hoped to meet up with Hasselborg. Instead, she learned that he had headed inland along the Anklin River on reports of an immense brown bear in the region. He had recently killed a large female that had been defending her cubs.[13] Alexander left word for the hunter to meet up with them on Hinchinbrook Island—and be sure to bring his canoe.

In 1908 Cordova was a small railroad town on the eastern side of Cordova Bay, along the eastern edge of Prince William Sound. Here Alexander's party spent four days rigging their skiff and stowing their gear before sailing to the head of the bay, approximately ten miles north of town. Camp, which consisted of two sleeping tents, a cook tent and a work tent, was established at the lower end of the valley floor, a vast expanse two miles wide by eight miles long and surrounded on three sides by sharp, jagged peaks. Occasional glaciers between the higher peaks gave rise to numerous streams that pitched down over the scarred slopes to form a river along the valley floor, which they traversed by boat. At low tide, sand flats extended for three-quarters of a mile on either side of the river. At night, arctic gusts would swoop off the mountains, making bear hides an essential component of their bedding.

The river valley around them was a mixture of willow, alder, and spruce thickets interspersed with bogs and swamps. A large salt meadow lay intersected by crooked sloughs. Land birds were not plentiful, except along the outer edges of the swamps, but the sloughs were a favorite haunt of geese, mergansers, and other ducks. Within a week Alexander had secured more bird specimens than she had taken during the entire expedition the previous year. The party also secured four porcupines and three hoary marmots, in addition to several of the smaller species of mammals.

From Cordova Bay the group spent a week collecting on Hawkins Island while awaiting Hasselborg's arrival. A week seemed to be about the right amount of time for each of their camps. Mornings were spent in the field checking trap lines or hunting birds. Afternoons were spent preparing the specimens as study skins. At the end of each day they would reset their traps and hunt for crepuscular and nocturnal species. Field notes were written up in the evenings and the following morning the whole process would begin again.

For the most part, the islands in Prince William Sound proved relatively poor in wildlife, being little more than mountain peaks shooting up from the ocean floor. With timberline often at 1,000 feet, the collectors found little habitat or species diversity among the islands. Some supported ground-dwelling ptarmigan and marmots, but others lacked the fairly ubiquitous tundra vole, a chubby, mouse-sized rodent with short ears, a short tail, and a blunt snout that advertised its presence by creating runways through the grass or other vegetation. Fox and song sparrows were common, as were hermit and varied thrushes, jays, crows, ravens, chickadees, and some species of warbler. Under these conditions, Alexander adopted Heller's habit of "squeaking" in order to draw in birds she heard but was at first unable to see. The sound often brought in not only the birds she heard but others seeking to investigate the attraction as well.

From Hawkins Island the group crossed the narrow strait to Hinchinbrook Island and dropped anchor in a bay on its northeastern shore. Hinchinbrook is much larger and more mountainous than Hawkins and its dense spruce forests teemed with bird life. Camp was made on a sandy spit at the mouth of a large stream and from there the group made several excursions about the bay and one to the top of a nearby peak. The trek to such high country was generally arduous and involved crossing vast melting snowfields that made walking in waterlogged boots treacherous. But Alexander loved to escape into this open country full of wildflowers that sprang up as soon as the snow had left the ground.

In Dixon's letters to Grinnell during this period he sounds like a homesick child writing home from summer camp. The gnats and mosquitoes he had experienced the previous year at Glacier Bay seemed inconsequential compared to the ravenously bloodthirsty insect pests he now had to battle. And the weather in the sound was decidedly miserable. The almost constant rain brought relief from the voracious insects, but the howling gales that swept down from the north meant that the party usually had to haul unwieldy gear off the storm-tossed boat and pitch wet tents in the driving rain. He even resented "carting around" the bird specimens he had prepared with hopes of their drying before they got moldy.

The boat itself was a wellspring of complaints. Often the party was becalmed, faced a serious headwind, or became caught in a gale. Under any of these circumstances the crew was forced to row the craft as the boat would not beat to windward. "This galley slave business is getting kind of ancient," Dixon grumbled to Grinnell. "Rowing a 30 ft. yacht is no Sunday School picnic."[14]

Heller's perspective on the unpredictable weather and treacherous waters was the exact opposite of Dixon's. At about the same time, he wrote to Grinnell that the party "enjoyed a good proportion of bright sunny days, days of extreme beauty and fascination peculiar to Alaska." Heller acknowledged the extraordinarily vicious insect life but did not allow it to prevent him from enjoying the occasionally beautiful days to their fullest.

Heller was equally pleased with what the party had accomplished to date. He remarked that Alexander continued to set the pace for the group. She was usually the first to rise in the morning and had accordingly appropriated the task of preparing breakfast for the group. He commented that her thoughts and attention were focused primarily on birds and that "she has a way of seeing a great deal of beauty in the rare ones regardless of their coloration."[15] He reported that she had not abandoned her interest in bears, though it was an unusual event even to find tracks less than a week old.

Heller noted that the islands within Prince William Sound lacked any significant individuality. Only about fifteen species were represented among the approximately two hundred mammals that had been captured thus far. This he attributed to the shallowness of the water in the sound and the similar topography and weather throughout the area, implying that there had been a great deal of faunal interchange among the islands with little selection pressure that might lead to taxonomic diversification.

It was July 4 when the expedition hoisted anchor and set off for Zaikof Bay on the northern shore of Montague Island. Its dark and mysterious gloom had haunted Alexander ever since her visit in 1906, and the expedition's greatest hopes for bears centered here. The party camped on a sandy spit with the mountains close at hand. Intersected by glaciers, the terrain proved rugged and the results of their collecting efforts poor and uninspiring. Disappointingly, they captured only a few of the same species of warblers, thrushes, and sparrows they had obtained at previous sites, but no bears.

In a driving rain, the party sailed to Latouche Island. *Edna*, as Alexander's 30-foot skiff had been christened, veritably flew the distance. Her "crew," who had spent so much of their time rowing rather than sailing the boat from destination to destination, rejoiced. Hasselborg, however, continued sewing a large "jigger sail," just in case.

At the first sign of good weather the party returned to Montague Island. This time they anchored in Hanning Bay near the southern end of the island. The site proved to be their longest and most pleasant camp, situated by a fine stream in a grassy meadow radiant with columbine, blue iris, and great stalks of wild celery. Alexander was both entranced and subdued by this portion of Alaska. Despite gloriously sunny days, the island remained a lonesome place for her. As she wandered by herself listening for birds, she cherished its solitude and beauty. Yet a short time later as she waited on the beach to meet up with her companions, that solitude seemed equally oppressive. Often the sparkling blue of the sea and the crystalline white of the snow would suddenly turn dark and ominous with an approaching storm.

The constancy of these colors in the landscape caused her to think fondly of vast stretches of green meadows that awaited her at home.

At this point her party had been collecting for a two full months and each of the participants had worked hard for the specimens captured, especially hard for the birds, Alexander felt, although they had made a reasonable collection. But bears were now their primary objective and within a week Hasselborg trapped two—a large male 8 feet in length with a 17-inch skull and a 14-inch hind foot and a much smaller female.

Having fulfilled their final quest, the party made its way back to Latouche Island where Kellogg was to catch the steamer for her return trip to California, a journey that would take two weeks. It was now early August and she needed to arrive in time for the opening of school. Heller wrote to Grinnell, "You have thus lost an indefatigable bird collector and we have parted company with one of the dearest girls on earth and an accomplished pie maker."[16]

Annie contemplated Louise's departure in a different light. She missed her companion sorely and confessed to Martha, "the greatest discovery of this trip was Louise. Surely I was fortunate but the more forlorn now to be left alone with the lapping of the waves on the beach and the rain on my tent roof."[17]

From this point on, written expressions that might offer insight into Alexander and Kellogg's personal feelings are sparse. Even in her letters to Beckwith, Alexander's comments about her new companion quickly give way to unenlightening narratives of their activities together. The tremendous volume of correspondence that Alexander and Grinnell conducted leaves the impression that Alexander dictated the women's schedules and assumed financial control of their affairs. The energy, enthusiasm, and physical support that Kellogg brought to the work balanced their relationship. And the partnership was an opportunity for travel and adventure that had inestimable value to both women.

Although Kellogg shared equally in all aspects of fieldwork that the women conducted, responsibility for management of the Museum of Vertebrate Zoology was Alexander's alone. While creation and support of the MVZ now occupied a large portion of her financial resources, it by no means accounted for the sum total of her benefaction to the university. At about the same time that Alexander began contemplating the establishment of the MVZ, she was already making regular donations to the university in support of her first love, research in vertebrate paleontology.

11

Support for Paleontology

At the end of October 1905 the success of Alexander's most recent fossil-hunting expedition to Humboldt County, Nevada, was still fresh in her mind when she wrote to President Wheeler: "In view of the possibilities for the advancement of science through the research work of the Department of Paleontology . . . I have decided to make the Board of Regents . . . the following proposition."[1]

What Alexander proposed was to donate $100 per month to the university for a period of five years, beginning January 1, 1906. The money was to constitute a special research fund "under the direction of the professor in immediate charge of the paleontological work."[2] Alexander's offer also stipulated that the department's annual budget not lessen in any way to offset her gift and that the university raise the salary of its vertebrate preparator.

But the monthly stipend was not the sum total of Alexander's benefaction to paleontology during this period. In April 1906, she wrote to John Merriam enclosing a check for $800. This she directed be spent on summer field research. An equivalent sum today would be roughly $13,800.[3] Yet before the introduction of income taxes in 1913, and charitable deductions much later, Alexander frequently presented similar sorts of gifts to individual researchers, no records of which exist in the university's ledgers. On one occasion when discrepancies were found in an audit of the MVZ's books, the items in questions turned out to be those that Alexander herself had purchased and subsequently donated to the museum.

Beginning that summer, and for several years thereafter, Merriam focused his attention on an asphalt bed that had recently been unearthed within the Los Angeles city limits. What emerged from those excavations were bones representing an extraordinary variety of large mammals that had roamed freely across North America 35,000 to 40,000 years ago, for example, lions whose skulls were greater in length than those of any cat now prowling the African plains, jaws of saber-toothed cats, limb bones of dire wolves and hyenas, as well as complete fossil skeletons of giant ground sloths, bison, mastodons, and camels. Tens of thousands of years ago, each of these animals had stumbled into this pit, only to become trapped in its highly viscous tar. Predators quickly followed hapless prey into the black abyss, ensuring death to all. The tar was known in Spanish as *la brea* and it exquisitely preserved its captives.[4]

Merriam's publications on the fossil mammals excavated at Rancho La Brea began to fire Alexander's imagination. Back in Oakland in the fall of 1908, she began to contemplate a paleontological expedition for the coming spring. After trapping birds and mammals in the cold, wet forests of Alaska for three summers, she was eager to ramble in hot sagebrush and chisel out long-extinct creatures with a pick and hammer. Her intended destination: Virgin Valley, a formation of volcanic ash situated in the northwestern corner of Humboldt County, Nevada.[5]

By late April Alexander was ready to set out, this time accompanied by the paleontologists Eustace Furlong and A. J. Heindl. The expedition was to last three months, with Kellogg joining the party in early June after her teaching obligations had ended for the year. Taking the train from Oakland to the farthest point served by the railroad in that portion of Nevada, the trio rented a wagon and horses to transport the enormous quantity of field equipment that they had brought with them to their final destination. On horseback they traveled to Quinn River Crossing, then west across the Pine Forest Mountains to Virgin Valley. Merriam arrived during the last week of their stay in the valley and remained with the group for several weeks thereafter.

Virgin Valley was an amphitheaterlike enclosure, eight miles long and hemmed in on all sides by steep lava walls. In the northwestern corner of the valley, the Table Mountains provided a splendid view of the region for anyone ambitious enough to scale their heights. A stream that entered the valley through a narrow canyon exited again between perpendicular cliffs in the northeastern corner, an area referred to as the "Gorge." Camp was made along Virgin Creek where a warm spring of water emerged from an adjacent ranch. After hours of chiseling rock in the hot sun, nothing seemed more restful than to bathe in the spring, an activity that quickly became a nightly ritual.

The surrounding countryside was composed more or less of volcanic ash, presumed to be Miocene in age. Some of these ash beds had become exposed when the overlying lava had broken down. In these mushy ashen hills and variously colored broken terraces, the researchers had most success at finding fossil bones. In order to determine whether or not an exposure actually contained bone, they had to trudge up and down this strange landscape, expending a great deal of physical energy as they examined each formation. Often they returned to camp tired and empty-handed. Scurrying around after his arrival, Merriam took immediate delight in the exposures of bone that Alexander and Kellogg had located, subsequently making pronouncements on the plausible factors responsible for the presence of this or that vertebrate assemblage.

The rhythm of the group's days varied little over the course of their six-week stay. The women would occasionally walk or ride several miles from camp in search of new material, their field notes detailing the list of teeth or bone fragments that they amassed and the site number where they had been collected. Once Merriam had arrived and surveyed their findings, he was eager to have the party move on and camp was relocated to the nearby Thousand Creek formation. The fossils found in these rocks were much younger than those in Virgin Valley and Merriam felt that, in all likelihood, the strata had been subjected to a great deal of uplift and disturbance. Bones of the Pliocene rhinoceros, *Titanotherium*, were found beside those of Quaternary horses, ungulates with curved horns like modern-day African kudu, and large unfamiliar rodents. The age and composition of these two formations made finding the large and spectacular fossil specimens that had been characteristic of Alexander's previous expeditions unlikely. Most frequently, the women discovered small teeth or foot bones of extinct species of deer, camels, pigs, dogs, rodents, and a rhinoceros but, at one locale, Kellogg uncovered an almost complete tusk of a mastodon.

Alexander did not intend to neglect her zoological interests during this trip. Not only did she fund a museum expedition to Alaska that summer headed by Swarth, but she also offered to cover all the expenses incurred by Taylor and Richardson if they would spend the summer collecting birds and mammals in the Pine Forest Range east of Virgin Valley. These expenses were over and above her regular monthly appropriation for the MVZ. She herself planned to do some small mammal and bird collecting and she had packed specimen traps and a shotgun among her paleontological field equipment. In addition, she took extensive notes on the animals she observed and directed a small bit of effort toward noosing lizards. The woman who a decade earlier had bemoaned her complete ignorance of bird identifications now confidently recorded all the species she observed and commented on the lack of others she had expected might be there.

As was her custom, Annie encapsulated her impressions of the trip in a letter to Martha. She summed up Merriam's presence by noting, "Really, to hear him talk you would think he was a little god walking in the pathway of light to whom all things are clear." Simultaneously, she noted astutely, "He is too cautious a man and too concerned over his reputation to run the risk of making a mistake."[6] Despite such sarcasm, Alexander felt that she and Merriam were the only two from the university with the expedition's interests truly at heart. Her impression of Furlong on this trip was that it had been a month too long for him now that he had a wife and young daughter waiting at home. Heindl, a Russian from Bohemia who was to act as the group's geologist, had behaved somewhat crazily. Alexander described him as a cross between a romantic and a "wild indian." His goal in life was to discover the North Pole, to which end he was inventing a power boat that would travel under the ice and every night he prayed that no one would reach the Pole before him. As a young boy he had read the novels of James Fenimore Cooper and now, in the American West, he fancied himself a Native American, a fantasy that he acted out by stripping naked to the waist under the blazing desert sun, festooning his cap with a feather and chasing antelope by moonlight. Despite his strangeness, Alexander admitted to finding him more mentally stimulating than any of the other young men on the expedition.

But she saved her most heartfelt comments for Louise, writing to Martha, "My friend Louise never failed me. She needs [to] go with me and see that I quit work when the long shadows began to creep down the hills. . . . Our daily rides were often over miles of country and when Furlong would be congratulating himself in fieldwork thoroughly done, Louise and I would push our search a little further and find truly exciting things."[7] The trip to Virgin Valley was the last one Alexander undertook as part of a university-sponsored expedition. From this point on, she and Kellogg would collect on their own.

Her clear view of others' abilities and foibles did not help Alexander form a positive intellectual image of herself. In matters based on personal experience she had no doubts of her competence. She continued to exude confidence in the field and aggressively supported her interests in paleontological research. But classes at the university were another matter. As 1909 drew to a close, she wrote to Beckwith that she and Kellogg had spent the fall semester auditing a course being given by Merriam. While she found the lectures "inspiring," she added, "I would become a pitiful object if the few friends I have didn't believe I was much account. We need so much encouragement. I declare—it is really pathetic! . . . I stand aghast when I am

Annie Montague Alexander, 1901, at the age of thirty-three. Courtesy Museum of Vertebrate Zoology archives.

A family portrait, date and location unknown. *Left to right:* Alexander's mother, Martha (Pattie); her brother, Wallace; sisters Martha (*seated*) and Juliette; and her father, Samuel. Alexander is standing behind her father. Courtesy Alexander & Baldwin Sugar Museum.

Annie Alexander (*left*) and Martha Beckwith, 1901. Courtesy archives, University and Jepson Herbaria, University of California, Berkeley.

Class photo, Lasell Seminary for Young Women, Auburndale, Massachusetts, 1889. Alexander stands on the far left, head tilted slightly, in front of the woman wearing the large white wide-brimmed hat. Courtesy Winslow Archives, Lasell College, Newton, Massachusetts.

Near the conclusion of their 1904 African safari, Alexander and her father were given an opportunity to shoot colobus monkeys; Annie declined but Samuel did not. Courtesy Bancroft Library, University of California, Berkeley.

Top: The 1904 party's first field camp was set on a broad plain in the vicinity of Nakuru, several hundred miles northwest of Mombasa. Courtesy Bancroft Library, University of California, Berkeley.

Near Kijabe, British East Africa, August 1904, near the end of Annie and Samuel's safari. Alexander is seated between her father (*left*) and Mr. Stauffacher, a missionary's assistant. Courtesy Bancroft Library, University of California, Berkeley.

On the first paleontological expedition Alexander sponsored—a three-month excursion to the Fossil Lake region of southern Oregon in 1901—she borrowed her brother's shotgun and supplemented the party's dry foodstuffs with fresh game. Courtesy University of California Museum of Paleontology, Berkeley.

Alexander on site, 1905 paleontological
expedition to Humboldt County, Nevada.
Courtesy University of California
Museum of Paleontology, Berkeley.

For two days, Alexander watched with fascination as university paleontologists excavated a saurian she had discovered in Humboldt County, Nevada, summer 1905. Courtesy University of California Museum of Paleontology, Berkeley.

Left: Alexander's mentor, John C. Merriam, professor of geology and paleontology at the University of California, Berkeley, in Humboldt County, Nevada, summer 1905. Courtesy University of California Museum of Paleontology, Berkeley.

Louise Kellogg, Alexander's partner and collecting companion for forty-two years; location and date of photo unknown. Courtesy Alice Q. Howard.

At the head of Yakutat Bay where Alexander's first expedition to Alaska camped for several weeks waiting for the ice to break up, summer 1906. Courtesy Museum of Vertebrate Zoology archives.

This mountain sheep, taken at Skilak Lake on the 1906 expedition to Alaska, became the centerpiece of a diorama in the Museum of Vertebrate Zoology. Courtesy Museum of Vertebrate Zoology archives.

Juneau, Alaska, a small town whose shops and stores provided Alexander's 1906, 1907, and 1908 field parties with last-minute supplies and provisions. Courtesy Alaska State Library, Allen Hasselborg papers, ms 2.

Campsite at Skilak Lake, Kenai peninsula, Alaska, September 1, 1906. Courtesy Museum of Vertebrate Zoology archives.

The Alexander party's field camp on Hinchinbrook Island in Prince William Sound, 1908. Courtesy Alaska State Library, Allen Hasselborg papers, ms 2.

On the 1908 Alexander expedition to Alaska (*left to right*), Louise Kellogg, Joseph Dixon, and Edmund Heller; location and date of photo unknown. Courtesy Museum of Vertebrate Zoology archives.

Louise Kellogg in Alaska, 1908.
Courtesy Alice Q. Howard.

brought face to face with what a real student accomplishes—what a trained mind with a large brain can do—and I feel perfectly hopeless."[8]

From 1901 through 1909 Alexander watched as friction and competition between faculty members within the Department of Geology escalated to an almost unbearable level. Historically, a schism within the discipline of paleontology has pitted those favoring its affiliation with geology against those who believe its natural affinity is to zoology and the life sciences. Most frequently, invertebrate paleontologists take the former position, arguing that to interpret fossil history one must understand the composition and depositional events of the rocks in which the organisms are embedded. Conversely, vertebrate paleontologists generally hold that to interpret fossil history one must study living organisms. Merriam's presence and aggressive research agenda gradually exacerbated the ideological division that would lie at the heart of the department's problems for decades to come.

Alexander was acutely aware of this growing divide. It politicized decisions concerning new faculty appointments and courses to be offered in a given semester. From her particular perspective, she feared for the life of a discipline in which she was now deeply invested, both personally and financially. Her fervent desire to see the development of a strong and independent research program in vertebrate paleontology led her, once again, to appeal personally to President Wheeler and the regents for intervention. Their initially tepid responses might have discouraged a less persistent individual, but Alexander's persuasiveness and commitment culminated in the creation of a separate Department of Paleontology in 1910 with John Merriam as chair.[9]

Merriam wrote to Alexander that September, enclosing a copy of his report to the dean of the college on the success of the summer school class that he had taught in the newly independent department. The paleontologist also pointed out that student enrollment in his course that fall was more than double what it had been the previous year, a fact that he obviously attributed to the formation of an independent department.

Pleased with these events, in December 1910 Alexander wrote to Wheeler offering to renew her support for paleontological research for an additional five years, this time doubling her financial commitment to $200 per month. Characteristically, her letter concluded with a postscript reminding the president that she wished the money to be recorded simply as from "a friend to the University."[10] She also reiterated her stipulation that the university's appropriation for the new Department of Paleontology not diminish in any way as a result of her gift. This time, however, she specified that the research fund that she was establishing was to be under Merriam's direct control.

From her earliest benefaction to the university, Wheeler regularly expressed gratitude to Alexander for her contributions to research; his letters indicate his continued awareness of the programmatic growth in paleontology and vertebrate zoology that her support enabled. In acknowledging her renewed pledge, he expressed his own, personal appreciation, writing, "I have seen during the years how steadfast your purposes are and how well-considered your plans."[11]

Because Wheeler had been out of town when Alexander wrote to renew her pledge, Merriam took it upon himself to express immediate appreciation for this latest gift. His two-page letter, written in longhand, was effusive and attributed the program's notable success to her assistance, both financially and in the field. He further credited her personal interest in the work with acting as a helpful and stimulating influence throughout the course of their association. He concluded with the expectation that her generosity would allow the collection to flourish, resulting in publications.

Merriam repeated these themes in subsequent years. In April 1913 he maintained that the campus had one of the largest concentrations of people engaged in paleontological research of any university in the country and he assured her of his complete sincerity when he stated that her assistance had been the most important factor ascribed to this statistic. Little more than a year later he would go one step further, asserting that the campus had the right to claim an equal rank among the best of American universities with respect to its paleontological contributions. He attributed this to her support, writing, "In whatever work is accomplished or advance made in this field I wish you to feel that credit belongs very largely to you."[12]

Such unabashed adulation gained little favor with Alexander and did nothing to mask Merriam's growing compulsion to maintain total control of the funds that she intended for support of departmental research, albeit under his direction. His desire to direct that money entirely to his personal research agenda became evident during a semester sabbatical in 1913. According to university guidelines, the individual serving as acting director in Merriam's stead retained the authority to administer the departmental budget. Merriam, however, made clear his wish to draw on Alexander's special fund (for purposes not specified) even during his absence from campus, claiming that it remained under his direct control. This audacity, coupled with Merriam's increasing involvement in administration, did little to impress Alexander, whose own objectives never wavered from the production of reputable published manuscripts.

Alexander did not yet fully comprehend the impact that Merriam's personal agenda would have on her plans, or on the future of paleontology at

Cal. His research output remained prodigious and his reputation grew apace. Of equal importance, the reputation of the collections and of the work being done at the university continued to grow nationally. Because these were the factors that motivated her benefaction, in December 1915, as her second agreement with the university drew to a close, Alexander wrote to Wheeler offering to renew her pledge for an additional five years.[13] Merriam was to direct the expenditure of her funds, and the yearly appropriation for the Department of Paleontology was not to change in any way as a result of her gift, she insisted. Alexander stressed that the moneys she proffered must be used to support research and that the university must remain responsible for routine academic functions such as salaries and the care and housing of the fossil collections. What Alexander failed to realize was that Merriam's desire to control her research fund, even as his productivity diminished and his personal agenda diverged from her own, would soon lead to a break in their relationship and to extended conflicts between herself and the university administration.

12

Hearst, Sather, Flood

From the issuance of her first formal check in support of paleontological research in 1903, to the scholarships that she created in 1948 for graduate students in the museums of vertebrate zoology and paleontology, Alexander's benefaction to the university was directed exclusively to programmatic concerns rather than to edifices bearing her name. Of the four women who were the university's great benefactors during this half-century—Phoebe Hearst, Jane Sather, Cora Flood, and Annie Alexander—Alexander is the only one whose name does not appear on a single structure on the Berkeley campus.

The size and shape of gifts from Hearst, Sather, or Flood interested Alexander only if they had anything to do with university programs that held her own interest. Demonstrating her practical head for business and the degree to which paleontological research had become the focus of her benefaction in less than a year's time, Alexander wrote to Beckwith in 1901, "Mrs. Hearst has given quite a sum of money towards investigations for ancient human remains . . . and Dr. [John] Merriam is put at the head of it. Now there are some caves in Shasta Co. that deserve attention from this point of view and must be explored but there is nothing to hinder Dr. Merriam slipping up between times to the "grey rocks" to look for the lost paddles of his Shastasaurus."[1]

The University of California was twenty-four years old in 1893 when Martin Kellogg assumed the presidency (1893–99), and its sole campus was located in the quiet community of Berkeley. Surprisingly, the university already stood sixth nationwide in annual income, seventh in enrollment, fifth in the size of its faculty, and was in a four-way tie for eighth place in the percentage of its students engaged in graduate studies.[2] In an attempt to keep pace with this burgeoning program, the physical campus had grown quickly, but without design or direction.

Just three years after Kellogg's appointment, Phoebe Apperson Hearst proposed to sponsor a $200,000 architectural competition to develop a comprehensive building scheme for the university. Widow of the businessman and senator George Hearst, and mother of William Randolph Hearst, publisher and editor of the *San Francisco Examiner*, Hearst felt that California's up-and-coming state university warranted a grand and cohesive physical plan. Implicit in her offer was the donation of additional funds for building construction—upon adoption of an acceptable blueprint for the campus—as well as various antiquities and works of art from her private collection.

Hearst was as good as her word. The next several decades saw construction of the Hearst Memorial Mining Building, Hearst Gymnasium (originally for women only), and an extension of Harmon Gymnasium (for men) on the Berkeley campus. She also contributed funds for the purchase of equipment for the departments of pathology, histology, and psychology; financed research expeditions; and created university scholarships and professorships in her areas of interest, primarily in the field of archaeology.

Hearst's philanthropy during the Kellogg era continued after Benjamin Ide Wheeler assumed the university presidency (1899–1919).[3] And she was not the only benefactor of note during this period. One of the hallmarks of the Wheeler administration was the tremendous growth that the university experienced both intellectually and physically. Increases in student enrollment were reflected not only in the construction of new buildings on the Berkeley campus, but also in the development of a second campus in Los Angeles and in the addition of several research stations throughout the state. Although Hearst's patronage may have persuaded other donors to step forward, Wheeler himself fostered an atmosphere on campus that led to unprecedented levels of benefaction, particularly from women: Jane Sather, Cora Flood, and Elizabeth Boalt, as well as Phoebe Hearst.[4]

Jane Sather used the accumulated wealth of her late husband, Peder Sather—a pioneer in the California banking industry and a founding trustee of the College of California (which later gave its land, buildings, and undeveloped property to the newly chartered state university in Berkeley)—to create a chair in classical literature, establish the Sather Law Library Fund, and provide money for the construction of two of the most famous landmarks on the Berkeley campus, Sather Gate, which graces its southern entrance, and the campanile, known as Sather Tower.

Like Alexander, Cora Jane Flood inherited her wealth from her father. James Flood, one of California's four "silver kings," and his partner, William

O'Brien, had owned a bar in San Francisco during the gold rush days. By eavesdropping on their patrons' conversations, the two staked an early claim in the Comstock Silver Lode, the largest and most valuable pocket of silver ever unearthed. Cora eventually donated her parents' Menlo Park estate, and a controlling interest in the Bear Creek Water Company, to the university as an endowment, creating in the process a foundation for the study of economics. Twenty-six years later, she also donated her San Francisco residence worth, at that time, approximately $250,000.

Elizabeth Boalt, widow of a California attorney, was less affluent than Hearst, Sather, or Flood, but notable for her enduring contribution to the campus. In 1901 she donated money to the university for the establishment of a law school and for the erection of Boalt Hall, the building in which it would be housed.

Because the catastrophic 1906 San Francisco earthquake occurred during the Wheeler presidency, the importance and timing of these gifts were enormous. The Berkeley campus itself suffered little physical damage during the quake but had staggering financial losses. The university owned a great deal of real estate in San Francisco; a major portion of its holdings was consumed by the fire that swept across the city in the aftermath of the quake. In 1907, the year in which Alexander proposed to establish a museum of natural history on the Berkeley campus, loss of income from university-held rental properties in San Francisco exceeded the salvage and insurance moneys that the university was able to collect on those properties.

Given the extreme degree to which the museums that Alexander founded and eventually endowed were dependent upon her monthly subsidy for their existence, it is likely that Alexander derived great pleasure from the assurance that other persons of means also valued academic and intellectual pursuits and were willing to support them financially. In later years, after the MVZ had become well established, Alexander occasionally would entreat Grinnell to solicit additional external sources of funding, seeming to resent the expectation that the responsibility for support of the museum should be hers forever. One example is a letter she wrote to Grinnell in 1911 stating that she had lost interest in supporting exhibition work and exhorting him to find other sources of funding if he wished to continue that venture. She noted, "As the scope of the Museum widens it will become necessary to encourage outside support while I pledge myself to the support of the main issues of the work.... It is much easier for you of course to turn over all the financial responsibility to me but I should like to see you go out of your way in behalf of the growing possibilities of our work."[5]

The "main issues" to which Alexander referred were research and publications. In the Museum of Vertebrate Zoology's formative years, she opposed Grinnell's teaching regular courses in the zoology department, arguing that all his time could be easily spent preparing articles for publication. "I don't believe in getting material unless something is done with it.... [A]ll the material we have collected is really unavailable without Grinnell to interpret it," she wrote to Beckwith, and she enforced such requirements through the conditions accompanying her benefaction.[6]

Alexander's hope that Grinnell would attract additional outside sources of funding for the museum was never realized. She continued to shoulder sole financial responsibility for its research program, apart from the university's limited contribution, throughout her life. Although she would have been aware of the many and varied donations being made to the university by Hearst, Sather, and Flood through articles in the local newspapers, she herself disdained such publicity. Each of her gifts was placed in the university's ledger with a notation indicating that it was not to be made public. This request was usually, but not always, adhered to. All unrestricted contributions to the university were reported regularly in the school's alumni magazine and occasionally her name appeared there as well, in all likelihood owing to oversight by a zealous but improperly trained employee in the comptroller's office.[7]

Given Alexander's upbringing and general reserve, her dislike of publicity is not surprising. One factor undoubtedly shaped her attitude in this regard. In the fall of 1905 the *San Francisco Chronicle* ran a full, front-page exposé in its Sunday edition on the Africa expedition that she and Samuel had undertaken the previous year. Its headline screamed "Big Game Bagged by a California Girl in Africa" and was subtitled "California Girl a-hunting in Africa." A montage of photographs, including one in which she, Samuel, and Stauffacher sat astride their dead elephant headed an article that, by today's standards, would have been worthy of the best of tabloid journalism. Somehow anticipating the worst, Alexander confided to John Merriam, "I had the misfortune to be interviewed by a reporter about the African hunting trip some days ago and I will appear as an Amazon probably in the next Sunday Chronicle. I hope it will not be so dreadful that my friends will all blush for me."[8]

Although that rather lurid article undoubtedly imbued Alexander with a distrust of the print media, her preference for anonymity probably reflected the conservative influence of her missionary roots. Both the Alexanders and Baldwins gave generously, but always without publicity or fanfare. Alexander was raised in an atmosphere where benevolence was expected but not touted. Many of Samuel and Henry's closest friends were not aware of either the nature or the magnitude of their many charitable contributions until after the partners' deaths.[9]

Like her father and uncle, Alexander seemed to derive a great deal of pleasure from assisting friends and associates. Her letters to Beckwith are filled with acknowledgment of her own financial good fortune and her fervent desire to share some fraction of her wealth with those she loved and about whom she cared. Annie frequently presented Martha with gifts including, on occasion, shares of stock in the Hawaiian Commercial & Sugar Company.

On a more substantive level, Alexander funded all Beckwith's research and fieldwork during her career as a cultural anthropologist. After Beckwith completed her doctorate in anthropology at Columbia, she accepted a research professorship at Vassar College that entailed only minimal teaching. The Vassar Folklore Foundation was created by Alexander to ensure the continuation of Beckwith's ground-breaking ethnographic studies and was established anonymously by "a friend to anthropology." Only Beckwith knew the identity of the donor, and its funds were available to her alone. The money permitted Beckwith to study the folklore of Hawaiians, Jamaicans, Native Americans, and the Portuguese residents of Goa, as well as folk festivals and college superstitions of Americans, unencumbered by the need to obtain outside sources of funding. Before establishing the fund, Annie wrote to Martha, "I always believed, if a person has the inclination, that getting knowledge first hand, is the real exciting and worthwhile thing in life. . . . [I]f it's original research such as you have just been undertaking in Jamaica I'm with you and will give you all the backing possible." Alexander's admission that "getting knowledge first hand, is the real exciting and worthwhile thing in life" provides yet another key to her motivation in conducting fieldwork and to her happiness. In 1951, a year after Alexander's death, Beckwith dedicated one of her best-known works, *The Kumulipo*, "to the memory of Annie M. Alexander, lifelong friend and comrade from early days in Hawaii," noting that Alexander would not permit such recognition during her lifetime.[10]

In the fall of 1901, when Annie's love for Martha flowed freely and exuberantly, she sent her a mink coat and muff, "to help keep you warm when the cold winter comes [on the East Coast] and ease my heart of the trouble that thoughts of a white landscape swept by heedless winds bring it." Her reply to Beckwith's embarrassed letter of thanks acknowledges one of the complexities attendant on wealth and at the same time characterizes Alexander's outlook on philanthropy, which Beckwith's decision to keep their friendship platonic did not change—

I'm glad you liked the coat and muff, awfully glad for it was a great pleasure to give it to you—and now please don't worry any more about it will you! It's a pity life isn't more simple. Why shouldn't it be as natural to share our big toys as we shared our small toys when we were children? Shouldn't the same rule hold good?[11]

Letters written by former researchers in the museums of Vertebrate Zoology and Paleontology at Berkeley also make mention of unsolicited checks received from "Miss Alexander" after the birth of a baby or in some other moment of financial need. Although Joseph Dixon resigned his position at the MVZ on more than one occasion to try his hand at farming, each time he requested reinstatement to cover his financial misfortunes Alexander provided the necessary salary. In all these circumstances, the money seems to have been proffered with a minimum of ceremony and with no expectation of reciprocity or obligation on the part of the recipient.[12]

These explanations notwithstanding, Alexander's dislike of publicity relating to the conduct of her personal life seems obvious. Her family's business connections and their involvement in community affairs undoubtedly placed her in contact with other affluent individuals jostling and jockeying for positions of prominence and influence in Bay Area society. Even within an environment as removed from social convention as many of the day felt that California was, it is understandable that a soft-spoken and reserved individual like Alexander would dislike public notice. On the one hand, by remaining single, Alexander had no need to promote herself as a means of enhancing the futures and fortunes of a husband and children. On the other, by taking care to keep out of the public eye, she deflected attention from her unconventional lifestyle. Making her disdain for publicity unmistakably clear, Alexander protected the objects of her philanthropy and herself as well.

13

Innisfail Ranch

Following their return from Nevada in 1909, Alexander and Kellogg did not confine their lives solely to the interests of the museum or to the unrelenting machinations of university politics. In the fall of 1911 Annie purchased 525 acres of undeveloped land on Grizzly Island in Suisun Bay, approximately 40 miles northeast of Oakland, intending that she and Louise would create a viable farm from its wet tangle of tule rushes and salt grass wilderness. A cousin had pronounced the land the "richest soil in the state," a mixture of peat and sediment that had been deposited over millennia by the Sacramento and San Joaquin rivers.[1]

Alexander viewed the undertaking as a business venture, prompted first and foremost by her desire to ensure Kellogg's financial security and independence in later life. While the Kellogg family could not be looked upon as impoverished, their estimated worth and, in turn, Kellogg's expected inheritance, would never approach that of the Alexander children.[2] With the opening up of the Sacramento and San Joaquin valleys through new methods of irrigation, the opportunity to gain a livelihood from farming was now a realistic possibility. Undoubtedly, there were other considerations. Alexander and Kellogg made their family homes in Oakland their primary residences. On occasion, evenings might be spent together at one house or the other but—absent the traditional sanction of matrimony—the women would certainly not set up housekeeping together in town while their parents were still alive.

Like her father, Alexander saw potential in an undertaking that others might have dismissed as improbable. Whereas Samuel Alexander and Henry Baldwin had made their fortunes converting the arid wastelands of eastern Maui into lush green fields of sugarcane, Alexander and Kellogg now confronted the task of reclaiming a portion of the Sacramento River delta subject to seasonal flooding and grossly fluctuating levels of salinity. Arguably, farming and all its attendant issues, from land and water rights to product marketing and delivery, was the business most familiar to Alexander. Growing up, she had witnessed what determination, intelligent planning, hard work, and some degree of luck could accomplish in this realm. It was also an enterprise for individuals who were drawn to live outdoors, an enterprise that could reward the spirit as well as the pocketbook. For Annie and Louise, the remoteness of the Sacramento River delta had an additional benefit—the opportunity to live in a home of their own. The farm was a haven where they could live and work together during interludes between collecting trips.

The delta region was certainly not unfamiliar to the women at the time of Alexander's purchase. Louise's father had been a founder and charter member of the Cordelia Shooting Club, "California's First Duck Club," and he had presided over its inaugural meeting in the parlor of the Occidental Hotel in downtown San Francisco in 1880. After calling the meeting to order and dispensing with the perfunctory opening remarks, Kellogg plunged into the first order of business—proposing that the new organization assume financial obligation for the portion of the Sacramento River delta that ran through Suisun Marsh, an area known then as the Chamberlain Tract. With the hearty approval of those in attendance, the necessary leases were soon entered into and a yacht hired. From that point on, the marshes in the delta became a regular haunt for club members and their families. In December 1910 Alexander recounted to Beckwith,

> Mr. Kellogg takes us once a year duck hunting in the Suisun marshes and it is always a treat to go. . . . The slough winds in and out among the tules full of red-winged black birds, tule wrens, song sparrows and yellow-throats—quite a little world by itself. . . . [W]e stayed overnight and set out some traps catching one interesting little mouse that is confined entirely to the salt marsh grass.[3]

In 1911 Grizzly Island was remote and undeveloped. It might well be said that the beauty Alexander saw in this soggy piece of real estate was comparable to the intrinsic beauty she seemed to find in all nature, from the stark, rocky outcrops that she and Louise scoured for fossils, to the plain-colored birds that had gained Heller's admiration of her in Alaska. Before the gold rush of 1848, Suisun Marsh had abounded with wildlife. The delta was a paradise for ducks, geese, swans, and sandhill cranes. Furbearers such as beaver, river otter, mink, fox, and weasel shared the region with game mammals such as tule elk, black-tailed deer, and pronghorn antelope. Predators such as mountain lion, black bear, and grizzly bear coexisted in this wilderness. The island derived its name from these ferocious creatures. One hundred fifty years ago, bears would regularly amble north across the relatively shallow bay and intervening slough from Mount Diablo to Grizzly Island to nibble on the rose hips and wild blackberries growing there in great abundance. As always, it was Martha to whom Annie intimated her feelings about this new venture, and what she felt to be the almost ethereal nature of the surrounding landscape:

> we are starting off to engineer a new enterprise—namely the reclamation of a small strip of land on Grizzly Island for farming purposes—near the junction of the Sacramento and San Joaquin Rivers. . . . The new conditions and new set of people we have met might in themselves compensate for all the trouble we are going to, if we weren't really sure we were going to make a profitable thing out of this rich tule land. . . . We need a poet to express what we feel to be a real beauty in the tule marshes with their winding sloughs subject to the influence of the tide.[4]

The Reclamation Act of 1902 not only set aside funds for irrigation projects that were designed to create farmland out of often arid tracts of ground but also, perhaps as important, fed its promoters' hopes for "development of the West," often through conversion of marshy areas into "reclaimed" acreage suitable for farming. As as result of the act, tidal wetlands that once surrounded San Francisco Bay and encircled the Sacramento–San Joaquin River delta were drained and diked over for cultivation. Farming and hunting, livestock and salt production, even the installation of telegraph wires, all encroached on the habitats that the region's once-abundant wildlife had depended on for breeding and feeding.

It is difficult to reconcile the ease with which Alexander accepted the notion of such perturbation and even contributed to it, given her conservation bias, but it would be equally erroneous to judge her behavior and actions by present-day perceptions and mores. In 1911 it was taken for granted that the state would be developed, that its population would increase, and that a pristine wilderness would eventually vanish. Grinnell's activism in the establishment of the National Park system was recognition of the need to set aside at least an iota of this wilderness to preserve for future generations "the conditions as we find them today."[5]

Given this view of impending destruction, it is not surprising that the Alexander and Kellogg farm quickly became a field site for MVZ staff and students who wished to collect specimens on the premises or who used the farm as a jumping-off point for trips to the islands of the bay. Alexander routinely recorded interesting natural history observations in her letters to Grinnell—the song of a common yellowthroat, the sight of ten kites circling about the tops of the eucalyptus trees at the corner of their property, the recent visit of a flock of tree swallows. Upon discovering a mockingbird nest in one of their crabapple trees, Kellogg began making regular observations. By climbing a ladder she was able to see at least one egg in the nest, which she watched repeatedly for several weeks. To aid the adult birds, she began leaving raisins in a dish on the front porch and added bluestone to their drinking water because she felt that the female looked soiled, unwell, and she suspected it suffered from intestinal trouble.

Alexander sent Grinnell any specimens trapped on the property. A female river otter caught by her neighbor, Mr. Luscomb, became the first record for that species from south of Humboldt and Shasta counties. Grinnell immediately described it as a new subspecies and asked if she would be able to acquire additional specimens. Luscomb eventually did. Two specimens of mink that Alexander sent to the museum became the only adult specimens from the Bay Area in the museum's collection. "The opportunity to secure such mammals is well worth taking advantage of. With reclamation and continued trapping, they will become less and less common; some may disappear altogether, especially the larger species like the otter," Grinnell wrote realistically yet soberly in his acknowledgment.[6] Within a decade, six new subspecies of birds and mammals had been described from the area (see appendix).

In 1911 milk, cream, cattle, pigs, and grain for livestock were the mainstay of Grizzly Island's economy.[7] Water was pumped from adjacent Montezuma Slough to wash dairy equipment and clothes. There were no roads during the winter months, no electricity, and no phones. The cream boat provided the primary means of access to and from the town of Suisun during the women's early years on the island, passing the Alexander and Kellogg farm on Friday afternoons. To make major purchases, island residents traveled to San Francisco, generally taking a commercial vessel from Suisun Bay through the Carquinez Strait into San Pablo Bay and then south to the city. There they would spend the night, returning the next day.

Alexander and Kellogg preferred to make the trip between the East Bay and the delta by rail, a no less lengthy and arduous journey. Train service ran from Oakland to the town of Fairfield. From there the women would drive the short distance to Suisun where they would catch a "gasoline launch" for the two-hour boat trip through Suisun Slough, around the southwestern corner of Joice Island, and back up Montezuma Slough to their farm, Innisfail Ranch, on the northern side of Grizzly Island (see Map 3).[8]

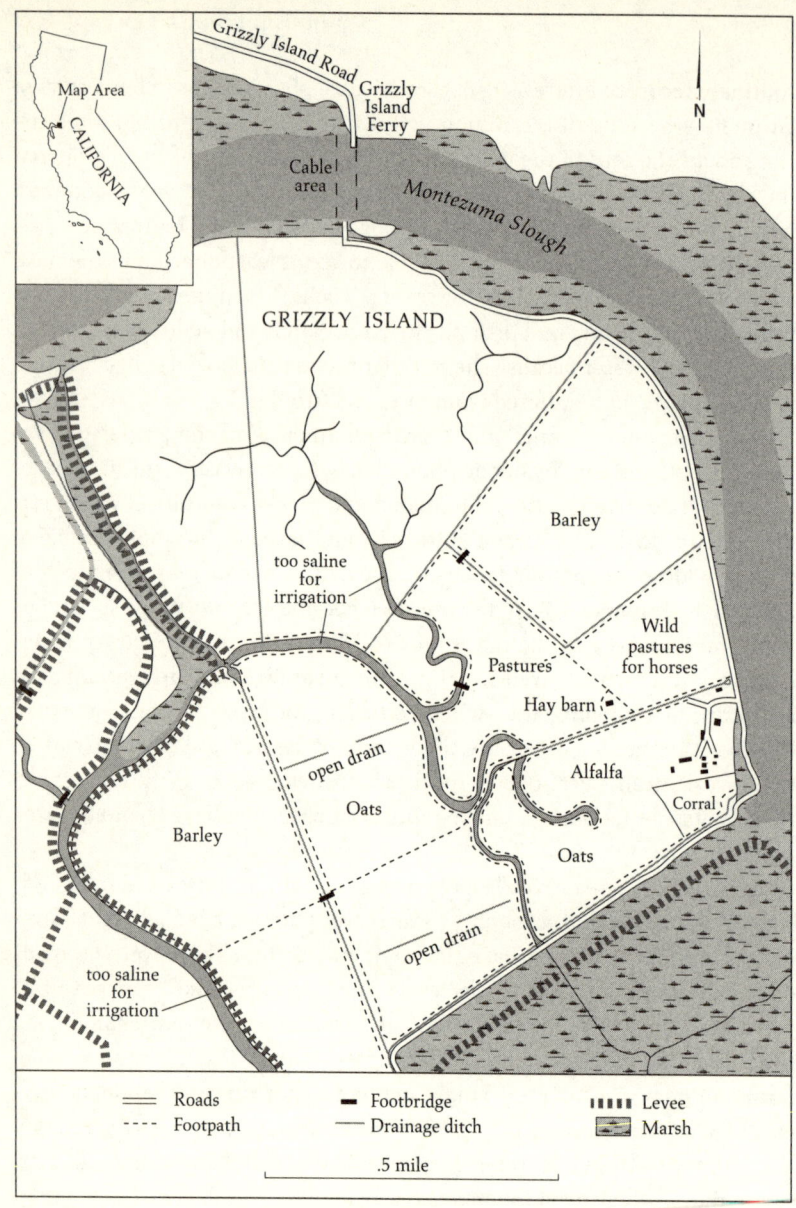

Map 3. Innisfail Ranch, 525 acres of undeveloped land on Grizzly Island in Suisun Bay that Alexander purchased in 1911. Over the years there, Alexander and Kellogg grew hay, bred milking shorthorn cattle, and raised asparagus (reconstructed from USGS Map 1941; R. L. Adams and C. M. Haring, "Grisly [sic] Island Project," College of Agriculture, University of California, Berkeley, 1921; and Alexander and Kellogg letters and Diary).

The farm, or ranch, as they frequently referred to it, was operated as an equal partnership between the women. They soon adopted a mud hen (American coot) as the official coat-of-arms but gave no clue to this seemingly unusual choice of an emblem. At one time mud hens were the most common species of rail in California, particularly in and around the state's tule marshes, but they are otherwise relatively undistinguished birds and traditionally disdained by hunters. Obviously undeterred, Alexander commissioned Canadian wildlife artist, Allan Brooks, to paint a pair for them. Several years later, Grinnell asked to borrow the illustration for publication in his volume on California game birds and the women graciously assented.[9]

Innisfail farm was an irregularly shaped piece of property cut up by natural sloughs. Approximately 4,000 feet of land fronted on Montezuma Slough, with Island Slough forming the major portion of the remainder of the property boundary. The water level in the sloughs fluctuated greatly and, like much of the delta region in the Sacramento and San Joaquin rivers, flooded the island during high tides. To reclaim this acreage and prepare the land for plowing, the women's first order of business was to construct an elaborate system of levees and drainage ways across the plot. One difficulty in building a levee on the deep sloughs was that of holding the mud in check; planks had to be drawn in on either side, a time-consuming task that required an enormous quantity of lumber. A neighbor who owned a gasoline launch often endured the four-hour round trip to Fairfield, making such purchases for them and running other necessary errands, but an additional obstacle presented itself on the return trip. At low tide even small, lightweight boats became stuck in the mud. When the water proved too shallow for docking, the women had to don tall rubber boots and wade out across the slough to retrieve their goods from the launch.

Alexander rented a dredger to dig the desired drainage ditches along their property lines where natural sloughs were not already present. Upon completion of this task, around 4 miles of open ditches and 4½ miles of levees crisscrossed their 525 acres, essentially carving the property up into numerous small fields. This made plowing more complicated, with the attendant difficulty of getting the plow across each of the open trenches. Alexander's solution to this problem was to buy six 40-foot planks that, placed side by side, would serve as a movable bridge. In a necessary but tiresome choreography, no sooner would the plow cross one slough or ditch than the planks—carried and laid across the next—would reposition the bridge ahead of the oncoming plow.

Because the land had never been broken, Alexander purchased a fat-wheeled tractor, the first of its kind on the delta. Her thinking was that its enormous wheels, reminiscent of those on a military vehicle, would keep the heavy piece of machinery from sinking and becoming mired in the soft soil. She also acquired a special sod plow and wagon and hired two of the neighbor's horses to pull the wagon.[10]

New problems seemed to crop up daily, well before any seeds went into the ground, and the solutions to each of these problems always seemed to require the expenditure of large sums of money and a great deal of physical exertion. For instance, at the same time that the women were intimately involved in dredging the land to regulate water levels during high tides, provision had to be made for irrigating their crops when sufficient fresh water was not available. Winter rains could usually be depended upon to start grain crops but, if there was insufficient rain in a given season, by May it would be necessary to irrigate in order to carry the crop through to a successful harvest. Alfalfa, for instance, had to be irrigated after each cutting.

The women conceived the idea of using the tidal capabilities of the sloughs, that is, the natural tendency of the sloughs to flood during high tides, as a means to irrigate their fields. Their plan entailed installing three floodgates, one each at the ends of the long ditches that led from the middle of their property out to Montezuma Slough. To regulate water levels, the floodgate installed in the southeastern corner of the tract was equipped with a centrifugal pump capable of draining 4,000 gallons of surplus water per minute. Initially the pump was operated by a 30-horsepower gasoline engine that had to be manually started and stopped during each operation, but it was eventually replaced by a self-starting and -stopping electric motor. By opening the flumes and tide gates at high tide and closing them during low tides, the women were able to trap sufficient water in the sloughs and ditches to cover the fields for irrigation. It usually required two high tides to irrigate the fields contiguous with the Montezuma Slough and as many as ten tides to irrigate the barley and oats that lay remote from this water supply at the back of their property.

This ingenious and comparatively cheap method of irrigation sufficed as long as the water in the slough remained relatively fresh, that is, when the Sacramento and San Joaquin rivers were running full after the winter rains and while they carried accumulated runoff from melting mountain snow. However, when the water became too saline, as it did during low-water stages of the Sacramento River when an influx of water from the bay sharply raised the salt concentration, the women could no longer irrigate in this manner. This might happen as early as July 1 or as late as September 1, depending on the intensity of the winter rains. Alexander and Kellogg tracked annual precipitation closely, usually recording daily temperature and rainfall in their personal diary and maintaining a typed list of accumulated rainfall by month between 1913 and 1923.

For all its ingenuity, the system that they devised to irrigate their crops was less than perfect. To begin with, it did not work throughout the growing season. In addition, there were continual mechanical problems and equipment failures. At one point a floodgate developed a large hole that the water's force continued to widen despite the women's best efforts. Their immediate solution was to fill sandbags and place them against the gate to relieve pressure on it and to keep the tides out, but this job took several days to complete, during which time several fields flooded.

Another problem was that their 30-horsepower pump did not have sufficient power to lower the water level quickly enough to prevent the crops in low-lying areas from rotting, while those on high spots got insufficient water, particularly in the more remote areas of the farm. To enhance the flow of water from the soil into the channels that they had dug, the women eventually installed an extensive system of terra cotta drainage tiles at a depth of $3^{1}/_{2}$ to 4 feet throughout 350 of their 525 acres. An added benefit of this system was that it helped leach an abundance of salt and alkali from the soil as well. The installation of the tiles, however, was a tremendous physical and monetary undertaking and necessitated the surveying of property lines, the purchase of a ditchdigger, and the hauling in of tons of tiles, the equivalent of 47 miles of tile once they were laid out in place.

Yet at times even this elaborate effort was not enough. During one several-week period of high tides, a portion of their fields not provided with this additional drainage mechanism risked ruin from flooding. The women were compelled to lay down additional irrigation pipe by hand from the affected area through their levee and on to another three-acre section of the property. However, since high tide advanced by about three-quarters of an hour each evening, after several days they were forced to abandon the effort and let the water run at will rather than to stay up around the clock moving pipe.

From the outset, Dixon kept Hasselborg apprised of the activities of their favorite female collecting partners. Early in 1912 he wrote, "They have had a dredger dike off and drain about 200 acres of it [the farm] and have a sixty-horsepower tractor in plowing it up getting it read[y] for oats [and] for hay. Of course they had to be right on the job so they have put up a tent and live there. Both are brown as a couple of Indians, but I'll bet they make good for hay won't likely go below $15 a ton this year and its [sic] above $20 now."[11]

Camping was perhaps the most apt metaphor Dixon could muster. Living conditions at the farm during the first several years were primitive at best. At first the women set up two small tents on platforms, one for themselves, the second for their tractor driver and plowman. Rain caught in large cisterns was their only source of water for drinking and bathing. Between the building of dams and the laying of tile, the women planted oats in a plot in front of their tent. Of their first crop, some of the oats were ruined from flooding, but one neighbor brought his team across the slough on a barge to help them cut that portion of the crop that could be salvaged.

Each succeeding year saw the expansion of their enterprise as well as the addition of fresh problems. In the spring of 1913 they had to put a new gasket on their pump. Their dredger sat idle in a ditch full of water because of a poor cable, and two of their workmen quit. The purchase of their own gasoline launch for making the trip to town helped them address the unrelenting stream of problems.

The women reserved a portion of their land adjacent to Montezuma Slough, plus a piece that was heavily bisected by smaller sloughs and ditches, as pasture for cattle. Their first crops were alfalfa, barley, and oats but, early on, they owned several milking cows and started purchasing Berkshire pigs. Although the women made a satisfactory profit from their hay, the composition of the soil and the salinity of the water in the sloughs precipitated a switch to breeding dairy cattle. The transition was gradual, with consideration also being given to simply renting out some of their property to other farmers rather than switching to cattle. But in 1914 Alexander purchased a foundation herd of milking shorthorns.

Alexander may not have been able to manage close, detailed work given the problems with her eyes, but pivotal to the women's success in cattle breeding was her eye for quality hoof stock. Of the original animals, she bought some from English breeders, the remainder from the best herds in the East and Midwest. She knew nothing about the field when they began, but as with each venture Alexander undertook, she made it her business to learn all she could about it.

The purchase of cattle precipitated a building boom on the farm that resulted in a proliferation of outbuildings in the northeastern corner of the property. A milking barn, a hay barn, an 18-foot silo, a nine-stall horse barn, an L-shaped, twelve-room bunkhouse with kitchen and dining facilities for farmhands, and a four-room rustic cottage 24 × 30 feet complete with toilet and bath for themselves all appeared south of the floodgate along Montezuma Slough over the next several years. In 1915 their first phone was installed, a party line that remained in existence until 1931, when they talked about purchasing a private phone line. Over time a farmhouse 30 × 30 feet, similar in construction to the bunkhouse and sporting a wrap-around porch on the northern and eastern sides, a calf barn, a barn for bulls, a chicken coop, and a seven-colony hog house also sprang up. An implement shed, a tool shed, a tank house, and a windmill rounded out the complement of improvements they made to the property. A 500-foot well attached to a windmill and tank supplied their drinking water and three auxiliary wells were drilled in the fields.

If the farm was a partnership wherein Alexander supplied all the financial capital, Kellogg expended most of the physical labor to support it: plowing and planting the fields, setting chickens and turkeys, weeding the vegetables, doing the wash, doing the cooking, or scrubbing the floors. Through cold winter rains and hot, humid summers when the temperature frequently rose to over 100° F, one or both of the women labored alongside the field hands Alexander had hired. Louise suffered more in the heat than Annie did and its unpleasant effects were exacerbated by the barrage of annoying flies that accompanied the onset of summer. During their first few months on the farm, the women had worked steadily to clear the land by cutting and burning the vast fields of tule, while two men walked ahead of their plow filling up the small sloughs with grass sods and covering the larger ditches with the wooden planks so that there would be no delay in crossing them. Opening a window onto the secret of their success, Dixon reported to Hasselborg, "They are making good on the ranch but they work like they didn't have a cent in the world and was [sic] afraid of spending their last days in the poorhouse."[12]

From the beginning, the women's search for reliable help was a persistent and exhausting problem. Often the contract laborers they hired failed to appear or would show up days late. This was a serious problem if a crop was ready for harvest as the delay might cost them the entire yield. Among those men who did show up for work, the women had to contend with drunkenness and generally unruly behavior. On one occasion, a cook they had hired to feed the fieldworkers threatened to quit, claiming the men were eating like hogs and refusing to clean up after themselves. On another occasion, a fight broke out between a field hand and the cook, with the latter brandishing a knife and the former waiting to seize him and drown him in the slough. Alexander wryly noted after the fact that the cook got quite irritable when he didn't get his whiskey. Once they had to break off a field expedition, hurrying back to stop a fight between their foreman and the engineer.

Alexander became adept at firing her employees and frequently recounted such trials in letters to her mother. In March 1915 she fired their foreman because she felt he was the least useful of all their employees. She did not plan to hire a replacement; she and Kellogg had tried three individuals in that position over the course of ten months and none had worked out. At that point, the women would have happily settled for someone who would merely milk the cows and care for the pigs since effective management seemed out of the question. Dixon apparently did not grasp the scope of the problem, focusing only on their monetary success when he wrote to Hasselborg, "Miss Alexander and Miss Kellogg stick to the ranch pretty close. They have had lots of trouble getting help and Miss Kellogg drove the horses on the hay baler last summer. . . . They seem to thrive on hard work and I guess have some money coming in from the ranch now."[13] Dixon's myopia probably reflects his own failed efforts, having been rehired by the museum after each of his unsuccessful attempts to make a living from farming in southern California.

Alexander and Kellogg remained on the ranch throughout most of World War I in the absence of competent and reliable laborers who had gone off to fight in France. Performing what she considered to be men's labor made Alexander rebellious (obviously fieldwork for the museums did not fall within this category), and her letters during this period sounded increasingly morose. She was not a quitter, she wrote to Grinnell, but "my efforts don't seem to count for much in this agricultural venture no matter how hard I work and I surely was never more exhausted than I am now."[14] Her desire to transfer the burden of the place onto someone else's shoulders seemed ever more fervent with the passing years.

Although pleased with their herd of milking shorthorns, the women found that the constant attention the cattle demanded tied them to the farm in an increasingly intolerable manner. Desperate to be in the field, either trapping mammals or collecting fossils, but unable to escape the responsibilities of the ranch during its formative years, the women channeled a portion of their energies into a kitchen garden. Over time, this horticultural effort included more than forty varieties of vegetables: two kinds of beets, five types of corn and three of peas, plus carrots, onions, potatoes, peppers, and tomatoes. Kale, kohlrabi, parsnips, and artichokes competed with asparagus, cauliflower, okra, and chard. They also planted fruit trees and flowers, an activity reminiscent of Samuel's activities on Maui. Peaches, pears, and nectarines cast long shadows across bushes of raspberries, stalks of rhubarb, and an array of red, green, and orange melons all generally yielded much greater quantities than either they or their farmhands could consume.

From pansies to petunias, dahlias to delphiniums, sunflowers, tulips, hollyhocks, roses, snapdragons, chrysanthemums, zinnias, gladioli, and narcissus, all created a riot of color and emitted delightful fragrances almost year-round.

When not working in the garden or in the fields, the women learned how to make smoked hams and sausage from the pork scraps. One neighbor boiled the pigs' heads for headcheese. What they could not consume themselves of all these products, they sold to other island residents.

From the outset, Alexander insisted that all their farm animals be of good stock, be they cows, pigs, or chickens. The women paid particular attention to the health of their animals and adhered to a specified breeding program. Their records were meticulous, no doubt a combination of Alexander's manner of conducting business and her adoption of the Grinnellian methodology.

From the moment the women began exhibiting their cattle at state, county, and local fairs in 1915, they won prizes, both within the industry and from the animal science faculty at the university. Here is perhaps the clearest indication of just how quickly Alexander had mastered her chosen subject matter. Dixon was frequently called upon to photograph their animals, and the women found good demand for their registered, prize-winning cattle through his pictures. But Alexander continued to lament the never-ending barrage of tasks and responsibilities that tied her and Kellogg to the farm. While raising shorthorns proved to be a successful financial venture, their purebred livestock demanded care and attention year-round. Living as they did on an out-of-the-way island where prospective buyers would be unlikely to find them easily or even know of their existence, the women felt they had to invest a good deal of time in promoting their cattle through exhibition. The preparation of purebred livestock for show and their transport to the exhibition venue both became additional time-consuming tasks.

The women were not at all hesitant to avail themselves of the expertise and resources offered to local farmers at the university's instructional farm operated in the town of Davis just west of Sacramento, and they quickly became known to the staff there. Not only did they meet with irrigation experts and grain experts, but at one point the assistant superintendent of the farm even agreed to look for a good foreman for them from among the school's students, an effort that, in the end, proved unsuccessful. Rather than keep trying to find a competent ranch manager, the women hit on a new scheme in which they would hire a farmer and his family to take charge of the farm on a share basis. At the same time, Alexander thought more and more about passing on the herd to "some intelligent, ambitious young man" who could build up a great dynasty with the foundation stock that she and

Kellogg had amassed. If both plans proved successful, the women would be free to come and go with ease and reestablish a regular schedule for fieldwork.[15]

Alexander's second wish was realized in 1919 when the women leased their herd to John Rowe in Davis, California. Rowe later bought the herd, carrying on some of the bloodlines and becoming one of the state's most successful breeders and exhibitors of milking shorthorns. But it took the women several more years to find the farmer, a trusted individual with a family, to take over the day-to-day responsibilities of running their ranch.

To celebrate their tenth anniversary on the island, Annie and Louise invited some of their neighbors to join them for turkey, cranberry sauce, and huckleberry pie topped off with a bottle of Cresta Blanca. Alexander reminisced to Beckwith on the eve of the celebration, her words evoking the beauty and harmony of the landscape that had drawn them to Suisun initially:

> It is hard to believe that ten years ago we pitched our tent here in a tule fog and made coffee out of a briny seepage from the levee. I remember that the red-winged black birds were outraged at our invasion and had a great gathering in the tules to discuss the matter, but they didn't need feel so abused, they like what we've done now. The variety appeals to them. They are in and out of our garden, perch on our trees tops, scratch in the manure pile, feed with the hens and make themselves perfectly at home.[16]

Annie vacillated between her delight with the farm and her desire to escape its confines and rigors. The following spring she confided to Martha, "I don't know what there is about this place that holds one like a magnet. . . . We seem to be continually wrestling with the problems as to what to do next with the farm—Just now everyone is excited over the possibility of growing asparagus on Grizzly Island which would certainly be more remunerative than hay or grain crops."[17]

At Alexander and Kellogg's request, members of the College of Agriculture at the university made a study of the ranch in 1920 to review the women's plans for further reclaiming their property and to recommend possible ways and means to realize more successful and economical handling of the land.[18] An extensive report suggested that improvements in irrigation were warranted, particularly for the acreage not adjacent to Montezuma Slough, that is, the fields that lay to the south and west. Not only was the present system of irrigation inadequate for these portions of the tract, but water from the sloughs adjacent to that acreage contained levels of sodium chloride and sodium phosphate ten to twenty times higher than that of well water considered suitable and safe for irrigation. The report went on to note that the cost of making improvements to the irrigation system might be prohibitive, given the margin of profit that could be expected, after taking into account what the women had already invested in the land, the outbuildings, and the farm equipment. Because of the presence of a considerable amount of natural pasture, grain stubble, alfalfa, hay, and an unusually well-built infrastructure on the farm, the committee recommended that the land be used to raise dairy cattle, with profits coming from both production of butter fat and the sale of livestock. They noted that the availability of skim milk would justify keeping hogs that could, in turn, be sold for pork. The women could increase the acreage devoted to raising alfalfa, which could then be used as feed for the cattle and pigs. If, in the future, surface irrigation procedures capable of removing excess salts became available, the report then suggested that it would be possible to grow asparagus, pink or large white beans, and possibly potatoes. The report also recommended that with the introduction of surface irrigation a limited amount of acreage along Montezuma Slough might prove promising for dwarf pears.

Alexander and Kellogg were only too well aware of the responsibilities of raising dairy cattle. For them the appeal of raising asparagus, or "grass" as they perpetually referred to it in their letters and diary, lay in its being a seasonally restricted crop, one that would afford them time during the winter to escape to the field. Because asparagus does not produce edible stalks until its second or third year, once the roots had been planted, the couple envisioned taking immediate advantage of the hiatus and leaving for the field.

Preparing their acreage for asparagus entailed leveling the ground, contracting to purchase the asparagus roots, planting them, and then cultivating the soil until the fields were in production. Once all these preparatory steps had been taken, new shoots would begin to sprout in late winter, roughly in the neighborhood of Lincoln's and Washington's birthdays. By the time Alexander returned from her annual visit to Hawaii in mid- to late April, they would be ready for harvest.

Young asparagus plants put out nice, fat stalks and the women began by hiring seasonal laborers, often Filipinos, to cut them. Using sharp, long-handled knives, the workers would carefully lay the cut stalks in boxes. As the boxes filled, they placed them at the end of each row in the field. A truck would then skirt the ends of the rows, picking up the boxes and transporting them to the packing house.

Initially Alexander and Kellogg converted their horse barn into a packing shed and, with the aid of Chinese laborers, packed their asparagus in wooden crates lined with moss. The moss not only cradled the stalks and protected them from damage during shipping but, watered thoroughly beforehand, kept the asparagus fresh and succulent until delivery. Cool weather and clear skies made the task easier (among other things, rain forced the workers to put on heavy rubber boots). Parchment paper emblazoned with the Alexander Kellogg label encased the individual bundles of asparagus. The AK label, which Alexander hired a San Francisco firm to design, had an orange square bearing the likeness of a grizzly bear standing atop a boulder.[19] Bear and boulder stood silhouetted against a backdrop of tule marsh and high Sierra peaks, behind which a bright orange sun peeked into a yellow sky. Against the base of this image was printed the AK brand logo, followed by the words, "Grizzly Island Asparagus, Grown & Packed by Alexander and Kellogg, Suisun, California." A larger version of the label adorned each end of the packing crates.

The women attacked this venture with the same zeal that had marked their earlier successes with hay and cattle. Kellogg proved to be an efficient manager and, within a relatively short time the AK brand of long, green asparagus had the reputation of being the best in the local market and became prized in the East. In 1927 while Alexander visited her family in Hawaii, Kellogg remained on the farm, planting approximately 80 acres in asparagus and "trying to see that the rows are kept straight."[20] By 1931 the farm was producing some 10,000 crates of asparagus, 8,000 in bunches and 2,000 in loose stalks. Once the women resumed fieldwork, they would often buy asparagus in local markets throughout the West and compare the texture and flavor of that produce to their own. Naturally, they were pleased when they felt that theirs surpassed whatever local variety they happened to try.

It was sometime during the farm's third incarnation, the one in which they switched from raising cattle to growing asparagus, that Alexander realized her third wish, that of contracting with one of her neighbors to take day-to-day charge of the farm. She wrote wryly of their decision to Beckwith, obviously equating the move in some small measure with moral failing. She observed, "Farming is an education in moral courage but we don't see much improvement in ourselves in that respect."[21]

With the arrival of Don Wilson and his family, the women were relieved of having to maintain a constant presence in Suisun. Whether it was to water the garden or supervise the laborers, repair a floodgate, replace the pump, await a delivery, or feed the chickens, Wilson was now on-site to attend to the multitude of daily tasks that had characterized their life on the farm for over a decade. Not that the women abdicated their responsibilities the moment that he came on board. The farm was theirs and their interest in turning a profit had in no way diminished. In the same way that Alexander had maintained oversight and directed the founding and establishment of the MVZ, so she remained in constant contact with the farm and directed its daily operations, whether from Oakland or from the field. Still an inveterate letterwriter, if the situation warranted a more immediate response than what could be achieved via the postal service, she would merely send a telegram. In many respects Wilson's role at the farm paralleled Grinnell's in the museum. In her quiet but unambiguous way, Alexander made clear that neither man was authorized to expend funds or commit the institution under his immediate direction to any new course of action without her express consent. In this regard, Wilson proved as compliant and agreeable as Grinnell had been.

The problems that had marked their early years in farming did not simply disappear with Don's hiring. The large issues, those concerning the availability of labor and a sufficient supply of potable water, persisted in one guise or another for as long as they owned the property. During the late 1920s, it became increasingly difficult to hire seasonal labor. In response, they ceased packaging asparagus on-site and began trucking their produce to commercial packing sheds up the river.

Fire was another persistent problem and a serious threat during the hot dry summer months in California. The farm experienced numerous fires over the years. Not only were these hard to extinguish but, perhaps of greater concern, threatened to spread to their neighbors' properties.

Letters that Alexander and Kellogg wrote to Wilson from the field continually inquired about the improvements he was making to the farm, the look of their current crop, and how their neighbors' crops were faring in comparison. It was now Don who fixed the leaky gate, relocated the windmill, and sub-irrigated various fields. It was Don who purchased and installed a new 15-horsepower discharge pump and a wooden gate after the salt water in the slough ate away at the galvanized metal screen at the end of their suction pump. When the screen rusted off, the pump sucked it up along with a block of wood, which stopped the pump from functioning and damaged the runners. Alexander felt that the manufacturers of the pump were culpable. They should have known about the salt water and they should have installed copper screens in the pump as they had with previous models, but there was little she could do from the field other than to express her relief that Don was now handling these problems.

A more serious incident occurred in 1930 while the women were traveling across the Great Basin and the Pacific Northwest collecting mammals for the museum. During their absence, which extended from late May through October, work was started on an illegal ditch along the periphery of their property. Mistakenly, the dredger hired to dig the ditch cut through the levee a short distance from their iron gate, precipitating a series of mishaps reminiscent of the women's early farming debacles. The break in the levee allowed an unusually large volume of water into their ditch. This, in turn, caused the ditch bank to collapse in several places, flooding the adjacent county road and making it unsafe for travel. Don wrote to them that until the situation could be corrected, he had directed all traffic to be detoured across the farm's lower field. Again, there was little the women could do from afar. Alexander was as supportive as possible and expressed to Wilson her relief in his handling of the situation. The women did not want a lawsuit on their hands and Alexander envisioned additional problems down the line if the situation was not dealt with properly at once. As the county had no funds for a new road—and putting it in would take several years—Alexander recommended that the current road be repaired.

Of all the changes on the island following purchase of the farm in 1911, perhaps none had a more profound impact on island life than the construction of a ferry across Montezuma Slough. Alexander championed the cause and single-handedly bore the cost of its construction. At the same time, she was a leading advocate for construction of a county road that would connect the ferry directly to Suisun, and she was largely responsible for persuading local supervisors to bear the ongoing costs and maintenance of the ferry once she had donated it to the county.[22] Suisun and the northern shore of Grizzly Island were only eight miles apart, but the road between them, by way of Dutton Ferry, covered a distance of thirty miles. What Alexander hoped to see was a direct connection between those two points.

Proposal for the ferry first came before the Solano County Board of Supervisors in 1915, but it was not until 1920 that construction was actually completed. As early as 1916 the county agreed to the location of the ferry but did not appropriate the $10,000 needed for construction until 1918, and it was 1919 before all the contractual agreements concerning land acquisition had been drawn up. While economic considerations clearly dominated negotiations during the early years, practical considerations entered into play as well. Before construction, much of Grizzly Island's trade had been going to the towns of Pittsburg to the south and Rio Vista to the east, river communities that maintained easy access to the island by boat. Installation of the ferry and construction of the county road leading to it opened up a territory of 25,000 acres of rich farmland for development and cut travel time between Suisun and Grizzly Island from two hours to forty-five minutes. Island life was changed forever.

Throughout the years, Alexander's musings about the farm and about the life that she and Kellogg had chosen to create there vacillated to the extremes. While her thoughts and expressions reflected an enduring love of the land, she yearned for a freedom that only life in the desert and in the mountains seemed to provide.

14

Vancouver Island and the Trinity Alps

Before their purchase of the property on Grizzly Island confined Alexander and Kellogg to the farm for several years, the women had conducted surveys of two regions in the Pacific Northwest—Vancouver Island, British Columbia in 1910 and the Trinity Alps in northern California in 1911. The specimens they collected not only increased the museum's holdings but also provided important material for Grinnell and his staff to study and interpret. Because Vancouver lies adjacent to the Canadian mainland, the origin and affinities of its vertebrate fauna presented an interesting evolutionary problem. Alexander and Kellogg hoped that intensive sampling of the area would reveal whether its present populations of birds and mammals were closely related to species on the Canadian mainland or uniquely evolved races. Coupled with the material that Alexander and her party had gathered in southeastern Alaska in 1907, and additional specimens that Harry Swarth and Allen Hasselborg had collected in the Sitkan region of Alaska in 1909, specimens gathered on Vancouver Island might extend the known distributions of taxa that existed to the north and south. Grinnell wrote to Alexander in the field, "The more I try to find out anything in the literature about Vancouver birds and mammals, the more apparent it becomes that mighty little is known of the region. You are right in the middle of about the least workt [sic] locality you could pick out in western North America."[1]

Upon their arrival in Vancouver, Alexander hired a local trapper, Edward Despard, to secure mink, marten, raccoons, otters, and beaver for the museum while she and Kellogg concentrated on collecting birds and small mammals. Despard did not remain with the women but instead met up with them periodically to deliver the animals he had captured. Because Swarth had spent the previous summer collecting in southeastern Alaska, Alexander offered to cover his salary, travel, and living expenses if he would bring his familiarity with that fauna to bear on Vancouver. While she and Kellogg worked the southern part of the island, she contemplated having Swarth collect in the north so that they might secure a greater representation of taxa. And if he wanted to, he could rent a room in a farmhouse or hire a second trapper to go after the large carnivores in that area—anything to save valuable time for collecting and specimen preparation.

Alexander and Kellogg spent two days collecting in the vicinity of Nanaimo in the southeast before heading to Parksville, a small settlement approximately twenty miles north. A good deal of this low country had been opened up to farming, but a large portion remained forested and relatively undisturbed. The women pitched their tent in a small clearing but took their meals in a nearby hotel to save time. They had brought a second tent with them in which to process specimens but, for the moment, it seemed easier to prepare study skins and write up field notes in their sleeping tent. Fieldwork is not an occupation that appeals to the fastidious, and in later years Alexander frequently commented in an offhand manner to family and friends about the disarray of a particular campsite that she and Kellogg were occupying—dirty traps heaped on the floor amongst food, dishes, piles of clothes, and, of course, specimens.

From the outset, their trapping success in the region was good and at the end of their first week they had amassed 137 specimens, including a cougar. Thanks to Despard, they quickly added three mink, a raccoon, an otter, and a good-sized black bear to their roster (Alexander purchased the bear from a local farmer). Alexander included a list of the birds that they had taken in a letter to Grinnell, knowing that he would be both pleased at their success and interested in the species that they had obtained. She confessed, "The special bird of this place seems to be the song sparrow but they are feeding their young now and it goes against the grain to shoot them though we have taken quite a number." The women obviously overcame their reluctance to take aim because two weeks later Alexander wrote, "Of song sparrows we already have about 75 but I suppose you put no limit on them!"[2] She and Kellogg had already learned that Grinnell's interest in geographic variation necessitated the collection of series of specimens across a broad geographic range (according to a prescribed and intelligent protocol).

If Grinnell was surprised by the number of sparrows that the women had collected, their success seemed merely to fire his enthusiasm and he did not miss a beat. He replied simply, "I am astonisht [sic] at the series of 75 song sparrows!! I would advise going a little slower on them now, making up on Red-winged Blackbirds, Yellowthroats, Woodpeckers, Vigors Wrens, and the like; but when you get to the *west* side of the Island, get as many more song sparrows. As you say, there must be local variation, at least from east to west, especially if the west side is wetter than the east. This may prove true of some other birds, too."[3]

From Parksville, the women moved their camp eight or nine miles northwest to Little Qualicum River and from there to French Creek, a small stream midway between the two previous locales. After four days of trapping along the river they relocated to an area known as Swain Ranch, about three miles west of Errington. The town itself was little more than a post office. Although grain fields and pastures lay scattered throughout the region, the greater portion of the area was composed of coniferous forest, interspersed with willows and alders in the low-lying regions, and wild cherries and madrona throughout the remainder. One of several dried swamps that now lay covered with *Spiraea* and willows offered a suitable site for their camp. From this vantage point they could look to the southwest where rolling hills slowly gave way to much higher peaks, the most conspicuous being Mount Arrowsmith, which they climbed.

On this particular trip, infrequent communication with Grinnell created a set of problems they had not previously encountered while doing fieldwork. The women were running exceedingly low on ammunition and they had been more spontaneous than anticipated in traveling about. Hence, when Alexander next wrote to Grinnell in late May it was from the town of Alberni, west of Parksville. "So now we are in a quandary as to where Mr. Swarth is. Our aux shells won't last more than two or three days longer from the way we hear the small birds singing around here."[4]

It was early June before Swarth caught up with Alexander and Kellogg near Beaver Creek, about fifteen miles northwest of Alberni, but his arrival did not elicit a warm or enthusiastic welcome. Alexander expressed an unusual level of irritation when she wrote to Grinnell two days later, venting her frustration over the fact that Swarth had arrived ahead of his baggage and field gear, without the badly needed ammunition and without traps. For such irresponsibility she held Grinnell accountable. The women were using all the field equipment that they had brought with them, and a collector without traps or shells was a virtually useless member of their party. There was little that Grinnell could say or do, either in his own defense or Swarth's. After a week's delay and additional expense, Swarth finally secured his gear.

Alberni lay on the western side of the divide from Parksville. One large stream and numerous small ones drained into the head of Alberni Canal from the broad, level valley in which the town was situated. Douglas fir and cedar were the dominant tree types on the valley floor, with the ubiquitous alders and willows scattered along the streambeds and around the periphery of the swamps. A dense undergrowth composed partially of devil's club, skunk cabbage, and thickets of salmonberry added to the vegetative diversity. Standing dead stumps amidst rotting logs and fallen trees attested to the fact that the region had been burned at one time. Near camp there were large areas of grassy meadows and, on all sides, the construction of beaver dams had created swamps. This diversity of habitat types reflected a welcome abundance of animal life and in late June, just before the women's departure, Swarth wrote to Grinnell, "I believe so far Miss Alexander and Miss Kellogg have put up over a thousand specimens! They are wonders. You never saw anything like the way they work; I'm not in their class at all."[5]

In the same letter, Swarth made reference to a manuscript that Grinnell had sent to Alexander for comment. The paper described several new species of birds and mammals discovered on the museum's 1909 expedition to southeastern Alaska. With reference to one such species, Swarth noted, "Miss Alexander objects to [a new species of vole being named] *Microtus alexandrae*—says there are too many things named after her. I failed to induce her to change her mind, but perhaps you will have more success. I certainly hope so."[6]

Grinnell's powers of persuasion in this matter proved no greater than Swarth's and the newly described species of vole was renamed in the final version of the manuscript. Whether it was an aversion to publicity or, more fundamentally, to solicitous behavior on the part of researchers Alexander had hired, she no longer permitted new species or subspecies of plants or animals to be named in her honor.[7]

While the 1910 expedition did not gather representatives of all taxa known to occur on Vancouver Island, of the 111 species of birds that were observed, representatives of 89 were brought back to the museum, much to Grinnell's delight. To Alexander's great pleasure, less than two years after their return, a 124-page "Report on a collection of birds and mammals from Vancouver Island" by Swarth was published under the museum's byline.[8]

The Trinity Mountains, or Alps as they are also known, is a high and relatively isolated range in northern Trinity County, California, bounded geographically by the Coast Range to the west, the Sierra Nevada to the east, and the Cascade Range to the north. The evolutionary question that lay at the heart of Alexander and Kellogg's proposed study of this area was the relationship of the mammals of the moist coastal belt in northern California to those of the drier interior. The vertebrate faunas from the three surrounding mountain ranges differed from one another but—given the unusual proximity of the Trinity fauna to all three—biologists suspected that the region's fauna would reflect an intermingling of these disparate forms.

Alexander viewed the proposed survey as a means to increase the museum's specimen holdings from that portion of the state and collect detailed information about the region's flora and fauna. She clearly understood the value that such a survey would have for those, like Grinnell, who were attempting to address the important theoretical questions in evolutionary biology taking shape at the time. But there was additional impetus to undertake the work. Kellogg wished to prepare a manuscript for publication on the collections that the women planned to make. She had already published two scientific papers on fossil rodents.[9]

In writing to Grinnell about this new twist, Alexander acknowledged that Swarth and Taylor, men she recognized as professional biologists, had been hired to write scientific reports for the museum and stated clearly that she did not wish to interfere with his management of the MVZ's research program. She readily admitted that she was unfamiliar with the protocol in such matters and trusted Grinnell to inform her honestly. Perhaps most revealing of the nature of their relationship, and of the way she viewed her role as benefactor and amateur naturalist, Alexander wrote, "I should not want you to acquiesce to Miss Kellogg's doing the work referred to merely to please me."[10] Alexander's comments instantly consigned her partner to the same amateur status she herself held—regardless of Kellogg's earlier publications. At the same time, her inquiry shows the respect she accorded Grinnell and the museum, even in its infancy. Curiously, there is no evidence that Alexander ever consulted with John Merriam before Kellogg's decision to publish on fossil material.

Grinnell did not address Alexander's identification of herself and Kellogg as amateurs or express any misgivings about their proposal. Instead, he strongly endorsed the expedition; the specimens it obtained and the resulting manuscript would be valuable additions to the museum's collections and to its series of publications on Pacific Coast fauna. Citing the enormity of the field awaiting exploration and study, he assured Alexander that their contributions would be a welcome supplement to the general pool of knowledge. He argued that the more individuals working in the field and the more information gathered, the more comprehensive would be the data available for interpretation to all interested individuals.[11]

Alexander needed no further encouragement. Having long wanted to try winter camping and collecting, in early February she and Kellogg traveled north by train to the town of Redding in Shasta County. From there they boarded a stage to Lewiston, continuing on to Weaverville. Many pages in Alexander's field notes are devoted to describing this latter segment of the trip, naming creeks crossed, towns passed, and elevations traversed, keenly tracing the grim changes that had already been wrought to a previously beautiful landscape. In one instance she noted tersely, "The vegetation east of the Shasta Divide for many miles has been seriously injured by acid from the smelters on the Sacramento R. The majority were shut down last fall as the farmers complained that their grain and vegetables suffered ill effects from the acid. The extensive view from the Shasta Divide presents a scene of devastation."[12]

From Weaverville, their stage climbed to Oregon Gulch Mountain, then on to Junction City, a deserted mining town. It continued along Cañon Creek for a short distance, and then over a low divide before entering Trinity Valley. They followed the Trinity River, turbid at this time of year, to Helena. A light dusting of snow covered the trees above 2,500 feet, and at 6 A.M. Alexander recorded the temperature in their camp as 32° F.

The women pitched their tents on the valley floor in a clearing that provided sufficient browse and water to sustain their horses and pack mules for the duration of their stay. Once camp was established, their routine consisted of setting lines of small mammal traps and of then going out on foot to hunt birds, mammalian predators, and various species of game animals. Often they would walk several miles looking for animal signs—footprints, feces, nibbled grass, or broken twigs—in an effort to find a trapping locality that would prove successful.

In addition to snap traps, the women frequently used "live traps" to capture small mammals, rectangular metal boxes with a raised door that snapped shut the moment the treadle mechanism at the rear of the trap was tripped. The problem with these sorts of traps was that material, either bedding or food, placed near the rear of the trap often became wedged under the treadle, preventing its release and ensuring that the door of the trap remained raised. The lucky animal visitor thus secured for itself a protected winter shelter as it feasted on the rolled oats, bacon, cheese, or other tempting treats that the women provided as bait.

After several hours of hunting, Alexander and Kellogg would return to check their traps in the afternoon. They would reset or relocate the traps, depending on a post facto assessment of the suitability of their current placement. Evenings were spent preparing the specimens they had caught or shot, that is, skinning the animals and stuffing the skins with cotton and wire before pinning them to a board to dry in a uniform position. Skeletal material, that is, the skull and postcranial bones of the animals, was set aside and allowed to dry. This would be cleaned by the museum staff. Soft tissues, that is, the brains, muscles, and internal organs, were generally discarded or were used as bait to trap larger carnivores. On occasion, the women would remove reproductive organs or digestive tracts and preserve them in formalin for future study.

Equally time consuming and important were the meticulous transcription of their field observations and the recording of specimen measurements into field notebooks each evening. Because their collecting regime was thoughtful and purposeful, it is not surprising to find that on this trip Alexander was prone to conjecture about the significance of the topographic features they encountered and their relationship to the species captured. On February 18 she recorded, "It would be interesting to discover if the Trinity R. is at this point as [sic] least a barrier to some of the small rodent species." A few pages later she mused, "The rocky nature of the cañon does not argue in favor of the advent of *Dipodomys* [the kangaroo rat] into these parts by way of the Trinity R. from the west nor can we say that the river is a barrier to the south altho we did not find sign on the south bank for Knowles [a local trapper] tells us there are kangaroo rats at Hayfork."[13]

Additional details about vegetation near where the animals had been taken, climatic variables, and the extent of human disturbance of the habitat were usually recorded as well. If their traps were empty, that was recorded too. The absence of expected or hoped-for species might be as significant as their presence. (Unless animals are "trap shy," an area in which the trap, its bait, and the immediate vicinity lie undisturbed usually indicates the absence of animals. Not infrequently the trapper returns to an empty trap, only to realize that the bait is gone regardless of whether or not the trap has been sprung.)

The women often asked government officials for information on animal sightings or help in collecting specimens. In the area around Weaverville, Alexander interviewed a local judge, the chief forester, and a local forest ranger concerning the types and abundance of mammals they might find and recommendations as to local trappers she might hire to secure some of the larger game species. Much to her frustration, no amount of explanation or cajoling could prevent such hunters from damaging the specimens they were collecting for her or cause them to record the necessary and valuable specimen information while collecting them. Somewhat exasperated she wrote to Grinnell, "You will notice that two of the skulls are broken. It seems next to impossible to persuade a trapper to kill an animal without whacking him on the head.... Knowles is about as good as the ordinary run of trappers who can't see anything in a skin except its commercial value—and the little extra care in skinning that we demand frets them."[14]

From the Scott River, the women pushed on and traversed the north fork of Coffee Creek, climbing to the summit of the Saloon Creek Divide before camping in a grove of white pine and red fir. They then moved their camp to Summerville and on to Hunter's Camp over the divide on the western side of the Salmon River valley. The trail here was poor and steep; they had to cut it out in places to accommodate their pack animals and gear. When they later camped at 6,000 feet at the head of Grizzly Creek, Alexander recorded in her notes without prelude or further commentary, "Howard [a hunter she'd hired] shot a *bear* and I went with him to try and load it on a mule but the mule took offense and deserted us—so the *bear* was left over night with my coat over it to protect it from coyotes."[15]

In all, the women made three trips to the Trinity Alps in 1911, two that winter and one the following summer. Their first trip lasted a month. Some days rain fell, often turning to snow and obliterating their traps. When a stiff north wind cleared out the precipitation, it also caused the temperature to drop. At some points the river rose, flooding an area where they had set their traps the previous evening.

Returning for a week in late April to nearby Tehama and Colusa counties, the women focused their efforts on collecting pocket gophers and botanical specimens. From the town of Red Bluff they traveled by carriage through pastureland that Alexander judged had been cleared of manzanita brush. They visited several ranches north and east of Red Bluff before crossing a rocky plain and ascending the canyon to Tuscan Springs at 1,000 feet. The extensive alfalfa fields in these counties were prime habitat for pocket gophers, and local farmers eagerly granted them permission to trap the elusive, subterranean rodents that were destroying their crops. The ground in this northern region was still frozen and setting traps proved slow, arduous work.

Alexander and Kellogg returned in early June to find the snowbanks just beginning to melt near the base of the mountains, with deeper snow lingering at the higher elevations. This time they took the stage to Callahan, then hired a rig to take them down six miles into Scott Valley between Sugar and French creeks. There they camped on a slough along the Scott River in a thick growth of cottonwoods. They moved southwest to the Salmon Mountains and camped at 6,900 feet on Wildcat Peak.

The women climbed the 7,900-foot Thompson's Peak, which afforded them a view of the entire upper portion of the watershed of Rattlesnake and Stewart creeks, then pushed on to Rush Creek, Kangaroo Creek, and Bear Creek. In terms of species diversity their trapping success remained high, although Alexander lamented their lack of quantity of any given taxon. One day their list of small mammals included a "flying squirrel (caught by Miss K. at the head of our bed in a rat trap. It had been disturbing us at night)." From June through late August, they set four to ten dozen traps apiece, per day. Thus, it is not surprising that Annie would write to Martha in mid-July,

> We have gotten a fair representation of the mammals so far so that the work has lost a little of its novelty but must bear in mind that their distribution is an important factor. I sometimes feel that one's efforts to advance science by this sort of work are so feeble that it is almost ridiculous. This straining after slight differences of species as of significance is because we haven't the penetration to see the more serious aspects—but I suppose . . . we must do the best we can, and leave to future generations with bigger brains what is beyond us to master.[16]

Her letter went on to cite *Aplodontia*, the mountain beaver, as an example of how little was known about the habits of most of the animals they were trapping and how difficult it was, therefore, to assess the significance of slight morphological differences. The women undoubtedly knew more than most about this particular rodent, however, because just a few weeks earlier Alexander had written to Grinnell, "Have taken some large trout in the lake but our main meat diet is *Aplodontia*—have tried it fricassee, broiled, stewed, and as soup."[17]

In reality, Alexander did not have to wait even one generation to see the kind of evolutionary synthesis that she had predicted would result from their efforts. The thoroughness with which she and Kellogg conducted their survey of the Trinity region, and the completeness and accuracy of the data accompanying their specimens allowed Grinnell to publish a theoretical manuscript based on that material five years later.[18] Despite the fact that the Trinity fauna was geographically in closest proximity to the coast fauna, he postulated that the abrupt change in humidity from the western to the eastern side of the Coast Range was a more effective isolating mechanism and barrier to species dispersal than either distance from the mountains or elevation was in the case of the Cascade and Sierra faunas respectively.

The mammals and birds that the women collected also formed the basis for Kellogg's 1916 publication, "Report upon mammals and birds found in portions of Trinity, Siskiyou and Shasta counties, with description of a new *Dipodomys.*" Hers was a descriptive piece that complemented Grinnell's theoretical treatment of the fauna. Whether it was the time required to properly prepare the manuscript, the increasing demands of the farm, or merely a diminished desire to undertake such work, the Trinity paper was the last manuscript that she would publish.[19]

In the summer of 1948 Alexander and Kellogg returned one last time to the Trinity Mountains. Annie was eighty-one, Louise was sixty-eight. They spent nearly a month camping in what was then referred to as the "Primitive Area" in the Salmon-Trinity Alps, a region of glaciated peaks, sparkling lakes, and unspoiled meadows. They slept on the grass, lulled to sleep by the murmur of a stream that passed directly below their camp. While Louise fished, Annie collected plants. The area was still wild and without trails, the slopes too brushy for easy hikes. The volume of Alexander's correspondence had not diminished over the intervening years and from their camp in Morris Meadow she wrote to the mammalogist E. R. Hall of their wanderings and adventures. In the same matter-of-fact style that had characterized her letters from her earliest days in the field, she recounted with a trace of satisfaction:

> a bear did us out of some of our provisions in spite of precautions taken. He didn't get the bacon and we fared very well with Aunt Jemima's pancakes for bread, and bacon grease for butter. A bear has an enormous stretch we have learned.[20]

15
The Team of Alexander and Kellogg

Rarely a year passed in which Alexander and Kellogg did not conduct fieldwork for extended periods of time. Collecting gave purpose to Alexander's life. Without it, the freedom and pleasure that she experienced in the outdoors and found vital to her mental and physical well-being would have been little more than self-indulgence.

In letters to Beckwith, Alexander frequently alluded to the societal pressures from which she felt she was escaping through fieldwork. Limited further by the strain that close work placed upon her eyes, she also experienced a feeling of uselessness at home that ran counter to her temperament. In the last days of August 1910, shortly after returning from collecting birds and mammals on Vancouver Island, she wrote to Beckwith from her mother's home in Oakland, "As long as I am in the mountains too busy to do any thinking, I seem to be all right. Here the gates are down and a multitude of matters that an ordinary vigorous mind would dispose of beset me."[1]

Alexander felt keenly that the happiness she experienced in the field must be earned, and the potential contribution to science of each collecting expedition was never far from her mind. The quest for something new, something that would add to the understanding of vertebrate evolution and enrich the research programs that she had established and nurtured, was paramount. Her approach to fieldwork was almost missionary in its zeal, as though the pure enjoyment of what she was doing was insufficient or inappropriate as an end in itself. She rarely attended church and never made reference to religion or to a belief in God, yet the values imparted to her by her father and grandfather seemed to provide an overriding context for her life's work.

But Alexander had inherited more than a value system from her father and grandfather; she shared their peripatetic tendencies. Fieldwork allowed her to indulge this passion. After a two-week trip to Sequoia National Park in the summer of 1918, Alexander wrote to Beckwith from the ranch, "It made me feel ill to think of coming back. I'm just a natural nomad but I suppose it is just as well to be tied down to a place and an occupation for a while at least. Otherwise what would become of me! But then, that doesn't worry me. I know a lot of interesting things to do besides farming!" This was true, and barely a month had elapsed when she wrote to Grinnell from the ranch, "I have a naturally roving disposition and I don't know how long I should fight that tendency."[2]

Among collectors who shared their enthusiasm for fieldwork and an appreciation of the scientific rewards to be reaped from such diligence, Alexander and Kellogg's skills and accomplishments as field biologists were greatly admired and respected. Among museum graduate students they were regarded as legendary even while they were still actively collecting. On museum field trips professors exhorted these students to speed up their collecting efforts, to prepare at least ten specimens a day, because that was the rate that "Miss Alexander and Miss Kellogg" maintained.[3]

It is difficult if not impossible to assess how Alexander's life, and thus her contributions to the university and to natural history, would have been different had she not met Kellogg. Relieved of the perpetual struggle of having to find a suitable companion each time she felt drawn to the out-of-doors or compelled to escape the confines of Oakland, Alexander was now free to travel at will with someone who shared her love of nature and fieldwork. Within two years of their meeting, Kellogg relinquished her teaching position and the women began their life together, centered around fieldwork and interrupted only by the demands of their farm. With the passing of years, they emphasized the collection of small mammals over that of birds and, with the exception of paleontological field trips in the early 1920s, their collection of fossils became an ancillary activity.

For forty years, the team of Alexander and Kellogg collected not only in California but also throughout western North America and abroad. Often they would collect in late fall or early winter, when crops demanded the least attention and members of the museum staff could not easily get away. The two women increasingly favored desert areas, both for the warm, dry climate that Alexander preferred, and because of the speed with which they were coming under cultivation and being perturbed. Alluding to collections that she had made shortly after the museum's founding, twenty-five years later Grinnell wrote to Alexander, "Mr. Benson showed me the card he got from you yesterday, picturing a trapping ground of yours which is destined to be submerged. What a transformation of fauna will occur—extreme arid to aquatic! It will surely be scientifically useful in future years to have in the Museum the representation of the mammal fauna that *was*."[4]

As early as 1910 even large, easily recognizable mammals such as the beaver were scarce in California and, in some parts of the state, had already become extinct. In a prescient and now classic paper published that year, Grinnell proclaimed, "At this point I wish to emphasize what I believe will ultimately prove to be the greatest value of our museum.—And this is that the student of the future will have access to the original record of faunal conditions in California and the west wherever we now work.—Right now are probably beginning changes to be wrought in the next few years vastly more conspicuous than those that have occurred in ten times that length of time preceding."[5]

Because she had been conducting fieldwork for almost a decade at this point, Alexander was one of only a handful of individuals who held this realistic yet sobering vision of the future. In establishing the MVZ, she had stood virtually alone as someone who intellectually and emotionally understood the need for immediate action and, at the same time, was financially able to address the urgency of that call. Over the years, impending dam construction or the bringing of extensive tracts of land under cultivation would motivate her to expend additional moneys for faunal investigations before the projects began. Her financial support of fieldwork in Baja California is a prime example. In 1925 she established a fund at the university, apart from the MVZ's endowment account, enabling an extensive survey of the terrestrial vertebrate fauna of the peninsula. She contributed to the fund for more than two decades, eventually expanding the scope of that work to include several other states in Mexico. As fieldwork in Baja was about to begin, Alexander wrote to Grinnell, "I agree with you that farming enterprises will change the complexion of that country before very long and it behooves us to make the most of our opportunities."[6]

Grinnell kept Alexander apprised of relevant proposals by the state and federal legislatures and the potential impact on wildlife conservation. In the early part of the century, the majority of such bills dealt with hunting, both of game species and of predators, but there were also efforts to regulate the import and export of animals and define the role of the state Fish and Game Commission. On occasion, Alexander would offer Grinnell her opinion about such legislation, generally based on personal experience in the field.

Alexander was keenly observant. Often she would return from the field with recommendations for where Grinnell might profitably send a collecting party, sometimes having viewed an area only from a distance through a pair of binoculars. Trusting that Grinnell would advise her of gaps in the museum's collection, at other times she would ask him to suggest areas from which additional series of specimens might be desirable.

On one occasion, Alexander and Kellogg decided to climb Clark Mountain, a peak almost 8,000 feet in elevation that rises up as a solitary cone from the desert floor of southeastern California. The mountain is biologically distinctive relative to its surroundings, supporting eight plant communities and a variety of interesting bird life. Museum staff and students had made several ascents, and by all accounts the climb was not an easy one. Adding to the difficulty was the lack of water on or near the summit. The women were not deterred, and the museum staff had no reason to doubt that they would reach the top. Upon their return, Alexander reported to Ward Russell, the museum's preparator for many years, that she had seen his initials carved in a tree at the summit. Russell later claimed that it was the only occasion in his life that he had ever carved his initials anyplace.[7]

Beginning with the 1908 expedition to Alaska, the quality of Alexander and Kellogg's specimen preparations continually elicited Grinnell's admiration. As a general rule, women of that generation learned to sew. Beyond that, Alexander's early training in art, and Kellogg's in manual arts, led them to create beautiful natural history specimens. Distinct from mounted taxidermied skins in lifelike poses, research specimens are uniform in shape and pose to allow detailed comparisons of color, structural variation, and size. Nonetheless, there is a definite skill associated with their creation. The more beautiful and uniformly prepared the specimens are, the greater their potential value and utility in research. But the task is not easy. The skin of mammals is thin and delicate, in birds even more so, and stretching or tearing it distorts a specimen's proportions. Every trace of fat and soft tissue must be carefully removed so that the specimen does not become greasy over time and rot or, worse yet, become subject to insect infestation. And from the moment the first incision is made in the body cavity, blood must be prevented from staining and matting the fur or feathers.

Not only were the specimens that Alexander and Kellogg collected meticulously prepared, but—in keeping with the protocol that Grinnell developed— maps, photographs of the specimens' habitats, and vegetation samples often accompanied their detailed notes. Frequently, the women preserved the contents of the animals' cheek pouches and stomachs or recorded this information in their notes. As specimens document the nature and presence of species at a given point in time, field notes simultaneously document the habitats and environmental conditions in which the specimens were collected at that same moment.

From Grinnell, the women learned the importance of obtaining not just one or two specimens of a given taxon, but of gathering a series of individuals that would reflect the complete range of natural variation within that group at a specified locale, for example, differences in age, sex, body size, and color patterns. They would then move on and collect similar series from additional localities within the animal's range so that the influence of climatic variables such as temperature and humidity on those parameters, and abiotic variables such as differences in soil color, could be properly assessed. To understand historical changes in the environment and their effect on species evolution, the women often went back to given localities in later years to obtain additional series.

Although the specimen collections made by Alexander and Kellogg were, and continue to be, invaluable as a permanent record of a vanishing fauna, within the hands of museum staff and students, these series of specimens took on even greater meaning. Examination of geographically and temporally diverse series generally led Grinnell and his students to proffer hypotheses about the evolution of a species, or groups of species, that differed markedly from the hypotheses proposed by researchers who had examined only a few individuals, usually representing a single locality. Grinnell based his classic papers on the taxonomy of pocket gophers on examination of the extensive series of these animals that Alexander and Kellogg collected from throughout the western United States. In what might be regarded as an early catalyst for those studies, more than a decade before their publication, Alexander wrote to Grinnell, "Have worked both sides of the river here for gophers . . . have thirty-three specimens. If Dr. [C. Hart] Merriam had had more material I think he would have given a different description of *Thomomys l. navus*."[8]

The specimens that Alexander and Kellogg collected, and the primary data associated with each one—for example, age, sex, reproductive condition, exact locality of capture, date of capture—were and are valuable not only to an understanding of speciation and evolution but also to the preservation of endangered species and the development of sustainable management plans for forests and wildlife. Alexander's belief that species preservation was critically important, and that public education was a vital component in achieving this objective, led her to approve and support Grinnell's involvement in a number of related arenas. Perhaps most notable was his involvement in development of the National Park system. Not only did he become a tireless advocate for its establishment, but he was also instrumental in shaping its philosophy and operation.[9] Upon returning from a trip to Yosemite in 1928, Alexander wrote to Beckwith, "We realized as we drove through the Park what a wonderful thing it is to have a reserve of the kind for the benefit of the people, where no shooting is allowed and the meadows and stream banks are not trampled by sheep and cattle. Nothing has distressed us more on our trips than to see the hideous results of unrestricted grazing."[10]

Overcoming her disdain of publicity, Alexander willingly lent her name to conservation efforts when either she or Grinnell felt they might yield a positive result, and she offered financial support to several of the prominent conservation organizations that sprouted up in the early decades of the century. From 1912–13, she sponsored the California Associated Societies for the Conservation of Wildlife, whose aim it was to influence legislation and reconstruct the Fish and Game Commission. Through her support, Alexander hoped to remove conservation from the realm of politics in which it was mired at the time and place it into the hands of the civil service, where it remains today. Beyond a donation of money, however, there is no record of her participation in the organization.

Similarly, Alexander supported the California Save-the-Redwoods League, founded in 1919. University comptroller Robert Sproul served as secretary-treasurer of the league and John Merriam as its chairman. Its executive committee included President Wheeler, several university regents, a number of eminent zoologists, and the director of the National Park Service. The league's purpose was to raise money to purchase redwood groves, establish a redwoods national park, and protect the trees growing along the state highways that were then newly under construction.[11]

The paving of roads and the construction of state highways that followed the introduction of the automobile accelerated habitat perturbation and species decline at an alarming rate, particularly in the developing West. The effect of these changes on fieldwork was considerable. On one hand, automobiles and roads increased access to remote localities, permitting more complete faunal and floral surveys. On the other hand, access to previously undisturbed habitats led to the disappearance of extensive tracts of wilderness and of the animals that had, for millennia, inhabited them. The presence of roads also meant that people could travel or live wherever they wished. A gasoline station every fifty or a hundred miles was all that was now needed.

By the 1930s, Alexander and Kellogg noted mining roads into almost every canyon that they entered in California and Nevada. Such roads, and the traffic they carried, invariably altered the faunal composition of those regions. Roads and highways also increasingly altered the behavior of some collectors by relegating their activities to locations within walking distance of their cars. Until backpacking came into vogue, one result of these changes was a skewed perception of the natural distributions of many species.

As paved roads began to redefine the contours of the country, place-names also changed. Often their meanings grew obscure or irrelevant as the cultural and topographic features to which they referred either disappeared or existed farther and farther from the beaten track. The exact location of sites in relation to the newly emerging traffic arteries became increasingly difficult to determine. With efforts worthy of the best detectives, Alexander and Kellogg sought out the localities of specimens described in the early literature so that Grinnell could compare individuals collected at these sites with later specimens that the women collected themselves. A favorite method of doing this was to interview old miners and ranchers, the owners of general stores, or any others who they hoped would recollect the name of a spring or the location of a wash. Meticulously annotated maps thus became an essential component of their fieldwork and, coupled with their collections, field notes, and photographs, contributed substantially to the museum's rapid rise in importance as a center for knowledge concerning the fauna of the western United States.

16
From "a Friend of the University"

As early as 1913 Alexander had begun contemplating an endowment for the museum, prompted by the steady growth of its collections and its increasing prominence as a research institution. Her pleasure in its success is apparent in her comment to Grinnell, "It is certainly gratifying that the Museum is more and more establishing itself a reputation as a center of learning for vertebrate zoology."[1]

Grinnell was largely responsible for the intellectual standing that the museum had attained, but he was quick to acknowledge that Alexander's financial support made such achievement possible. On many occasions, and in a variety of contexts, he tried to impress upon her the importance of her monetary contributions to the work:

> I am giving people to understand that whatever of standing and authority we possess . . . is due to our attainments in zoological research, and this in turn has been dependent upon the opportunities afforded by the founding and maintenance of the Museum. *You* have made the *research* possible. Scores of other people will give in varying amount to the cause of game-preservation, who won't give a cent for pure research; and yet it is the research which properly precedes correct handling of game problems. *You* have already done your share![2]

While Alexander recognized that her financial support was essential to the museum's development, for many years she remained chagrined that the university's president, its regents, and even the faculty of the zoology department failed to comprehend the true value of the collections or of the research programs that she had created. Grinnell attempted to reassure her:

> I think the letter from President Wheeler is fine. You must consider his limitations (and those of the regents) in forming any conception of the methods and aims of such an institution as the Museum. It seems nothing more than natural that these men should measure your work for the University in terms of the dollars involved. Money is the common standard, and, too, it *is* the money that makes the major part of our work possible. You deserve all the credit expressed, and more, on this score alone. It is nothing to be ashamed of, or to resent, if their appreciation seems to be prompted only by a recognition of the money cost of the Museum. They don't know any better, and the intrinsic value of the Museum and your work for it remain the same.[3]

Unlike many patrons, Alexander did not wish to have buildings or monuments on campus erected in her name. She believed fervently that the moneys she proffered in support of research had far greater value to the university than any physical structures that she might provide. Accordingly, she often tried to persuade the administration to acknowledge this value through additional financial commitments of its own to paleontology and vertebrate zoology. Understandably, she remained disappointed when such pledges were not forthcoming.

Grinnell also attributed the museum's growing success to the academic freedom it enjoyed under Alexander's sponsorship and to her unwavering commitment to long-term research projects. And because Alexander remained impartial to the outcomes of the research being undertaken by museum personnel, the results that he and his staff obtained and the positions that they came to advocate, on topics ranging from game management and habitat conservation to predator control and species preservation, were viewed credibly. Studies that culminated in books such as *Fur-Bearing Mammals of California* went forward unhampered by external pressure from disparate factions with varying economic interests such as farmers, ranchers, fur traders, or preservationists.[4]

Owing in large measure to this academic freedom, when Grinnell was invited to apply for a position at the Biological Survey in Washington, D.C., in the fall of 1915, just eight years after the museum's founding, he declined. He preferred to remain in Berkeley. In the process of reaching this decision he negotiated nothing for himself but simply requested that Alexander maintain the museum's annual appropriation of $7,000 for at least five more years.

Alexander had originally agreed to support the museum for seven years beginning March 23, 1908, and she continued that level of support in the months following the expiration of her original contract. At the time of Grinnell's request in 1915, however, she had not made any additional long-term financial commitments to the university with respect to the MVZ. Early in 1914 she had written to Grinnell advising him that she had redrawn her will, leaving approximately one-third of her estate to the museum (but had she died at that time, the income from her estate would have covered nothing more than his salary and Taylor's). Alexander anticipated that her financial fortunes would brighten before her actual demise and she viewed this measure simply as tentative and precautionary.

Museum expenses that Alexander assumed outside the scope of its annual budget, such as the hiring of additional staff, hinged on economic conditions and, in particular, on the price of sugar and on the extent to which tariffs were or were not imposed by Congress. Because sugar was Alexander's chief source of income, a reduction in the sugar tariff in any given year meant increasing competition from foreign growers and lower prices for Hawaiian sugar (and for her, a lower annual income). When Swarth announced in 1912 that he was leaving the MVZ to take a position at the Los Angeles County Museum of Natural History, Alexander wrote to Grinnell that she would prefer to leave his position vacant "in view of the fact of the very probable reduction of the tariff on sugar at the extra session of congress next spring.... I regret exceedingly to put any damper on the work of the Museum but do not think it wise to commit myself too much until the crisis is over."[5]

When Swarth asked to return to the MVZ three years later, both Grinnell and Alexander were delighted given his skill as a collector, preparator, and writer. But the timing worried Alexander as she wrote to Grinnell, "[I] do not want to make him an offer until we can make him an offer that would be worth his while accepting. Congress meets Dec. 4 and I trust the matter of Tariff will be taken up without delay, and be settled favorable to our interests!"[6]

Swarth's desire to return to the museum coincided with Grinnell's request for a new five-year commitment from Alexander. Two months later, the tariff issue was resolved satisfactorily in her favor and she was able to fulfill both requests. In her six-page reply to Grinnell concerning his offer from Washington, D.C., Alexander warmly acknowledged the magnitude of the museum's accomplishments in just eight years under his direction, noting:

> The [MVZ's] work so far has all been [more scholarly] than any output from the Biological Survey and is already recognized as such by those whose opinions are worth having. That they want you shows that it is recognized in the survey itself and is very gratifying indeed.[7]

In a relatively rare expression of personal feeling, she concluded, "Have written you a long letter but haven't begun to tell you how much we think of you!"

It is likely that the intrusion of World War I postponed Alexander's immediate desire to endow the museum, as the subject did not come up again until 1919. In that year she presented the regents with $137,000 in sugar stock and $63,000 in cash as a "perpetual fund" for the museum. The agreement accompanying her gift directed the regents to incorporate the moneys into the university's general endowment pool with the hope that it would generate a fixed, annual return of 5.5 percent. The terms of the endowment were straightforward, in essence ensuring the continuation of policies that had been in place since the museum's founding. The income from the endowment was to be used for staff salaries, fieldwork, curatorial expenses, and research. For its part, the university was obligated to continue adequately housing the collections and to provide utilities and janitorial services for the museum. All staff members would continue to be appointed by the regents— "upon the recommendation of the Director, the latter to consult Miss Annie M. Alexander throughout her lifetime upon these and all other important matters relative to the conduct of the Museum." In its final form, this last point in the agreement was rendered as a "desire" rather than as a condition of her gift, as Alexander feared that imposing too many restrictions might jeopardize her true intentions.[8]

The agreement also specified the duties of the staff. First and foremost they were to care for the collections. Following this, they were to conduct research and fieldwork "relative to the distribution, speciation, life histories, and economic relations of the higher vertebrates (mammals, birds, reptiles, and amphibians) of western North America, particularly California." Last, they were directed to increase and spread scientific and popular knowledge about vertebrates and their natural history. This third duty echoed Alexander's hope that public education would lead to the conservation of species and their habitats. Although the MVZ's exhibit program had ended in 1911, its commitment to public education through lecture series, teaching, publications, and collaborations with state and federal agencies had not.

The regents gratefully accepted Alexander's offer and readily assented to the conditions as stipulated. Grinnell announced the event with obvious pleasure by placing a notice in *Science* magazine, but the true measure of his feeling was demonstrated by the donation of his personal collection of bird specimens to the museum. He had selectively gathered its 8,312 individuals between 1893 and 1907 in locations ranging from arctic Alaska to the most southern reaches of the California desert.⁹

Apart from Alexander's regular allocation to the museum, her pattern of subsidizing research expeditions continued unabated following creation of the endowment. In addition, at the end of 1920 she signified her intent to enter into another five-year agreement for support of the museum. As first proposed, the arrangement specified a subscription of $11,000 annually for five years, contingent upon a matching provision to be furnished by the regents. This sum represented a decided increase from the $7,000 annually that Grinnell had requested in 1915, but it was far below what the endowment returned in 1921 when the value of Hawaiian sugar stocks soared and the endowment increased in value approximately $26,000 over what it had been worth at the time of her donation.

In 1922 Alexander wrote to the university, this time offering to set her annual contribution for the next five years at $8,100, beginning July 1, 1923, provided that the university assume responsibility for salary increases for the museum staff. This proposal not only represented an increase over her previous annual appropriation of $7,000 but also marked the first time that the university became liable for a portion of the museum's business other than maintenance of its physical infrastructure. While Alexander viewed the increased appropriation as necessary to maintain the MVZ's level of research productivity, she was becoming increasingly thoughtful and concerned about the future of her museum should something happen to Grinnell or to herself. She was also becoming increasingly sensitive to the university's expectations of future donations that she might make in the wake of her recent gift. In that concise way she had of rendering an understated yet thoughtful opinion, she wrote to Grinnell, "I don't want to be under the urge to give more and more. It is proverbial I guess that the more one gives, the more is expected of one, and while I won't say it is less appreciated, it isn't any more."¹⁰ In this context, her insistence that the university cover all future salary increases seems prompted more by good business sense and a desire to have the university assume some responsibility for the intellectual contributions being made by the museum than by any limitation on her own financial resources at the time.

Within just a few years of its founding, the studies emerging from the MVZ began to inspire researchers nationwide, who were quick to adopt the methods that had been developed and implemented by Grinnell and his staff. The studies' results also involved the museum in a growing number of conservation issues. In 1914 U.S. Secretary of the Interior Franklin Lane formally approved and authorized a multiyear natural history survey of

Yosemite National Park to be conducted under Grinnell's direction. By Grinnell's own admission, volumes such as *Animal Life in the Yosemite,* which were the culmination of that work, would not have been possible without the continuous financial support for fieldwork that Alexander bestowed over and above her regular monthly appropriations, in this instance for the decade from the Yosemite project's inception through publication of its final report in book form.[11] Recognizing and acknowledging that Alexander's contribution to the success of the museum's program was more than just monetary, Grinnell wrote in the introduction to that volume, "Her unswerving faith in the worthiness of the undertaking served continually to encourage and energize those who were concerned with its conduct."

Similarly, in 1925 Alexander began financing fieldwork in Baja California. In this instance, her rationale stemmed from the realization that "farming enterprises will change the complexion of that country before very long and it behooves us to make the most of our opportunities." Likewise, when she wrote to Grinnell in 1934 proposing that the museum staff undertake fieldwork along the Colorado River at her expense, he was quick to concur, "In other words the river-bottom very likely will become changed in important respects. By making a careful record of conditions as they exist now, just before the impounding effects of the Hoover Dam, we will have something later for comparison. The section now to be worked, be it noticed, is *above* Needles, and hence is complementary to the work I did from Needles down, in 1910."[12]

In addition to field expeditions, Alexander provided funds for the purchase of rare or unusual specimens for the collections. In acknowledging one such check, Grinnell wrote,

> It is perhaps needless for me to refer to the fact that your continued performance of acts of this sort is resulting in the phenomenally rapid growth of our bird collection, which even now is the foremost reference collection west of Chicago. This is of course gratifying to me personally inasmuch as it brings a consciousness of adequate equipment for any line of research work where specimens are essential. Students of birds generally are realizing this fact, and looking to us more and more as an authoritative center for information concerning the Ornithology of the Pacific region.[13]

When the value of Hawaiian sugar stocks rose dramatically in the years immediately following the establishment of the MVZ's endowment, the university consulted Alexander about the advisability of retaining her stock or selling it and reinvesting the proceeds. Although Grinnell credits her with advising the university to keep the stock, in fact Alexander informed the comptroller's office that she assumed no responsibility for the management of the fund and directed the university to an individual in San Francisco with whom they might consult should they feel so inclined. She merely wished to see an annual yield of 5.5 percent accrue from her gift.[14]

What came to light through this inquiry, however, was that her gift had not been put into the university's general fund as originally intended and, by 1929, had accrued an additional $25,000. Learning of this surprising turn, Alexander asked that this sum be combined with the principal of the endowment and that the income generated be added to the museum's annual budget. But, rather than allowing the interest to accumulate indefinitely as originally proposed, Alexander decided that a safer strategy would be to apply any annual surplus from here on directly to the museum's budget as a reserve fund in the event that the endowment did poorly in a given year, and she directed the university to alter her original agreement accordingly.[15]

If the depression affected Alexander's financial position appreciably, she shielded her research programs from its impact. When Grinnell took a rare vacation in the summer of 1932, he wrote to Alexander from South Fork Mountain on the Humboldt–Trinity County line in northern California. His words convey the warm and sincere respect he had for Alexander. More than simply a patron or benefactor, she was a fellow visionary, committed as he was to the great experiment they had undertaken jointly. In four handwritten pages, Grinnell wishes Alexander to know that he personally, as well as the museum, is ready to make the necessary budget cuts if she is struggling financially to maintain her gifts to the university, and offers to take a salary reduction of up to 25 percent as a first step.[16]

In a reply that reflected a measure of her own affection for Grinnell, as well as her love of the manner in which she had chosen to live, Alexander responded simply, "Don't worry about my supporting the Museum in these hard times. Things can't get much worse can they! It is a good thing to have an absorbing interest and not to be devouring the papers for financial news all the time."[17]

Following a conference that the two held in the spring of 1933, Grinnell felt that Alexander was on the verge of modifying her regular payments to the university.[18] With other departments on campus facing budget cuts and salary reductions in the wake of the country's financial crisis, he now anticipated the same for the museum. While intimating that her gifts would not cease, Alexander had requested information about the university's appropriation for the museum for the coming year. In addition, she had not approved expenditures from the MVZ budget for a proposed installation of a beam in the museum from which large mammal skulls and antlers would hang.

In light of Grinnell's concern, Alexander's proposal to double the size of the MVZ's endowment must have come as a great surprise. This she did in the fall of 1935, formally donating an additional $225,000 to the university. While Alexander continued to fund MVZ field expeditions apart from this gift, she now proposed to deduct the amount earned by the endowment from her monthly remittance in support of the museum's daily operating expenses. In typical fashion, she closed her letter to the regents by requesting, "I ask that this gift, if accepted, be entered on the books of the University not in my name but in the name of 'A Friend of the University.'"[19]

Given the museum's committed and directed field program, and Alexander's own prodigious specimen collecting, within fifteen years of its founding the MVZ stood third nationally in the size of its mammal collection. Only the Biological Survey in Washington, D.C., and the American Museum of Natural History in New York had greater numbers of mammal specimens but their collections differed greatly in taxonomic scope and geographic representation from the MVZ's, whose holdings of mammals from the western United States had become the largest in the world.[20]

In August 1910 C. Hart Merriam made a second visit to the museum to examine those bear specimens that he did not already have on loan in Washington, D.C. The trip was prompted not only by his taxonomic revision of North American bears but also by his need to examine material of the western fauna for his monumental work on the mammals of North America. His presence on campus confirmed John Merriam's earlier assertion that Alexander's research collections would make the university a recognized center for the study of birds and mammals. In less than three years' time, the MVZ collections had grown sufficiently in size and scope that they could no longer be ignored by investigators on the East Coast or in the Midwest, including the chief of the Biological Survey.[21]

Merriam may have been the MVZ's first visitor of note, but he was not its last. After Theodore Roosevelt's two terms as president of the United States (1901–9), the transition to private citizen was not easy, and he turned his attention and energy to the outdoors. His first step was to mount a major expedition to British East Africa and the Sudan, an activity that he felt sure would sustain the level of adrenaline to which he had become accustomed during his years in politics.

Roosevelt wrote to Alexander following her return from Alaska in 1908, outlining his plans for the trip and requesting that Edmund Heller be permitted to accompany the expedition. Both Alexander and the mammalogist had apparently come to Roosevelt's attention through C. Hart Merriam, Roosevelt's lifelong friend and the individual responsible for selecting most of the expedition's personnel.[22]

Alexander now faced a difficult decision. The fact that a MVZ staff member should come to Roosevelt's attention only a year after the museum's founding was undoubtedly flattering. She knew that Heller had become fascinated with Africa while accompanying Carl Akeley to the continent just before the MVZ hired him and might not return to the museum if Roosevelt now chose to make Africa "his country." Yet Alexander's own safari experiences four years earlier made her sympathetic to Roosevelt's request and to Heller's desire to join him. She graciously assented.[23]

Roosevelt's African exploits did nothing to diminish his appeal with the public. Still a national and international figure after his return to the States, in 1911 he was invited to serve as the Charter Day speaker at Cal. In accepting the invitation, Roosevelt not only requested a meeting with Alexander and a tour of the MVZ but also gave the university, through the museum, the skin, skull, and skeleton of a large bull elephant that he had shot, an exploit that he vividly described in *African Game Trails*.[24]

Fully cognizant of Alexander's dislike of publicity, Grinnell struggled masterfully to avert a scene while orchestrating the former president's visit. He wrote to Alexander several days in advance of the occasion, "In attempting to arrange Roosevelt's visit, I have tried my best to avoid a crowd. I planned to have *just* President Wheeler, Dr. J. C. Merriam, yourself and your friends. But it is surprising how importunate many people can be, when it comes to forcing their presence, when there is a conspicuous person to meet."[25] Wishing to provide Alexander an opportunity to talk with Roosevelt undistracted, Grinnell proposed that they tour the museum's skin room. Unfortunately, no record of the afternoon survives.

As Alexander predicted, Heller never returned to the MVZ. He and Roosevelt continued to conduct fieldwork in Africa over the next several years and the young mammalogist eventually coauthored *Life-Histories of African Game Mammals* with the former president.[26]

Perhaps one of the highest compliments paid to the museum during Grinnell's lifetime occurred in 1939 on the eve of C. Hart Merriam's departure from the Biological Survey in Washington, D.C. The "dean of American mammalogists," as Grinnell referred to him, chose to donate his accumulation of more than seventy years of research materials—publications, photographs, clippings—on mammals to the MVZ. This gesture was prompted by his feeling that, in so doing, "this unique collection will be available to the largest number of present and future Zoologists."[27] The Biological Survey may have held the largest number of mammal specimens in the country, but Merriam recognized that its research productivity was sorely lagging. The museum that he had encouraged Alexander to establish more than thirty years earlier had, in that interval, grown in scope and reputation to become the "center of authority" for the study of vertebrate evolution on the West Coast that both she and Grinnell had envisioned.

17

Founding a Museum of Paleontology

Alexander's collaboration with Grinnell was in striking contrast to her relationship with John C. Merriam. While the MVZ continued to grow steadily and its reputation to blossom, the program in paleontology at Cal proceeded by fits and starts. Alexander's championship of vertebrate paleontology in 1910 had led to the creation of a separate Department of Paleontology and held out promise of a bright and successful future for research in that discipline on the Berkeley campus. The specimens that she and Kellogg contributed to the collection were coupled with generous funding on her part for fieldwork. The moneys that she proffered enabled Merriam to carry on his excavations in the Pleistocene caves of California, at Rancho La Brea, and in the Miocene Barstow formations in San Bernardino County. The fossils uncovered at these sites provided a wealth of material that served as the basis for many notable publications. Simultaneously, Alexander provided travel funds and encouraged Merriam to promote his findings, and the department's burgeoning program of research, at national meetings.

What Alexander had not predicted was that Merriam would slip away from research. In the decade following the founding of the Department of Paleontology, he began to devote an increasing proportion of his time and energy to administrative matters and self-promotion. Commensurately, the tension between himself and Alexander gradually increased. In contrast, Grinnell's involvement in public policy, at the state or national level, centered on his research agenda: to understand the mechanisms of evolution and preserve the natural world around him. Power outside that realm did not tempt or interest him.

While Merriam effusively acknowledged Alexander's contributions in his many publications, his letters to her did not mask his growing desire for recognition and influence outside the confines of Berkeley, a campus still regarded as somewhat of a rural outpost among academics at the time. His willingness to supplant research with administration was visible to Alexander as early as 1917 when she wrote to Beckwith,

> Grinnell keeps to his course with a single-mindedness that is most comforting. Some of our good scientists have gone service-mad like my Dr. Merriam who thinks he is bearing the whole burden of national defense on his shoulders. Paleontological research has gone to the winds for the present although he asserts he is still doing some work. I also assert that a man cannot serve two masters. Destracting [sic] duties "when their minds need their keenest edge to cut their path to the elusive truth."[1]

Alexander hoped to redirect Merriam's interests toward research but found her efforts thwarted by an administration that strongly supported his heightened visibility at the national level. Wheeler viewed Merriam's appointment as chairman of the National Research Council in 1919 as evidence of his prominent standing in the scientific community and, by extension, of the rise in stature of the University of California under his own presidency. For this reason Wheeler took up Alexander's campaign to ward off outside offers and keep the now nationally recognized paleontologist on campus.

Despite Wheeler's efforts, in 1920 Merriam left Berkeley to become president of the Carnegie Institute of Washington. He did not completely sever his ties with the university but asked to retain a research appointment in the Department of Paleontology. His request was granted by an administration courting exactly these sorts of affiliations.

For her part, Alexander felt angry and betrayed. In writing to Grinnell about Merriam's departure, only a thin veneer of restraint covers her thoughts:

> Dr. Merriam's disaffection and desertion of his Department which has owed much of its development to my support of research work since 1900 . . . has made me in a way lose my sense of direction. . . . You all seem interested in your work and I am more than ever convinced that it is quiet research that brings results—not two, three or four months a year but every month of the year. I have never heard it said that any branch of science could be exhausted and when a man begins to feel he is too big for his special line of research—to my opinion he ceases to be a scientist and becomes a politician. So I hope in your case you will find your work interesting enough to always claim your best endeavors.[2]

Three months later, still upset at this turn of events, Alexander wrote to Beckwith of Merriam's departure. He purportedly planned to spend three or four months a year pursuing paleontological research, but she had no faith in his ability to accomplish anything of import under this new arrangement:

> To me it is the case of Browning's Lost Leader—
> "Just for a handful of silver he left us—
> Just for a riband to stick in his coat"—
> I have supported the research work of the department for twenty years and it has really become a part of my life—to see the work open up from year to year, and the great possibilities. To me it seems incredible that a man professedly a scientist with opportunities scarcely equaled anywhere, should leave his work for an administrative position, honorable and influential though it is.[3]

Merriam's departure left a hole in the department's leadership. In the vacuum, the paleontologists who remained made an uneasy affiliation with the researchers and faculty in the Department of Geology. Merriam had suggested no one in-house to replace himself as department chair. Still hoping to have a say in departmental affairs and in the use of Alexander's research fund, he began to lobby for the hiring of Ralph Chaney, a paleobotanist from the University of Iowa, as his replacement.

During this uneasy period of transition, Alexander kept her focus on research. It was the discipline, not the individual, that inspired her commitment. Accordingly, in December 1920 she renewed her five-year pledge of support for the paleontology program. On this occasion, however, she made two additional stipulations—that the university bear an increased share of the financial responsibility and that the Department of Paleontology remain autonomous, separate from the Department of Geology.[4]

Merriam's departure from Berkeley coincided with high-level administrative changes in the university. Wheeler's retirement in July 1919 had marked the end of an era. During his presidency, the University of California had grown from a single campus with several hundred students to a multicampus complex spread across the state. Because the regents felt that no single individual could master all aspects of the university in order to govern it successfully, they appointed an "administrative board" to replace him.[5] The result was disastrous. Wheeler had been an academic. The decisions he had made and the priorities he had set for the university's growth and development reflected this background. The faculty refused to accept the sudden change in direction that the newly appointed board portended. Within six months, the regents reconsidered their action and elected David

P. Barrows (1919–23), former dean of the academic faculties, to succeed Wheeler.

Barrows was also an academic, yet his appointment marked a first step in the transition from the appointment of academicians to the appointment of administrators to the university's highest position. As incoming president, Barrows could appoint, promote, demote, and dismiss faculty members—but only after he had consulted with the Academic Senate. And although he could make recommendations—after consultation with the senate—the appointment of deans and directors was now in the hands of the regents.

It was Barrows who responded diplomatically, and somewhat evasively, to the stipulations outlined in Alexander's proposal for a continuing five-year agreement with the university. In thanking her for her generous offer, he responded that the regents agreed to appropriate the moneys necessary for additional support of the department as she had requested, "in order to make the most effective use of the fine sum which you propose to supply." However, reflecting his diminished administrative capacity and foreshadowing future conflicts that she would have with the regents, he assured her only that the Department of Paleontology would not be consolidated without "the general support of the men composing it."[6]

Throughout 1920 Alexander remained frustrated with Merriam's attempts to direct the department from a distance and manipulate the research funds that she was providing. What underlay the issue of money was her belief that her former mentor had abandoned the ideals that she strove to foster through her financial support of research. In contrast, Grinnell implemented them. In a lengthy reply to the MVZ director, acknowledging the copy of a manuscript that he had sent to her for comment, entitled "The Museum Conscience," Alexander applauded his emphasis on "morale fibre" and expressed her indignation with Merriam. She critically contrasted what she felt was the significance of Grinnell's analysis with comments by Merriam that she had read the previous year in his paper on "constructive thinking." With the issue of "moral fibre" obviously much on her mind, Alexander wrote to Beckwith that she had been reading Thomas Huxley, "Now there's a man that I admire—one that has real moral fibre! I wish I could have been associated with a type like that all these years!"[7]

Grinnell's reply to Alexander is further evidence of the unusual relationship that these two individuals maintained. He thanked her for her generous commendation, noting:

Your own personal standards are extraordinarily, almost uniquely high, and you naturally want other people to approach them in their own conduct. This situation keeps me, for example, continually on trial. *But*, I grant freely, this is an *excellent thing*. A function which you can, and do perform, and have done from the start, is to exercise a vigorous influence for high standards in all that pertains to this Museum and to the individuals which are associated with this institution. That is a big factor in our existence and in whatever grade of achievement the Museum attains to.[8]

Because Alexander was unable to counter Merriam's continued interference in the Department of Paleontology, she chose to exert her own authority by circumventing his sphere of influence. In the spring of 1921 she wrote to the university offering to create a museum of paleontology separate from the department whose creation she had helped foster more than a decade earlier. In her proposal, Alexander signified her intention not only to continue her support for paleontological research but also to increase her annual appropriation for research—*if* the university would provide the new museum with a budget for equipment and laboratory expenses separate from that of the department. She proposed that the museum be placed under the direction of a single individual who would be responsible for its budget and for her continuing donation of funds. She also stipulated that all paleontological collections that belonged to the department would now become property of the new museum. In essence, Alexander's creation of a paleontology museum was an attempt to replicate the success of the MVZ, which had been established as an autonomous research unit within the university under her direct control. Merriam might retain his affiliation with the department, but he would not be a member of the new museum, and he would no longer have access to the moneys that she was providing for research.[9]

The regents were quick to accept Alexander's proposal, envisioning in its development an institution of academic excellence comparable to the MVZ. In so doing they agreed to assume the expenses for an assistant to the director and a museum curator to look after the collections, as well as to provide an annual stipend for expenses associated with running the new museum and alterations to the existing building that Alexander felt were needed. But in their eagerness to take advantage of Alexander's offer, the regents seem to have given relatively little thought to the administrative repercussions and schism within the department that might follow the birth of the new museum.

Almost immediately upon creation of the University of California Museum of Paleontology (UCMP) in 1921, the Department of Paleontology once again became subsumed within the Department of Geology. Bruce Clark, an invertebrate paleontologist and biostratigrapher, was appointed director of the new museum. Clark had earned a master's degree in geology at Cal in 1909. While working toward his doctorate he had been appointed an instructor in the newly formed Department of Paleontology. On receipt of the degree he became an assistant professor of paleontology. Although Clark's research focused on invertebrate paleontology—a discipline frequently allied with geology rather than vertebrate paleontology—the quality of his research had earned him Alexander's respect, and his attempts to create harmony among the disparate factions within the department had made him her choice to head the new museum. Joining Clark was Charles Camp, a vertebrate paleontologist who was first introduced to reptiles and amphibians by Grinnell as an undergraduate at Throop Institute. Having completed a doctorate at Columbia, Camp had returned to California to pursue his career. Both paleontologists would retain faculty appointments in the newly reorganized geology department, in addition to their museum affiliation. The vertebrate preparator, Eustace Furlong, was also given an appointment in the new museum but kept no affiliation with the reorganized Department of Geology.[10]

Alexander's establishment of the UCMP did not deter Merriam from asserting what Grinnell referred to as "retention of his prerogatives in paleontological matters"[11] at the university. He continually tried to influence faculty appointments in the Department of Geology and challenged Clark's authority over the museum. By the end of 1921 a rift of such magnitude had developed between these two scientists relating to details in organization of the new museum that it threatened Clark's ability to direct the research program.

During this difficult period of transition in paleontology, Barrows offered little support or guidance to Alexander. He was struggling to maintain control of the presidency, but in 1923, after only three and a half years, William W. Campbell, director of the university's Lick Observatory, replaced Barrows as president (1923–30).

Campbell had the faculty's respect and brought to his new position commendable experience both as a scholar and as an administrator. Aware of the difficulties that Barrows had faced, Campbell wrote to the regents before accepting the post to demand certain conditions that he felt would strengthen the office of the president. Aware of Campbell's greater authority, Alexander was vexed and disappointed when no action followed her letter to him in 1924 about her displeasure with Merriam's continued connection to, and interference with, the Museum of Paleontology.

In Alexander's mind, the most egregious of Merriam's offenses centered on research papers that he and his Berkeley colleague Chester Stock were publishing under the byline of the Carnegie Institute, papers based on university specimens that both she and her research funds had helped procure but for which neither the university nor the Museum of Paleontology received any acknowledgment. Although disdaining publicity herself, Alexander wanted her museums to receive proper credit for the fieldwork, specimens, and research programs that had developed under her support. With uncompromising vision, she wrote to Beckwith,

> It is interesting to see how he [Merriam] has developed since he was a humble professor and is now in a position where he handles several millions of dollars and has great honor. He is getting so accustomed to running things that anything in a scientific way that looks at all promising he wants to have a finger in, in fact dominate. My private opinion of him is that he is a bad man, vindictive and unscrupulous. It seems to me that I can read in his face that he has incurred opposition and hostility by his policy. Science has become a political game with him, nothing more. I don't believe we can ever get together and maybe I shall be glad to withdraw when my pledge is up. It is a humiliating situation we are in and I see no way out of it but I mustn't bother you with it.[12]

Whereas Merriam's error in failing to acknowledge the university in his publications accentuated the rift that had developed between Alexander and her former mentor, Campbell's neglect in addressing Alexander's concerns reflected the increasingly unresponsive administrative bureaucracy with which she was negotiating. By 1926 the UCMP's director could no longer abide the continued acrimony between paleontologists and geologists in the department and the tension it created in the museum. Clark resigned his position. Merriam and Stock were now both nationally recognized authorities in the field of vertebrate paleontology but, given their disruptive influence on the museum's research program, Alexander knew it would be difficult to attract another individual of stature to assume the directorship of the UCMP. Once again she wrote to Campbell, this time inquiring as to how the university wished to resolve what she felt was a blatant conflict of interest—that of Merriam's agenda versus her own, a conflict she felt threatened the very core of a flourishing research program.

Alexander interpreted Campbell's lack of swift and decisive action in response to her letter as a failure on the part of the administration and the regents to grasp the paleontological collections' inestimable value. She regarded their hesitancy to name a successor for Clark as forfeiture of the discipline's research preeminence at the university and further evidence of Merriam's inappropriate influence on museum affairs. When the university finally appointed Ralph Chaney to succeed Clark as director of the UCMP, Alexander began to question the administration's very integrity. From her perspective, Chaney's selection was a triumph for Merriam over the concerns she had been voicing.

Alexander's response to the university's action was calculated and decisive. She immediately withdrew her financial support for the museum—until such time as the administration secured a satisfactory director for the museum and articulated its understanding of a clearly defined and creditable program of research. These measures gained the attention she desired. The university reversed its decision and replaced Chaney as the UCMP's director.

Within a year of Clark's resignation as director of the UCMP, paleontology again split from geology to form a separate department, and William Diller Matthew arrived from the American Museum of Natural History in New York to assume both chairmanship of the new Department of Paleontology and directorship of the museum.[13] When Matthew accepted the appointment at Berkeley he was already well known for his publications on fossil mammals and biostratigraphy, and he brought to the UCMP the sort of leadership and research focus that Alexander desired. His arrival marked the turning point for paleontology at Cal that Alexander had long envisioned, and for which she held out great hope.

Alexander revealed the depth of her feeling about Matthew's appointment, and the extent of the hope that she pinned on his leadership, in both word and deed. To Beckwith she wrote, "Dr. Matthew comes out to California July 1—it is the salvation of the work at the University to have such a man in charge." At the same time she wrote to Campbell offering to renew her support for research in paleontology for an additional year by pledging the sum of $900 monthly for a period of twelve months. At the end of that year—"or at such time as the work becomes well organized"—she tentatively proposed to donate $100,000 to the university as a permanent endowment for the UCMP. Finances permitting, she further offered to supplement the income from the endowment for at least five years so that the museum's budget would remain at $900 per month. "It is my ambition to see the Department of Paleontology one of the strongest research departments of its kind in the country," she wrote.[14]

For the next several years, research in paleontology at Cal proceeded according to Alexander's vision and she generously maintained her annual financial support. Under Matthew's leadership, fieldwork in paleontology expanded eastward to the Rocky Mountains and as far as the Great Plains. Enrollment by both graduate students and undergraduates in the department increased dramatically. Then in 1930, just as the program was gaining momentum, Matthew became seriously ill. His kidneys failed and, after a brief hospitalization, he died.

Alexander was stunned. The far-reaching plans that Matthew had set in motion might have transformed paleontology at the university. But four years had not been sufficient to implement them all and, once again, the lack of strong leadership in the department and a change in the university administration threatened the program's future.

The year of Matthew's death also marked the appointment of Robert Gordon Sproul as president of the university (1930–58). Generally regarded as a man of unusual ability and personality, Sproul was well respected by the faculty and regents. He had been an undergraduate at Cal and had served as assistant comptroller for the university during the Wheeler presidency. Sproul had been promoted to comptroller in 1920, the year after Wheeler's retirement. In that position Sproul interpreted his responsibilities broadly, garnering many of the administrative responsibilities that Wheeler had maintained as president but had eluded both Barrows and Campbell. In 1930 when the regents accepted Campbell's resignation after seven years as president, Sproul's immediate appointment as his successor came as no surprise to the faculty.[15]

Sproul's years of service as comptroller had made him intimately familiar with Alexander's benefaction to the university and its contingencies. Owing perhaps to this fact, he arranged to have Grinnell chair a subcommittee to look into the matter of Matthew's successor. Based on Grinnell's years of service as director of the MVZ, he took it upon himself to write to the president, "As regards the nominations from your Committee on the Paleontology situation, may I make the unofficial, personal suggestion that it might be well to inform Miss Alexander (and to seek her reaction) in advance of making an offer to any nominee."[16] While there was no administrative reason for the university to involve Alexander in its decision, her opinion was still considered worthy of respect and consideration, given her substantial level of ongoing support and the priority that she placed on research excellence.

Alexander had hoped that Charles Camp, the vertebrate paleontologist, would be named both chairman of the department and director of the museum following Matthew's death. In an attempt to further this end, she met personally with the chair of the geology department and the dean of the college. Despite her efforts, this option was not one of the committee's four recommendations presented to her by the administration. Her immediate reaction to the university's proposals was to reply that she was not yet ready to commit herself. Her lack of enthusiasm for their suggestions resulted in a meeting with Sproul in which she and he agreed that Camp would become director of the UCMP if Alexander would continue her financial support of the museum. Against her wishes and much to her displeasure, Chaney was to become chairman of the Department of Paleontology.

Chaney immediately began trying to wrest control of the museum away from Camp. In light of Chaney's actions, Alexander wrote again to Sproul, on this occasion unambiguously defining the relationship of the museum vis-à-vis the department that she felt must now be actively implemented. Included in her letter was the stipulation that there be "complete separation" of museum and department funds, with Camp retaining sole charge of the museum funds and of its collections. She also wished to see complete physical separation of the two units, with separate staffs for each. Finally, she specified that an additional line in the university's budget be added for support of the museum, thus reflecting its status as an independent research unit on campus. If the president approved these suggestions, Alexander stated that she would continue her support of the museum and would increase her annual donation for the coming year. If not—the implication was clear—she would withdraw her support. Whereas she was requesting only $700 per year in matching funds from the state for the UCMP, her own contribution would be increased to $11,800 through this arrangement. The university agreed to meet her demands.[17]

The complete separation of the UCMP and the Department of Paleontology having now been disposed of, Alexander's next battle with Sproul was over the acquisition of new and adequate space to house the fossil collections. Incredibly overcrowded in the basement of Bacon Hall, one of the first buildings erected on the Berkeley campus, the specimens were barely accessible and at risk from fire.[18] She had raised this issue with Campbell in 1926, stating at that time that it had been eleven years since a request had first been made for a fireproof building. Apparently no action had been taken in the interim. With Grinnell's assistance, Alexander had recently succeeded in negotiating the move of the MVZ into the newly constructed Life Sciences Building on campus, vacating the original aging and cramped galva-

Top: Alexander bought "Blundie," a six-wheel Ford with detachable tractor treads (designed for use on snow), because she envisioned it as an ideal vehicle for driving across desert sand; location and date of photo unknown. Courtesy Alice Q. Howard.

"Birdie," the women's much-beloved Franklin, had an air-cooled engine and a bumper immediately behind the rear wheels to make the car favorable for desert travel as well; Ebbetts Pass Highway, August 31, 1945. Courtesy archives, University and Jepson Herbaria, University of California, Berkeley.

Left: Joseph Grinnell, the Museum of Vertebrate Zoology's first director; University of California, Berkeley, December 1930. Courtesy Museum of Vertebrate Zoology archives.

The original Museum of Vertebrate Zoology was a corrugated iron building situated in Faculty Glade. It was razed after 1930, when the museum moved into the newly constructed Life Sciences Building; University of California, Berkeley, January 27, 1928. Courtesy Museum of Vertebrate Zoology archives.

Steller sea lion in one of the three dioramas Alexander commissioned as exhibits for her new museum (the others were of California sea lions and mountain sheep). Courtesy Museum of Vertebrate Zoology archives.

Top: The library on the second floor of the original Museum of Vertebrate Zoology served for both specimen examination and the writing up of reports. Courtesy Museum of Vertebrate Zoology archives.

The large atrium on the first floor of the museum was intended as an exhibition hall but came to house specimen cases as the museum's collections grew. Courtesy Museum of Vertebrate Zoology archives.

Alexander (*left*) and Kellogg at an exhibit of their prize-winning milking shorthorn cattle; location and date of photo unknown. The banner behind their cows reads "Innisfail Farm," the name of their ranch. Courtesy Museum of Vertebrate Zoology archives.

Alexander and Kellogg's AK brand of asparagus, with its distinctive orange and yellow grizzly bear logo, soon became recognized as the best in the local market. Courtesy Museum of Vertebrate Zoology archives.

Tidal flooding of fields on the women's farm on Grizzly Island; date unknown. Courtesy Museum of Vertebrate Zoology archives.

Top: AK brand asparagus on the ferry Alexander paid for and named for her farm; the ferry connected Grizzly Island to the town of Suisun. Courtesy Museum of Vertebrate Zoology archives.

The unusual fat-wheeled tractor that Alexander purchased for the farm (a forerunner of the caterpillar tractor), intended for use on the farm's soft soil. Courtesy Museum of Vertebrate Zoology archives.

Top: In 1923 Alexander (*center, wearing her usual field garb*), Kellogg (*not pictured*), and a party from their hotel visited the grotte de Niaux, a cave in southern France, shortly after it had been opened to the public. Courtesy Museum of Vertebrate Zoology archives.

In 1924 Alexander (*left*) and Kellogg traveled by camel to explore mammalian fossil deposits in the Faiyûm, on their guided tour of Egypt and Palestine. Courtesy Alice Q. Howard.

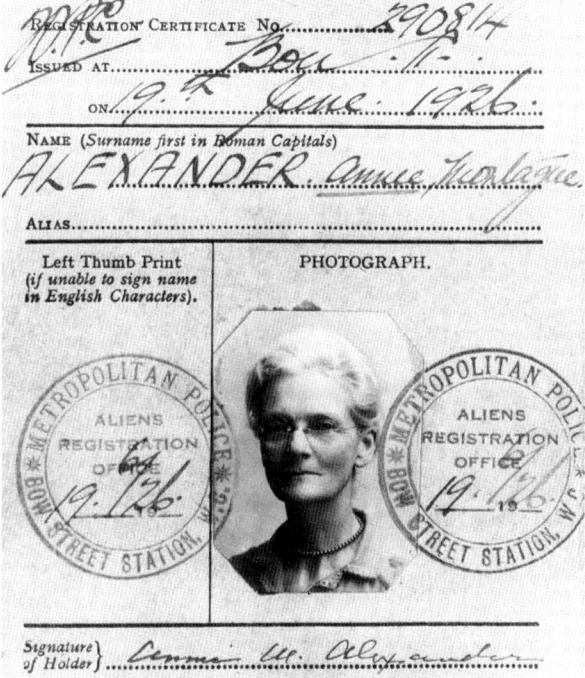

Top: Among the other members of Temple Tour no. 126 at the great pyramids at Gîza, Alexander and Kellogg are in the center of the photo, front row, wearing large, floppy hats—Alexander to the left on a light-colored donkey, Kellogg to her right on a dark-colored one. Courtesy Alice Q. Howard.

First page of Alexander's passport, issued in 1926, when she sailed to England for diagnosis and treatment at the Hospital for Tropical Diseases in London. Courtesy Museum of Vertebrate Zoology archives.

Kellogg in the field, wearing the practical, comfortable clothes the women preferred to skirts and dresses; location and date of photo unknown. Courtesy Alice Q. Howard.

Top: The large circular Arab tent Alexander bought in Egypt for conducting fieldwork in the States; it dwarfs their standard camp tent and truck; locality and date of photo unknown. Courtesy Alice Q. Howard.

Campsite in Saline Valley on the eastern side of the Sierra Nevada, late December 1936; $3^{1}/_{2}$ miles east of Lead Canyon, Inyo County, California. Courtesy Museum of Vertebrate Zoology archives.

Top: The Franklin unable to defeat the depths of snow that clogged the entrance to Saline Valley; January 2 or 3, 1937; 1 mile northeast of Waucoba Spring, Inyo County, California. Courtesy Museum of Vertebrate Zoology archives.

Kellogg collecting desert plants along the road west of Burro Creek, Mohave County, Arizona, June 24, 1932. Courtesy Museum of Vertebrate Zoology archives.

At the base of sharp, conglomerate rocks 12 miles northeast of Cedarville, Modoc County, California, the women set up camp; the morning of September 22, 1948, the thermometer had registered 34° F. Courtesy archives, University and Jepson Herbaria, University of California, Berkeley.

Top: Alexander's Dodge Power Wagon, purchased for a three-month expedition to Baja California in the winter of 1947–48, has a steel mesh shell constructed over the bed; location and date of photo unknown. Courtesy Alice Q. Howard.

Mary Erickson, the first female graduate student in the Museum of Vertebrate Zoology, who went with Alexander and Kellogg on a two-month collecting trip through southern California and western Arizona in the winter of 1933–34; Quitobaquito, Puma County, Arizona, January 23–29, 1934. Courtesy Museum of Vertebrate Zoology archives.

Left to right: Jack Waterhouse, Louise Kellogg, and two Samoan friends in the summer of 1934 seated on the steps of a *fale,* a Samoan guest house. Courtesy Alice Q. Howard.

Alexander (*left*) and Kellogg, location and date of photo unknown. Courtesy archives, University and Jepson Herbaria, University of California, Berkeley.

The photograph Alexander had taken at Grinnell's request in the fall of 1935 (and, she stipulated, not to be hung in the museum during her lifetime). Courtesy Museum of Vertebrate Zoology archives.

nized iron building that had served as the MVZ's home since its founding in 1908.

Sproul was not Alexander's only avenue of recourse in this matter. She approached the task of gaining support for a new building by meeting individually with several of the regents on the university's Finance Committee in the fall of 1930. She related the outcome of this strategy to Beckwith, noting that of the six individuals with whom she had met, "four were sympathetic but dubious that funds could be secured" and two seemed "quite indifferent. Even the possibility of my terminating support has little weight—I guess the subject of paleontology doesn't interest business men. Hell."[19]

The next several years represented a period of relative tranquillity for the paleontology museum, but the issue of a new building for the UCMP remained open. Alexander's continued support of its research program enabled faculty and students to produce a continuous stream of creditable publications under its byline. For the first time since Matthew's death, Alexander again contemplated the museum's long-term future. For reasons never specified, Alexander had not endowed the UCMP as she had proposed in a 1927 letter to Campbell. In December 1933, hoping to ensure the viability of the institution and safeguard its collections in perpetuity, she wrote to Sproul once again outlining her thoughts in this vein and indicating a sum of $25,000 to inaugurate the plan. Her offer was contingent on the regents' commitment of financial support for the museum, with housing for its staff and collections.[20]

In the spring of 1934 Alexander again wrote to Sproul. Her letter was similar to that written three months earlier but with two subtle differences. The first was that she now offered to donate $30,000 as an endowment for the UCMP rather than $25,000. The second was that her gift was contingent, not only on the regents continued support of the museum but also on their agreement to support the UCMP "as a separate administrative unit for twenty-five years, at least."[21] Alexander hoped that the assurance both of support and of autonomy for the UCMP would ensure a bright future for research and education in vertebrate paleontology, free from external interference.

At their meeting three weeks later, the regents voted to accept Alexander's offer. Upon Sproul's formal announcement of Alexander's gift, Grinnell wrote to her of his personal pleasure in the matter. He believed, as she did, in the intrinsic value of research in all areas of "pure science," and he reiterated his gratitude for all that her vision and financial support had provided researchers on the Berkeley campus. He wrote,

> I believe you are wise thus to scatter your "investments" in men and in fostering scientific work. . . . Being able to do this *now* means that you can *watch* progress and get the concurrent satisfaction from successes, which is the real reward for planning and effort.[22]

In August 1934 Alexander increased the UCMP endowment by an additional $60,000, part of which was given as cash, the majority represented by the transfer of 1,000 shares of common stock to the university.[23]

Grinnell had envisioned Alexander gaining satisfaction from watching the UCMP's progress and success after creation of its endowment, but by 1938 any such feelings had turned to frustration as friction between the department and the museum intensified once again. Even though the department no longer had access to Alexander's funds, it did influence the museum's research program through control of the teaching curriculum and, thus, of student training in vertebrate paleontology. Hence in April 1938, when Sproul wrote confidentially to Alexander that he, the faculty, and the regents had voted unanimously to invite her to receive an honorary doctorate, she declined. Her reply shows her disappointment and displeasure over the university's failure to intervene decisively in departmental affairs on behalf of the museum:

> I assume, of course, that this honor is in recognition of the many years I have supported the research work of the Museums of Vertebrate Zoology and Paleontology but, President Sproul, as you know and have known, I am not happy in my relations with paleontology. What would mean far more to me than any degree the university might confer would be to see Dr. Charles L. Camp made head of the Department of Paleontology, and on his own terms, that the dignity and usefulness of the department may become established beyond question and a bright future assured. No degree could compensate for the disheartening situation which exists and which I need not discuss here.[24]

Neither Sproul nor Alexander made further reference to this exchange, and its effect on the deteriorating situation in paleontology can only be surmised. Because of the paleontology department's historical relationship to the geology department, courses in the paleontology department emphasized training in invertebrate paleontology. Camp felt strongly that students majoring in vertebrate paleontology and zoology were not receiving sufficient preparation for upper-division courses. He wrote to Sproul in May 1938, requesting that courses in vertebrate paleontology be considered a function of the museum and placed under the UCMP's control. He also asked that museum staff be allowed to enhance the curriculum by offering a course in vertebrate paleontology of the Tertiary. And, Camp mentioned, factions in the museum and the department often gave students conflicting advice and used the students as pawns during doctoral examinations—unacceptable behavior within academia. In echoing the original terms of Alexander's agreement with the university, he suggested that the office of the department chairman be removed from the vicinity of the museum, preferably to another building on campus.[25]

Shortly after writing to Sproul, Camp left to conduct fieldwork for the summer. The gravity with which Alexander regarded the university's pending response to his letter, however, is clear. As soon as Camp returned in mid-August, she wrote asking him if the administration had acted upon his request. In concluding, she stated unequivocally, "Upon its attitude in this matter for or against,—the establishment of a greatly improved teaching program, or rejection on purely political grounds, depends my whole future connection with the University. It seems to me the time has been ample in which to inaugurate the new program for the semester."[26]

A response from the administration was not immediately forthcoming. The fall semester came and went. Early in 1939 Alexander drew up a list of her financial contributions to the paleontology program at Cal, from 1899 when she wrote her first check, through 1938, the year of her most recent support of Camp's fieldwork. The total was $300,622.12. A "Statement of Progress of Work in Paleontology under the Alexander funds" was also drawn up and sent to the administration.[27]

The university remained unresponsive. On May 1, 1939, UCMP Director Bruce Clark wrote to Camp about the deteriorating situation, reviewing the history of Alexander's relationship with the Museum of Paleontology. His letter highlights the real significance of her contributions from an academic perspective, and it points out what severing her relationship with the university would undoubtedly mean for the future of the museum. Alexander had supported practically all the work on which Merriam based his now renowned reputation, he noted, both in money and through the specimens her fieldwork had contributed to the collections. He stated,

> I have a very high regard for Miss Alexander. She is a woman of high ideals. She has been trying to do something to help science and she has been doing it in a most intelligent way. She knows what she wants to do and if she is given encouragement will undoubtedly go even farther

than she has in the past. I want to emphasize the point that she is not merely a rich woman giving money to the University but she is giving it with a decided purpose, a purpose that is backed by her own training in zoology and paleontology....

I hope that things can be straightened out so that work in the Museum will not be hampered.... There is no institution in the country that is doing any better type of work than is being done in vertebrate zoology with the funds available. It is a great pity that anything should come up to hinder the future of this work. It seems to me that the University owes a great debt of gratitude to Miss Alexander for her many years of support to this institution and others on the campus and it would be a great pity to shake her faith in the University.[28]

As follow-up to a personal meeting between Camp and Sproul, the paleontologist forwarded Clark's letter to the president along with his own of the same date, reiterating the argument that "her [Alexander's] chief concern is to protect her interests in paleontology from control by Doctor Merriam and those whom he directs." Camp concluded, "It would be most unfortunate if we lost her support through failure to understand her point of view."[29] In between these remarks, Camp substantiated Clark's premise of Chaney's intention, despite promises to the contrary, to secure control of the museum for Merriam.

The contents of Camp's letter proved to Alexander that the university had not lived up to the spirit of the agreement with her, though it gratefully accepted her donations. When Sproul did not respond immediately to this latest appeal for intervention, Alexander felt that the situation in paleontology was untenable and unresolvable. One week later she wrote to Camp of her intention to withdraw her financial support for paleontology beginning June 1, stating that she felt she had no other avenue of recourse.

In actuality, Alexander did not withdraw her support until July 1, the start of the university's fiscal year. Within two months, Camp wrote to her that Sproul had offered him the chairmanship of the department. With characteristic optimism, Alexander once again renewed her support. But Chaney's interference continued; a year later the situation in paleontology remained in decline. Alexander now took her case directly to the regents, explaining her concerns and asking them to intervene on behalf of the program in paleontology. Briefly outlining the history of her association with the department, the extent of her monetary donations to the university as a whole (more than $1 million by this time), and the administration's unresponsiveness, she went on to describe Sproul's inherent conflict of interest in the matter.[30]

The issue of a patron's influence on university policy is still a delicate subject. But Alexander raised it in order to fight for the research, the publications, and the training of students that she rightly recognized as the true source of value and virtue in a university. For Alexander, paleontology was a "foundation science," fundamental to any understanding of the evolution of life.[31] It was this imperative that drove her, and this perspective that she felt the university administration never adequately understood or appreciated. She was trying to protect the nonmonetary value of her contributions to paleontological research, that is, the ideas that emanated from her resources.

No written response to Alexander's appeal is available. Because the regents held Sproul in high regard, there was little likelihood of their interference with his handling of the problem.

Small fires periodically flared up over the next few years, yet the situation in paleontology came sufficiently under control so that in April 1945 Alexander wrote to Camp proposing to increase the UCMP endowment to $100,000. This she did in December 1945 (through shares in utilities and mining stock), hoping to net the museum an annual income of $11,000.[32] Alexander's strategy in presenting the university with stock while she was living was to ensure the museum sufficient income without worrying about the degree to which taxes might diminish her gift after her death. Because Alexander believed that the university should bear primary responsibility for teaching and for the collections' upkeep, her agreement further stipulated that the income from the endowment should go primarily to research and fieldwork and that the university must provide at least $8,000 annually to support these other museum functions.

Alexander's wish to provide for future support of the UCMP and its research program long after she would be able to fight for them personally caused her to double the size of the UCMP endowment in 1948. Her contributions to the endowments of the MVZ and the UCMP were now virtually equal.[33]

As Grinnell had done after Alexander's initial endowment to the UCMP, Alden Miller, director of the MVZ following Grinnell's death in 1939, wrote to her in the wake of this most recent gift. In his letter he emphasized his strong, personal conviction about the importance of an endowment to the life and future of an institution such as the UCMP, owing both to the stability of purpose and to the security that it guaranteed for the collections over the course of many years and throughout the unpredictable shifts that are an inevitable consequence of changing university administrations. Although the MVZ's endowment was now contributing proportionately less than it had to the museum's support than were other sources of income, Miller stated unequivocally that without the endowment the MVZ would not have thrived or had the assurance that "changing fortunes of the University would not disrupt our work and alter the ideals of the institution."[34]

Although relations between Alexander and the administration relative to the situation in paleontology were never satisfactorily resolved during her lifetime, Miller's letter mirrored Alexander's own thinking and reflected the principles upon which her actions were based. Six months earlier she had written to him somewhat bitterly, or perhaps it was merely with a deep sense of resignation and disappointment, "Paleontology is a foundation science yet it takes a rather spectacular quest . . . to arouse the interest of the administration."[35]

18

A Restless Decade

For several years following their purchase of the farm on Grizzly Island, its unrelenting demands had severely curtailed Alexander and Kellogg's ability to conduct fieldwork. World War I had further interrupted their collecting schedule and created within Alexander an almost intolerable feeling of confinement. During this period, she began to conceive of new projects into which she could invest her seemingly boundless energy and her money. Merriam's decision to accept the presidency of the Carnegie Institute in 1920 coincided with her restlessness and may have precipitated the brief flurry of paleontological activity that followed.

In February 1917 Alexander had written to Beckwith proposing that the two undertake publication of a magazine about Alaska. Annie was willing to fund the venture if Martha would supply the prose necessary to make the magazine a success. She suggested that they embark on a three-month cruise of the Alaska Panhandle so that Martha could see the beauty of the region through her own eyes and begin to comprehend the wealth of opportunities that Alexander felt it afforded for literary exposition. "The scenery with its high peaks and glaciers is well worth describing as well as the queer characters one meets in the out-of-the-way places," Annie wrote by way of enticement.[1]

Although Alexander professed seriousness about the venture, nothing seems to have come of it. A year later, however, Beckwith countered with her own publishing scheme. Alexander gladly agreed to back it, but no further details of the plan were revealed and it, too, seems to have been abandoned.[2]

Martha next proposed to furnish, and then lease, an apartment in New York City as a money-making enterprise for herself. Annie held little enthusiasm for the plan but nonetheless offered to guarantee Martha the

$5,000 to $10,000 that she would need to get started. But realizing her friend's lack of practical experience in such matters, and revealing her own insecurities, she warned,

> When it comes right down to competing in a business way with other people a woman is up against a whole lot of things—chiefly man! For a man will bluff a woman every time if he gets a chance. He is a better student of character. He knows your weak points in a flash and before you realize it you are feeling like two cents and he is getting the best of the bargain.[3]

While their correspondence continued unabated, during the first half of 1918 Alexander's thoughts remained closer to home. For weeks her mother had been too weak to move her limbs, turn her head, or even open her eyes. Finally, in early July, Annie wrote to Martha that her mother had died, likening death to "an ever encroaching tide that finally submerges the last stand of vital forces." She confessed, "It has never seemed to me as if I could lose my mother—she has been so intimate a part of my life."[4]

Alexander's admission is difficult to reconcile with the slight evidence of Pattie's role in Annie's life. The daughter's diary and the letters to Beckwith scarcely mention Pattie. Other than Annie's letters from Africa, and a series of letters written while she and Louise were conducting fieldwork on Vancouver Island, the only correspondence that has survived between mother and daughter was written in the years following Alexander's purchase of the farm on Grizzly Island. In these letters, Annie would recount the litany of crises she faced in developing and managing the ranch. Perhaps because Pattie herself had been married to a plantation owner, she could empathize with this aspect of her daughter's life. Alexander's reflection, "it is sad to lose one's loved ones just when we are growing to appreciate each other's society," suggests that mother and daughter did come to understand, and perhaps value, the other's choices.[5]

The conclusion of the World War I coincided with a period of tranquillity on the farm sufficient for the women to go back to fieldwork. While confined to the ranch, Alexander confessed that she had begun to dream about collecting fossils—in the hot, dry sands of Nevada or in the more temperate regions farther north. A brief trip to the Sequoias in the summer of 1919 filled her "with ecstasy. . . . The Sequoias are so peaceful and stately in their growth—it is a rebuke to us little mortals who race over the world looking for we hardly know what," Annie mused. "I'm just hungry for the wilds—there's not one percent of the people I meet who know what it is to really live."[6]

After this excursion came a quick trip to the Southwest that fall, after which Alexander declared that she was "a new person," revitalized in a way that only time spent in the outdoors managed to accomplish.[7] These outings prefaced a spate of paleontological expeditions that Alexander and Kellogg undertook during the 1920s.

From early May through early July 1920 the women collected in southern Oregon and throughout northeastern California and northwestern Nevada.[8] Their field vehicle was a much beloved Franklin named "Birdie." Pirie Davidson, a paleontology graduate student, along with Chester Stock, Ward Russell, and a Mr. White, joined them about a week into the work. With the arrival of their field companions, camp was moved to Scott Ranch, and from there to High Rock Canyon. As they crossed the mountains, climbing to an elevation of 6,950 feet, snow still capped the highest peaks and ledges. Their first night was spent in a small canyon beneath a grove of aspen trees. Alexander slept under the stars, only to awaken and find that a layer of ice coated their drinking water.

High Rock Canyon extended approximately seventeen miles through solid lava rock. Its long slopes of red talus lay beneath a horizon of solid basalt. Interspersed among these layers were large exposures of unusually shaped and weathered tufa, varying in color from white to orange. Most of the fossil material that the women recovered from the site was embedded in white tufa, an exposure that was clearly delineated from the yellow horizons above and below it. Teeth and jaw fragments were common, along with miscellaneous foot and toe bones. While the women scoured the landscape, the men searched for other promising localities. They returned with tales of yellow tufa bluffs, replete with projecting bone fragments. The new bluffs lay directly beneath the lava capping of the canyon walls and contained a wealth of horse and rhinoceros teeth, as well as bone fragments of deer, large cats, and camels.

The group moved back and forth across this landscape, carefully plotting each locality with annotations on the geology of the surrounding region. Sometimes they proceeded on foot. At other times they would take their vehicles and park at the entrance to what looked liked a promising canyon or streambed, often hiking miles with a canteen and backpack in search of valuable material. Annie loved such hikes and the views that the climb to a ridge top usually afforded, perhaps a snow-covered mountain range against the horizon or a lush valley that seemed to extend endlessly in a carpet of wildflowers.

Davidson parted company with the group at Denio, Nevada, and the remaining members of the party went on to the Thousand Creek Formation.

Although the beds in this region had been well worked over the years, Louise made a superlative find—the complete skull of an *Hipparion*, a slender-limbed precursor to the modern horse, in addition to a hoof, a limb bone, and a splint. Their work continued unabated and to good purpose until the evening of June 28 when Ralph Chaney strolled into camp. Conflicts that Alexander would have with Chaney still lay several years in the future and she simply recorded in her field notes, "He brought letters which contained news that decided us to go home, leaving the rest of the party in the field."[9]

The remainder of Alexander's notes sketch the country through which she and Kellogg passed in the three days after leaving Logan Butte but do not mention the cause of their hasty departure. Within days of her return to Oakland, however, Alexander sailed for Hawaii. From here she wrote to Grinnell about the death of her niece, Elizabeth, who had been killed in an automobile accident on June 20, 1920, just three months shy of her seventeenth birthday.

Annie remained in Hawaii until late October, returning to Suisun in time to cast her vote for Harding for president. Regardless of their schedule, she and Louise always made sure to return to the farm in time for the November elections. The Alexanders were Republicans, Samuel referring to them in his youth as Fremonters after John Frémont, the first Republican candidate for president. Alexander's interest in politics seemed to complement her head for business, and she monitored the ebb and flow of the political tides, both international and domestic, with similar enthusiasm and equanimity. Upon returning from Crater Lake in 1899, she had written to Beckwith,

> Don't think then that this matter of my eyes [her failing eyesight] is preying on my mind. I take more interest in the Dreyfus trial than ever. Am up in the morning as soon as the first bell rings so as to get a look at the paper before breakfast, see whether Labor has expired during the night and how lame a stand the friends of Dreyfus are making.[10]

Three years later, she wrote with visible relief of recent election results, "The overwhelming Republican victory was a fine thing for us and shows the world after all that we are not governed by demi-gods and irresponsible natives."[11]

Other than stuffing envelopes for the Republican cause in various elections, Alexander made only one foray into politics that were not university-related. In the fall of 1940 while driving back to Oakland from the East Coast in a newly purchased Oldsmobile, she and Kellogg had the "good luck" to hear Wilkie speak at a rally in Harrisburg, Pennsylvania. Inspired by the candidate's rhetoric, and sensing a graveness in the current political situation, Alexander tried to muster local community interest in Wilkie's cause. Attacking this problem as she did all others—head-on—she had a platform constructed on a vacant lot in the town of Fairfield, the Solano County seat, and erected a large, circular tent with gaily colored, embroidered panels. Inside she placed a table and chairs, as well as a cot and blankets. From this command post she distributed Wilkie literature and addressed letters and pamphlets to the Democratic voters of Suisun concerning what she saw as the worsening crisis facing the country and the opportunity for a change to Republican leadership. Much to her chagrin, the country did not share her enthusiasm for change. On the eve of Roosevelt's reelection, Annie simply recorded in her diary, "very depressing."[12]

In December 1921 Alexander returned to Hawaii. Mauna Loa continued to erupt and she spent Christmas on the Big Island, fascinated with the geologic events unfolding before her. While she remained unequivocal in her professional pursuits, on a personal level she had conflicting feelings. She loved Hawaii and at times felt homesick for the sense of family and belonging that she experienced there. But family feeling warred with her peripatetic tendencies. From the field she confessed to Beckwith, "the 'resident aunt' is not my vocation—a little visit, then off again."[13]

True to her words, in early April 1921 she and Louise boarded the Santa Fe line for the desert regions of northern Arizona. Although it was snowing when they arrived, the women spent six days at the Grand Canyon seeking out Indian trails, archaeological sites, and promising fossil locales before Louise returned to Oakland.

Following Louise's departure, Annie rented a car and drove to Adamana for ten days to explore the Petrified Forest and Painted Desert. Between intermittent snowfalls she spent several days in the Blue Forest where the fossil formations were similar to those that the women had worked at Logan Buttes. She camped in the cold, windy landscape, cooking her meals in a hollow in the hillside. One day she explored Billing's Gap on horseback, attempting to locate the area in which paleontologists from Stanford had previously collected. She gathered some fossil material, but most of the specimens she unearthed were miscellaneous and unimportant fragments (a good proportion of the material that Alexander donated to the UCMP in later years fits this description, a delicate point that its staff never belabored).

Alexander found the fossil fragments that she had uncovered in the Blue Forest so compelling that less than two weeks after returning to Oakland she and Kellogg packed the Franklin and headed south on a more intensive six-week collecting trip. Past Bakersfield they crossed a rough, barren landscape of lava flows; west of Needles, an eroded floodplain. By the time they reached Williams, Arizona, the Franklin required the attention of a trained mechanic to repair a loose connecting rod and bad bearings.

The women made their first camp $2^{1}/_{2}$ miles south of Adamana on the western edge of the Petrified Forest. There they erected a newly purchased 9 × 12-foot tent and added cots, a table and chairs, and a small stove to their standard retinue of field equipment. The stove sat immediately outside their tent and baked biscuits beautifully.

In the light-colored exposures several miles from camp, Annie found shells, teeth from Triassic reptiles, and a few bone fragments. At the same time, she and Louise could not resist setting a few small mammal traps in the sandy landscape. On their first night out the traps yielded a fair catch—three pocket gophers, five antelope ground squirrels, four deer mice, and three kangaroo rats.

The Indian lookout above their camp was littered with pottery shards and red stone, and it offered the women a fine view of the surrounding desert. More bits of pottery and Indian inscriptions lay scattered in the general vicinity of their camp, but of greater interest were the fossil remains of a large reptilian snout that Louise unearthed.

From the Blue Forest, the women continued on to Zuni Well and the Painted Desert before backtracking to Holbrook, where they had a spring replaced on the Franklin. From Holbrook they moved on to Seligman and from there took a roundabout road to Nelson Canyon, finally camping in a ravine beyond Kingman. They had driven no farther than Oatman the next day when they suffered a tire puncture, the last of their vehicle's malfunctions before the trip ended.

In the spring of 1922 Alexander and Kellogg conducted what would be their most important fossil-hunting expedition during this period, an excursion to the Miocene formations in the Mojave Desert near Barstow.[14] Their first collecting locale, however, was Mint Canyon, a site just north of the San Fernando Valley and east of Saugus. These fossil beds were purportedly Pliocene in age and predominantly red, consisting of gravel and sand with some clay. Farther up the canyon the beds appeared pink and the formation was composed almost entirely of clay. They also discovered greenish beds and layers of cemented pebbles and sand, beautiful to look at but devoid of bone fragments.

Interspersed throughout Alexander's field notes are details of the surrounding landscape. Her observations reflect the women's perpetual preoccupation with farming and crop success and document the tremendous amount of agricultural development taking place in the greater Los Angeles area in the early 1920s as irrigation was introduced in an ever-widening radius throughout the valley. On March 5 she wrote, "We took the road to the southeast through a fertile valley of walnut, orange and lemon groves. We were impressed with the depth and productivity of the soil, apparently a light sandy loam and by the almost universal concrete systems of irrigation. Within fifteen miles of Saugus the country takes on a more arid aspect." The following day she noted, "Where the Los Angeles aqueduct crosses the highway there is a large field of fine looking alfalfa."[15]

The majority of collecting localities marked on their map contained shells and shell fragments, rather than vertebrate bones, and their collecting proved relatively unproductive. A terrific east wind blew much of the time and snow still capped the peaks of the San Gabriel Mountains to the east. Then, as they were preparing to leave Mint Canyon for a more promising locality farther south, they received a telegram from Pirie Davidson that caused them to alter their plans. Davidson had received a wire from Chester Stock informing her that Childs Frick of the American Museum considered this new locality near Hemet, southeast of San Bernardino in the San Jacinto Mountains, "his territory" and he preferred to work the area himself.[16] Believing, like Grinnell, that the field available for study was enormous, the women conceded Frick's priority and, instead, arranged for Davidson to meet them in Barstow. From there they would explore Miocene beds that lay just north of the city.

Finding nothing of interest directly north, they backtracked and followed the road toward Hinkley and what they hoped would be the western end of the Barstow formation. The blacksmith in the Barstow garage had given them explicit directions to a mining camp fourteen miles from the edge of town. There the women pitched their tent, cut wood, and had just finished unpacking the Franklin as drops began to fall from the sky. The rain continued steadily for several hours. As they watched, the dry streambed beside their tent became a rushing torrent. By morning the rain had turned to sleet and a thin coating of snow covered their table and stove.

The weather kept the women in bed until after 8 A.M. but, once up, Annie immediately set out to survey the site. By afternoon the sky had cleared and the ground was no longer slippery, although the temperature remained cold. Each of the women found bone fragments or teeth, but it was not until three days later that Davidson made the find of the trip: at the base of the canyon she spotted a narrow shelf of hard material projecting from the wall of tuff about four feet above the streambed. It proved to be crowded with bone.

The women worked the entire formation for over a month, collecting a variety of carnivore and ungulate remains—horns, teeth, jaws, skull fragments, and toe bones. Between her annotations of fossil finds, Alexander noted plants and animals that she and Kellogg observed in the area—a large flock of buzzards that passed overhead at sunset one evening, as well as linnets, canyon wrens, and flycatchers that foraged in the vicinity of their camp. As the weather warmed, bats swooped and dove across the evening sky like kites caught in a strong breeze. One day she recorded the presence of tracks in the sand that she thought must belong to a tortoise, and farther along in the canyon bed she saw fresh footprints that were unmistakably those of a panther. Signs of kangaroo rats were everywhere. Near the strata that Davidson had discovered, Annie noticed a large owl nesting in a cliff hole, as well as a species of yellow flower that she did not recognize. But the fossils remained her primary focus. More than a month after their arrival at the site she wrote to Martha,

> We have been such busy mortals! Wouldn't you think you were busy if you had been working for half a month with pick and shovel digging a camel out of the ground? More than that we have been on our knees with jackknife and chisel following the bone along with a caution and patience I didn't know we possessed. The excitement of the chase never quite equaled this. First we found the forelimbs with vertebrae, limbs, and small bones mixed up with it. Then six neck vertebrae beginning with the atlas and axis. These half then encircled the skull, the greatest prize of all and in front of this the lower jaw complete and the scapula with more vertebrae and ribs.... [T]he camel isn't our only specimen I can boastfully say. We never go out looking but what we find something. Just this morning Louise picked up part of a horse jaw and a good deer jaw with teeth, the little *Merycodus*, in places we have crossed several times so a fossil hunter's work is never finished.... It is strange how absorbing this work is. We forget the outside world. I was walking along one day with a light step feeling young and happy when the wind blew a wisp of white hair across my eyes and reminded me![17]

The material about which Alexander wrote so lovingly proved of great import, so much so in fact that, following their departure, Frick subsequently moved his field party into the region and continued to remove many valuable specimens. Alexander was perturbed. Frick was plundering a site whose material she believed rightfully belonged to the University of California, a site that the women had discovered only after vacating and relinquishing a locality in southern California that he had claimed for his own institution.

Never one to deal with intermediaries, Alexander wrote directly to Henry Fairfield Osborn, president of the American Museum. Osborn replied promptly and sympathetically. He presumed that Frick had acted unintentionally but promised to take up the matter personally with his collector when Frick returned. He then reiterated what Alexander had believed from the outset, and what she undoubtedly had hoped to hear, "We have all of us felt that important parts of our collections from the Coast should go to the Museum of Paleontology at the University of California, and I will be very glad to take up with Curator Matthew and Mr. Frick your suggestion."[18] The Barstow Formation eventually yielded large quantities of important material for the UCMP.

Following their return to the Bay Area, Annie wrote almost predictably to Martha, "I hated to come back. Some day I'll cut myself loose from everything and keep going until I dry up and blow away like the grass of the field."[19] It is a sentiment that echoes throughout her correspondence all her life. Unable to stay in one place or be tied to one project for very long, Alexander was in turn drawn to the farm, to her museums in Berkeley, and to her family in Hawaii. Yet after a relatively brief residence in each locale, she would be off again. In the summer of 1923, between field trips to southern California, she and Louise boarded a ship bound for Europe.

19
Europe, 1923

Alexander's restlessness continued throughout most of the 1920s. After leasing their herd of milking shorthorns to John Rowe, she and Kellogg were at liberty to travel as they had not done since buying the farm almost a decade earlier. "This will leave us free for fossil land or the wilds of British Columbia, or the still more remote Siwalik Hills of India, where the early ancestor of man once trod the jungles and left his bones in the wash of the rivers," Annie wrote exuberantly from Suisun to her cousin Mary Charlotte. "At all events a career of adventure is mapped out for us, redeemed by its scientific prospects."[1]

Throughout the first two decades of the twentieth century, remarkable discoveries of early human remains were being brought to light in Europe and Asia. During this period, Alexander began to take an increasing interest in human evolution, reading Henry Fairfield Osborn's *Men of the Old Stone Age* and *The Origin and Evolution of Life* shortly after their publication.[2] She even went so far as to express the belief that a comparative human osteological study would prove extremely useful to researchers and that peoples of different cultures should donate their skeletons to science.

But this new interest was only part of the impetus for Alexander's trip abroad in the summer of 1923. She and Kellogg had built their herd of milking shorthorns around prize-winning livestock that originated in England, and the women wished to compare the two groups of cattle. At Alexander's invitation and expense, Beckwith and the Keys sisters, friends from Annie's childhood days in Hawaii, accompanied the women on this excursion, although the five frequently split up.[3]

As John Merriam had done twenty years earlier, UCMP Director Bruce Clark now took it upon himself to have a letter of introduction written for Alexander. Knowing that Osborn was well acquainted with most of the prominent vertebrate paleontologists in Europe, Clark thought that a letter written by Osborn as a professional courtesy would give Alexander and Kellogg entrée to those individuals and to the most important sites where human fossil remains had recently been excavated. Clark's letter to Osborn assumes that the paleontologist-cum-museum president was already familiar with Alexander and with her benefaction to the university, although his letter predates Alexander's own correspondence with Osborn concerning Frick by more than a year. Osborn willingly agreed. But Alexander responded critically to Clark's good intentions.[4]

It is likely that Alexander felt uncomfortable about what she believed was an imposition on Osborn's kindness. She did not like to call attention to herself. And she did not like to feel obligated to others. If at least a portion of her concern stemmed from anticipation of a reciprocal request by the American Museum, it was not misplaced. In 1926, just three years after her trip to Europe and only a year after her donation of $500 to that institution, Alexander received a plaintive appeal for money from Roy Chapman Andrews, leader of the Central Asiatic expeditions that were being sponsored by the American Museum.[5]

When Andrews inaugurated his work in Asia, the aspect of it that garnered the most public attention was his intended search for man's earliest fossil remains, the so-called "missing link." Andrews fervently believed that the bones to support this contention lay buried in Mongolia. But the worsening political situation in China, coupled with rising anti-foreign sentiment, had forced him to cancel his ongoing explorations for at least two years. He had managed to salvage his field equipment and the specimens that his party had collected to date, but the war in China had cost the expedition dearly in terms of both time and money. In order to resume the work once the political crisis ended, he had to raise additional funds. Thus, he wrote directly to Alexander, "We must go there and read its story concealed in the rocks and beneath the marching sands. But it is so remote, so inaccessible, that we must have time and money. . . . We believe that you are interested and ask you to help with a contribution of any amount, either singly or on a basis of three years; your support will be greatly appreciated."[6] Regrettably, Alexander's reply cannot be found, but Andrews's appeal reveals the degree to which word about her philanthropy and her interest in human origins had spread outside the confines of Berkeley.

The women arrived in New York in June 1923, several days ahead of their ship's scheduled departure. At least two days were spent at the American Museum examining relevant artifacts in the anthropology collection, during which time they also collected their letters of introduction. Once on shipboard, their week at sea passed quickly. Between games of deck tennis and bridge, Annie read the remarkable saga surrounding the discovery of the Piltdown fossils. At tea, she and Louise pored over a map of England and for amusement, as well as edification, drew up a list of English place-name endings and their meanings, for example, -*chester* = cattle camp; -*wick* or -*wich* = bay. Then, almost before they had settled into a routine, a wooded landscape dotted with chalky white cliffs sporting aging fortifications announced their arrival at Cherbourg. As their ship made its way across the channel toward Southampton, the scenery changed from neatly plowed fields separated by rows of poplar trees to densely wooded hillsides dotted with small villages, the church spire in each punctuating a landscape that was otherwise uniformly green.

Alexander immediately purchased a copy of Muirhead's *Blue Book of England* and a geologic map of Britain. Their first archaeological destination was Stonehenge, that amazing grouping of enormous upright pillars and horizontal lintels dominating the Salisbury Plain in southern England. As their train headed west toward Salisbury, gates in hedgerows and along brick walls afforded glimpses of English lawns and country gardens interspersed with thatched-roof cottages and potting sheds. Annie noted the flowers—delphiniums, fox glove, Canterbury bells, pansies, calceolarias, white lilies, snapdragons, and rhododendron. Nearly every windowsill had pots of brightly colored geraniums. As their train continued west, such scenes gave way to fields where farmers could be seen cutting, raking, and stacking hay, weeding root crops, or tending to vegetable gardens.

Stonehenge itself was located eight miles north of Salisbury and at nine in the morning, in a drizzling rain, the women set off on foot to explore the site. Unexpectedly, but much to their pleasure, the man in charge of the excavations was at the site when they arrived. Colonel Hawley believed that the unusual formation had been erected for defensive and domestic purposes, and he based his assumption on a circle of holes that he had uncovered outside the ring of monolithic stones. Charred human remains had been found in the holes and a number of deer horn picks had been recovered nearby, suggesting the manner in which the holes had been dug. Intrigued by Hawley's discourse, the women spent two hours wandering and surveying the site in the steady drizzle.

The following morning they headed southwest by train toward the town of Exeter. As they lunched on fish, roast lamb with potatoes and cabbage, and a sweet cheese accompanied by crackers, radishes, and watercress, the passing landscape became increasingly hilly, dotted with herds of milking shorthorns grazing in lush green pastures. In Exeter, Alexander hired a car to convey them to Sampford Courtney, a village constructed around a narrow main street with picturesque thatched-roof houses where they asked directions to the farm of Mr. Thomas Lang. A jolly, thickset man of about fifty, sporting sandy brown hair and merry blue eyes, Lang proudly showed them his hoof stock on hand—a dozen Devon bullocks in one paddock, forty long-wool Devon sheep in another, several Devon yearlings, and three calves. His purebred Devon bull stood apart from the others in a small corral attached to one of his many stone barns. In addition to several pastures, Lang kept fields of meadow hay, rye, and clover, which he alternated with oats and other grains. Alexander was impressed with the cattle, and more so after Lang confided to her that he never went to market with his stock—the butcher always came to him. If he and the butcher could not agree on a good lump sum for his animals, Lang merely sold his meat by the pound.

For the next several days, the women proceeded at a leisurely pace across the English countryside. They picnicked by the side of the road, enjoying the heather just coming into bloom, as well as the gorse and wildflowers that blanketed the moors. They stopped along the shore at Tintagel to admire the views of rocky cliffs up and down the coast, the white foam of the surf, the crisp salt air, and the brilliant sunshine along the beach.

In Clovelly, Annie and Louise exchanged their rented car for bus transport. After traveling north up the coast to Bideford, the motorbus turned east, heading inland and stopping in Barnstable before continuing on past Taunton to Bath. The countryside now became quite flat, cut up into small fields by long rows of low-lying hedges. The soil here looked less fertile and had been planted to hay, oats, barley, and wheat—or occasionally, cabbage or horse beans. Far less land was devoted to pasture.

From the western coast of England, the women headed northeast to Newcastle upon Tyne. Early July signaled the start of an annual fair and the two headed for the show grounds where cattle judging would take place. Long a veteran of such events where their own animals were concerned, Louise recorded the results in detail. Milking shorthorns were judged first and this took the entire morning.

> At two the judges began on the bulls. They took a long time deciding between a Conjuror bull bred by Hobbs and Sons and [the] Bessborough Polonius [bull] but the latter finally went to the head and the decision agreed with the opinion of the ringside to judge by remarks and some applause. The Conjuror bull was a little more stylish in color and appearance but Bessborough was the better bull throughout and we were glad to see him win as he is related to our Bess Blonde, dam of Betty.[7]

Annie took photos of three of the best Dairy Shorthorns and, after the judging, she and Louise looked hastily at the different breeds of sheep and pigs, Dexter and Kerry cattle, the small black cattle, and the Blue Albions, cows that resembled a cross between a Shorthorn and a Holstein.

Back in London, the women were ready to begin their examination of the artifacts and fossil sites that dated from Early Man. Paleolithic cave paintings that had only recently been opened to the public were generating tremendous interest, both nationally and internationally. The women met first with Dr. A. Smith Woodward, a distinguished (and later famous) geologist at the museum in South Kensington. Woodward showed the women bits of the museum's paleontological collection, including the London specimen of *Archaeopteryx*, considered the ancestor of living birds; bones of the moa, a large, flightless bird from New Zealand; a ground sloth from South America with portions of its hair and skin still intact; the perfect head of an arctic mammoth found in Siberia that bore unusually slender tusks; a fine specimen of Irish elk; and a giant Egyptian mammal, *Arsinoitherium*, which had two large, straight horns projecting forward from its skull. They were also shown perfect fossil ichthyosaur, plesiosaur, and amphibian specimens, all of which had been collected in England. But it was not until after lunch that Woodward showed them the highly prized fragments of the skull and lower jaw of Piltdown man, fossils with which his name had already become synonymous. While Kellogg noted "several ape resemblances" (she specifically recorded in her journal, "width of skull across the back, widening of border of symphysis of lower jaw, interlocking canine tooth"), she felt that the Piltdown skull was "distinctly human." He also showed them the "Rhodesian skull," a photograph of which Annie had previously found intriguing. "It presents a gorilla like face with its heavy brow ridges but Dr. Woodward thinks it much later than the Piltdown," Louise recorded. "It is not mineralized like the Piltdown, the latter, in his opinion being the oldest human fossil yet found, possibly of Late Pliocene age."[8]

By 1923 the human skull fragments collectively referred to as the Piltdown fossils had been famous for more than a decade. Another three decades would pass before they would be officially pronounced a hoax.[9] Mr. Charles Dawson, lawyer, amateur geologist, and antiquarian, had announced their discovery in December 1912 and had named the fossils for the locality at which they had been found, Piltdown Common in Sussex, England. Within the anthropological community, the bones were considered extraordinary. The cranium of Piltdown man was essentially human in character, whereas the mandible appeared apelike. Based on this specimen, Dawson boldly proposed that during the Early Pleistocene the "missing link" in human evolution had lived and walked across the plains of southeastern England.

As a respected amateur, Dawson had often worked at the British Museum, using its collections and consulting with its curators. Upon discovery of the fossils at Piltdown, Dawson had immediately sought out Woodward for confirmation of their importance. Woodward became convinced of their authenticity, and in December 1912 the two men presented a scientific paper on the specimen at a meeting of the Geological Society of London. The unusual combination of human and apelike characters in a single individual led them to propose that in western Europe man had already differentiated into widely divergent groups by the end of the Pliocene. This, in turn, led to the suggestion that Recent man might well have arisen in Europe rather than in Asia or Africa, as was the prevailing theory. The fossils became an overnight sensation.

The evidence presented in 1912 did not win over everyone in the scientific community, and Dawson died in 1916 with controversy and speculation about the fossils and their place in human evolution still swirling. In 1923, however, Woodward himself showed the fragments to Alexander and Kellogg, and they listened intently to his discussion and description of them. When Alexander spoke of wanting to see the site from which the fossils had been excavated, the famous geologist suggested that she and Kellogg join a party that was leaving the following day.

At nine the next morning, the women met their group at Victoria Station for a two-hour train ride to Uckfield. From there the party drove four miles to the Ouze River, stopping at Piltdown Common where they were shown a view of the north and south downs bisected by the eroded section of Wealden. The group walked from the bridge over the river to the site where the famous fossils had been found. The slope of eighty feet over a half-mile suggested the long period of erosion that had taken place since the Piltdown gravel had been deposited. This estimated erosion had been crucial to aging the fossils and, thus, to their importance evolutionarily. Woodward allowed the women to dig in the gravel, "not however in the most promising place where the gravel runs under the road," Louise recorded later.[10] If she and Annie had expected to find something of import they were disappointed, and they returned to their hotel that evening empty-handed.

The women visited the British Museum more than once to examine additional fossil material, particularly the collection of skulls representing different human races, and talk with the curators about archaeological work currently under way in France and Great Britain. On these occasions they were shown more of the museum's collection of artifacts, including primitive flint instruments, and they attended a special lecture on the differing periods of Early Man. From it they learned that during the Neolithic, circa 2000 B.C., England had been inhabited by lake dwellers. Finds of polished implements and arrowheads indicated that these people were able to grow wheat and barley, that they ate wild apples, plums, blackberries, and strawberries, and that peas and beans were a part of their diet, details that fascinated Annie and Louise.

In addition to time spent at the British Museum, the women made several side trips—to the local museum in Ipswich, to Grimes Graves, a series of chalk pits dug by Neolithic men mining for flint implements, to Norwich where they toured the cathedral and the fossil collection of the castle museum, and to Cromer where they looked for eoliths on the flint outcrop adjoining the beach that became exposed at low tide.

But their time in London was not strictly confined to the examination of artifacts and paleontological relics. They visited many of the sites that tourists generally frequent, from Westminster Abbey and the National Portrait Gallery, to Bread Street, where Milton was born, and the church of St. Giles, where he was buried. They spent a day at the zoo and they viewed the Elgin Marbles. Evenings were spent at King's Theatre or watching a summer repertory company perform at Hampstead.

One activity not on most tourist itineraries was Annie and Louise's visit to a Citroën dealership. Alexander went in search of the Kégresse, an automobile with caterpillar treads instead of rear wheels that had crossed the Sahara to Timbuktu.[11] She was quick to appreciate the implications of such a feat and had decided that this vehicle would be ideal for fieldwork in the deserts of the Southwest. The dealership in London did not have the car in stock but, from this point on, across Europe and North Africa, she would attempt to track down the elusive, almost mythical, vehicle.

Approximately a month after arriving in Britain, the women left England for France. In Paris they visited the museums and churches, the gardens and monuments for which the city is famous. They strolled along the Seine and down the Champs-Élysées. They dined at Restaurant Marguery on filet of sole served with a lettuce salad and dessert, the entire meal topped off by a bottle of champagne. Their "best meal yet," Louise proclaimed.[12] One evening they had tickets to Mozart's *Magic Flute,* on another occasion they attended a performance at the Opéra-Comique.

The women also visited the village of Saint-Germain-en-Laye, a forty-five-minute train ride west of Paris through some of the city's loveliest suburbs.

The object of their visit was an archaeological museum housed in a twelfth-century chateau rebuilt by Francis I. The castle was a favorite of Louis XIV until he completed Versailles, after which he abandoned it. From its terrace that seemed overrun with flowers and potted trees, they glimpsed the Seine winding back toward the city, where the gleaming white dome of the basilica of Sacré-Coeur was visible through the sunlit haze. The women carried with them a letter of introduction and the curator-in-charge, Dr. Hubert, spent an hour showing them the Piette collection from the cave of Brassempouy, including carved human figures regarded as exquisite examples of Cro-Magnon art, and pieces of bone and mammal teeth decorated with drawings of horses, reindeer, mammoths, boars, and elephants. Representations of birds with long necks, fish, and an occasional human figure could also be discerned. Two other rooms in the museum held implements and pieces of pottery from the Neolithic and later periods.

On the morning before their departure from Paris, the women sought out the local Citroën-Kégresse factory and asked to be given a demonstration of the vehicle on which Annie had pinned such high hopes. This was no mere car, with its powerful engine and tractor-tread wheels. She and Louise test-drove the vehicle, motoring up and down a steep embankment on the inside of the city wall—what looked to Louise like an almost perpendicular grade—and then out onto a parade ground piled high with mounds of rubbish, rocks, and sand. "The machine rides as easily as an ordinary car and we were delighted with the ride and what it could do. Just the thing for the desert," Louise noted gleefully.[13]

Why Alexander failed to order the vehicle once she had tested it is never intimated, but we can only assume that she regretted her decision. The following year, she and Kellogg vainly continued their search for the car amidst the port cities of North Africa.

Whereas the women enjoyed the scenic beauty or architectural history of each town or village in which they stopped, what directed their itinerary from this point on was the desire to view firsthand the renowned Paleolithic cave paintings in central and southern France and in northern Spain. Issues of the paintings' original meaning and purpose did not concern Alexander. She had been drawn to the skulls of Piltdown man and Rhodesian man in an attempt to understand human evolution, but she was no less taken with the "art" that the Cro-Magnon peoples had presumably painted several thousand years earlier.

From Paris, the women took the train south, passing first through Bordeaux, then Bayonne, before arriving in the coastal town of Biarritz famed for its beautiful sandy beaches and its tamarack trees whose feathery foliage had been trimmed to resemble giant beach umbrellas. During part of the journey their train passed through the moors, a countryside of sandy soil covered with brush and pine trees from which farmers were gathering resin to make turpentine. At other times they glimpsed fields of corn and acres of vineyards from their window seats.

The following morning their train crossed the border into Spain, continuing west along the northern coast to San Sebastián and on to Bilbao. They traversed numerous valleys that ran down to the sea, passing from one to the other through dark tunnels carved through the intervening mountains. A lush shade of grass-green carpeted the lower slopes while oak and chestnut woodlands adorned the upper reaches. The mountains gave way to vistas of narrow canyons and distant misty peaks. Each valley bore a river, and each river a village in which the tiled roofs of picturesque stone houses radiated out from a solitary church steeple. And all along the banks of the river, cultivated fields lay ready for harvest.

From Bilbao, the train continued west to Santander, then southwest one and a half hours into the mountains toward their final destination, Puente Viesgo. The town was famous not only for its caves but also for its mineral springs that welled up into hot baths and were popular with visitors from Madrid. An ancient stone bridge spanned the high, rocky banks of the river that ran through town. Behind the town was a limestone peak and, halfway up it, the entrance to Castillo Cave. The following morning a zigzag trail to the summit afforded the women a panoramic view of the surrounding countryside. From its height, they were able to look out over fields of corn in the valleys below, each valley separated by towering bluffs on either side. Rising up the slopes from the valleys were irregularly shaped pastures for sheep, cows, and donkeys, delineated by finely crafted stone walls that had stood for centuries.

A steep trail on the peak to the southwest led Annie and Louise to a dense forest of beech and oak. Here they saw and heard numerous birds, but most were new to them, their songs unfamiliar. The only exception was a shrike, which they recognized despite the bird's brown and gray coloring (those they were accustomed to seeing in California were black and gray). They also saw signs of small mammals and Louise spied a lizard. But, without traps, a shotgun, or noose, the women had to content themselves with merely recording their observations on paper.

Early one morning Annie hired a Ford to take them fourteen miles northwest to the town of Santillana, the entry point for access to the Altamira caves that are no longer open to the public. Louise described their visit this way:

At Santillana we looked at some quaint houses and sat under the oak trees in what seemed to be the village common while our driver got permission for us to go into the cave. It took nearly an hour and then a guide came along with an acetylene lantern and we started off across the fields to the cave, a walk of half an hour up on to a sort of down from which we got a fine view of the mountains around. The entrance to the cave lies in a small hollow and above it stands a monument, made of the rock of the country, in memory of the man who found the cave. You enter through an iron grilled door to the first chamber and a short distance in to the left is another door that leads into the chamber where the ceiling frescoes are. This is about fifty by a hundred feet in size with a low ceiling and the animals, mostly bison are painted in red, the line of the back, the limbs, tail and horns being outlined in black. Several natural [bosses] in the rock have [been] used for the body of the bison, giving the effect of a raised sculptured figure. The colors are well preserved. A[nnie] thinks they must have been burned in to have withstood the dampness. The whole ceiling was wet and water was dripping from it in many places. The guide told us that some of the paintings had been destroyed by dry air blowing in before the door was put in. They had peeled off. Beside the bison there were horses, deer, wild boar, and a hand.

After leaving the main chamber the guide took us further into the cave and showed us on the side walls engraved line drawings of a deer and horse and black outline drawings of a tapir. A bison and at the very end of the cave and at a considerably lower depth two longhorned antelope, running, and the head of a deer with the ears especially well done.[14]

The following afternoon, after the rain had subsided, their guide took them to see the paintings at Castillo Cave. Inside, a pit thirty feet deep had been dug to expose the successive layers of Paleolithic culture that had occupied the site, from Pre-Chellean to Magdalenian. Fragments of animal bones, presumably the remains of meals eaten over the millennia, could be seen protruding from the walls of the pit. By digging around in the loose dirt, the women managed to find several teeth that they recognized as horse, deer and, perhaps, bison.

The cave was composed of a series of chambers, some circular, others elongate, some with high ceilings, others in which they were forced to bend over to avoid hitting their heads. At the end of what seemed like a long, damp corridor they were rewarded with the red outline of an elephant. Immense stalagmites and stalactites attested to the age of the cave; from one an artist had carved a bison, making use of the natural shape of the formation. On the walls of the cave they observed drawings of deer and curious markings whose meaning they could not discern.

The paintings at Castillo were not as numerous and were more crude, less graceful, than those the women had seen at Altamira, but they included fascinating red stencils of human hands and two feet, as well as some nonconventional figures that have been interpreted as writing or as depictions of houses and villages. The following morning Annie and Louise returned unaccompanied to the cave and spent several hours digging for fossils at the base of the pit wall. Broken bones showing contact with fire were numerous and the women unearthed additional horse and deer teeth. Annie uncovered a bear tooth and a broken bone that she felt might have been used as an implement, while under an overhanging rock Louise discovered a layer of bird and rodent bones that she suspected might have been Recent, perhaps the droppings of an owl, or remnants from an eagle's nest. In the afternoon, they returned to Altamira for one last look, finding the skeletons outlined on the ceiling of that cave even more impressive on a second visit.

Three days after leaving Puente Viesgo, the women reached Lourdes in the southwestern corner of France. The narrow streets of the city were crowded with shops selling religious statues, pictures, and rosaries, in addition to the usual array of tourist souvenirs. Just as they arrived, a procession was forming for a march to the grotto, its participants caught up in song. They followed the group and, standing behind the railing that enclosed the space in front of the grotto, viewed the statue of the Virgin that hung above it. They watched as a bishop prayed for the sick to be healed, the crowd enthusiastically responding to each appeal.

The train that they caught that evening took them to Toulouse, where their letter of introduction to Comte Bégouen elicited a call to his friend and pupil, M. Jacques Estanove, who escorted them to the museum of Toulouse with its collection of Recent birds and mammals. In addition to a specimen of the rare Saiga antelope, the animal that most caught their fancy was the desman, *Myogalea* (= *Galemys pyrenaicus*), a semiaquatic mole found in the Pyrenees. With fur like a mole, a scaly tail like that of a rat, teeth most closely resembling those of a shrew, and a nose that might be more accurately referred to as a proboscis, the creature was unlike any other small mammal that the women had ever seen. Estanove also showed them the bones of cave bear, hyena, and bison that had been taken from the grottos. But it was the implements recovered from the caves near Toulouse that Alexander had come to see and, later that afternoon, the comte himself vividly recounted for them his discovery of the caves on his estate.

From Toulouse, the women boarded the train for Ussat, a famed summer resort in the narrow valley of the Ariège and a recognized center for Magdalenian culture, an Upper Paleolithic people distinguished by their ability as artists and for the importance that art must have played in their lives. Monsieur Cugullière, a friend of Comte Bégouen who was drawn to Ussat every summer to work in its caves, met the women at the train station. Many of the region's more than three hundred caves had never been examined, but the one Alexander and Kellogg wished to visit was the famed grotte de Niaux.

At their hotel, others expressed interest in visiting the cave as well. In the end, thirteen persons piled into a small wagon drawn by two horses for the trip, including a portly gentleman who introduced himself as the director of an orchestra in Toulouse and who sang and shouted to the group throughout the ride, in spite of the heat. Cugullière hung on to the step in the rear and proved to be a lively companion, laughing and talking the entire way, about an hour's ride from the village.

It was a short but rather steep climb from the road to the entrance of the cave. At the end of a long tunnel that opened into a larger chamber, the group was rewarded with spirited drawings of bison, horse, reindeer, and antelope. The cave itself consisted of a variety of formations, and red marks on the walls indicated the number of steps to be taken before making a turn to the left or right. Water could be heard running in the depths. The orchestra leader continually sang and made noise, checking the acoustics in each new locale, while Monsieur Cugullière sounded a conch shell that hung around his neck, an item that Louise noted served alternately as a drinking vessel should beer be at hand.

From Ussat the women continued by train to Foix, where they arranged for a car to take them to the cave at Mas-d'Azil twenty miles north. At the Hotel Avignon, they presented a letter of introduction to the proprietor, a friend of Monsieur Cugullière, who agreed to escort them. Located about a mile outside of town, the cave at Mas-d'Azil was an astounding natural phenomenon, as well as an important prehistoric site. A magnificent 213-foot arch marked the entrance to a much longer tunnel that had been hollowed through the mountains along the River Arize. The Azilian culture was intermediate between the Magdalenian and the Neolithic; a process of miniaturization eventually made its tools and implements too small and too fine for efficient use. Most of the artifacts that had been found in the cave were engraved bones and small painted stones. The cave was also filled with bats that Annie and Louise detected by the hissing sound the winged mammals emitted in the darkness. Once disturbed by the light that the women carried, the bats immediately took flight.

The women spent several days exploring the caves and villages around Mas-d'Azil before heading north to Les Eyzies and the valley of the Vézère in Dordogne.[15] More than two hundred prehistoric sites dot the region, and Alexander carried letters of introduction to some sites; others seemed unattended or required no prior notice to enter. At a spot along a road where repairs were being made, pieces of flint tools, scrapers the women judged to be from the Magdalenian, and pieces of bone with apparent evidence of use as implements, lay in the debris. They visited the national museum of Les Eyzies with its excellent collection of material from the region—including numerous examples of carved ivory and reindeer horn from the caves—and climbed amongst the rocks of Saint-Christophe, great cliffs along the Vézère inhabited during early Christian and medieval times.

In all, the women visited twenty-eight archaeological sites in the Vézère River district during the ten days they spent in the region. They toured the local museums and tried, to the greatest degree possible, to meet with the individuals who had done the actual excavations or who had curated the artifacts removed from each site. They noted the geography in each locale and seemingly left no implement unexamined. Strangely and inexplicably, they made no mention of visiting Lascaux. Finally, on the morning of August 24, two months to the day after they had arrived in England, they boarded a first-class compartment for the return trip to Paris, and the first leg of their trip home.

20

The Temple Tour

As the women traveled east in 1923 to begin their European excursion, Alexander was already contemplating a trip to Egypt and Palestine the following winter. Aside from restlessness, her interest in human history and her desire to add new specimens and taxa to her museum collections may have been enough to justify the trip. Keeping research in mind as she cruised gently down the Nile, Alexander wrote to UC paleontologist Chester Stock, "Excavations of ruins in Egypt are being carried on by several organizations, including Americans and it is a pity no effort is being made to salvage the mummies that are being found for the purpose of comparative human osteology. The opportunity for finding skeletal material is exceptionally good we are told."[1]

The official "discovery" of King Tutankhamen's tomb, announced publicly in February 1923, had drawn unprecedented interest. A public weary of war and its suffering eagerly embraced news of the young king and his crypt of fantastic treasures. The mystique of Tut's youth and the mystery surrounding his untimely death, coupled with descriptions of his dazzling jewels, controversy as to their rightful ownership (British vs. Egyptian), outrage over the exclusive contract given to the *Times* of London to cover the story, and intrigue over the deadly "curse" said to be associated with disturbing the tomb immediately drew wealthy Americans and Europeans to the site in numbers that threatened to overwhelm the country's modest tourist infrastructure.[2]

While in England, Alexander made inquiries about steamers that would be bound for Egypt the following winter. Oddly, she made no reference to King Tut or the remarkable discoveries in progress there while formulating her plans, and neither her letters nor her diary contain any mention of the frenetic events of the day. Her aversion to sensationalism and publicity, both of which clearly surrounded the excavations at Luxor, may explain her detachment, but her silence on the topic is remarkable considering that she and Louise stood just a few miles from the site on February 12, 1924, when the king's sarcophagus was finally opened.

Alexander's decision to visit the Middle East as part of Temple Tours' trip number 126 is perhaps the venture's oddest element. She and Louise were not accustomed to "taking tours." Up until this point, they had always traveled on their own, hiring guides or drivers as the need arose. Ever since Annie had had the good fortune to meet Louise, the two had not embarked on a trip with anyone else except family members or, occasionally, a close friend. Whatever induced her to select this particular tour—which she approached with some trepidation—at the journey's end Alexander willingly told the tour company that the experience had vastly exceeded her expectations.[3]

Annie and Louise sailed from New York on January 8, 1924, intending to join the other members of their party in Egypt. Their ship was scheduled to make brief stops at several port cities in Europe but had to skirt the coast of Madeira, owing to a fierce gale. As the boat neared Spain, the wind and waves gradually subsided and the ship docked at the entrance to the Mediterranean, on the rocky peninsula of Gibraltar. The following day it anchored at Algiers, a bustling port nestled at the base of a hilly city. Alexander and Kellogg walked through the Arab quarter, along narrow alleys wound between continuous blocks of houses whose stone steps had been worn smooth and hollow by centuries of use. The local women they passed were all veiled, clothed in baggy trousers and voluminous white robes; the men wore layer upon layer of caftans; beggars wore the remnants of filthy burlap sacks. Although the two visited the botanical gardens and the city's museum of antiquities, Alexander's primary objective was the local Renault agency. The automobile they now sought was a six-wheeled vehicle reported to have three axles, two in the rear, with two differentials to transmit power to both rear axles simultaneously. The company ads claimed that it could navigate almost any terrain, especially in sand, and that it could carry nine to ten people. Failing once again in her quest, Annie wrote disappointedly to Edna Wemple, "We were crazy to see one but they had none in stock as they had sent out the cars the day before."[4]

Shortly before they left New York, Annie and Louise had taken a New Year's holiday in Maine. While there, Alexander interviewed the staff of Herrick Bros., the local Ford dealership, in regard to what she termed "a belt device" that the company had invented for traveling across snow in an automobile. This she imagined could easily be converted to run on desert sand and would make the resulting truck an ideal field vehicle. The belt was actually an elaborate rigging of steel cables and crosspieces that increased the contact area between vehicle and road surface from 35 to 180 square inches. Having failed to test-drive or even examine the Renault she had hoped to purchase in Algiers, Alexander posted a letter to Herrick Bros. from Naples, placing her order for the unusual vehicle, "equipped with extra pair of wheels and the belt device such as you have built for traveling over snow."[5] As hers was to be a desert vehicle, she also ordered an extra water pump and a spare transmission, special chains for the front tires, and several unspecified options in accordance with her previous discussion with the salesman. She then wired Alexander & Baldwin to send a $500 deposit with the $500 balance payable upon notification from the company that the car was ready for shipment. She hoped that it would be waiting in Oakland upon her return.

From Naples, Annie and Louise sailed to Alexandria, stopping briefly en route to see the Greek and Roman ruins and the museum of antiquities at Siracusa on the eastern end of Sicily. In Alexandria, they caught the train for the three-and-one-half-hour ride to Cairo. From the window of their second-class compartment, they saw green fields sprouting young alfalfa, while peasants plowed fields with water buffalo, preparing it for cotton. A procession of farmers walking or riding donkeys, leading water buffalo, cattle, and camels, or simply herding sheep and goats, passed along the road that paralleled their tracks.

Reaching Cairo before the tour was scheduled to start, Alexander hired a guide to accompany them on a six-day camping trip into the desert. Their objective—to collect specimens for the museum, especially jerboas, and explore the Upper Eocene mammalian fossil deposits of the Faiyûm.[6] The women traveled by car as far as Gîza, site of the great pyramids. In a small village, squatting along a wall by the side of the road and tightly clutching cloth bags, several Arab trappers awaited their arrival. Tipped off by Alexander's guide, the price they demanded for their specimens was high, the bargaining tough. Once the negotiations were completed, one of the men was immediately dispatched back to Cairo to purchase chloroform, formaldehyde, and jars so that the women could preserve the items they had bought.

The fossil localities that Alexander and Kellogg hoped to explore still lay several hours across the desert past Gîza. A caravan of five pack animals carrying their camp gear and provisions had already been dispatched to the site. Each woman carefully mounted a camel, grabbing the wooden pommels both fore and aft that served as the only evidence of a saddle. A broad cushion was provided to sit upon, enabling them to cross their legs in front of them or leave them dangling to the sides, whichever proved more comfortable. The symmetry of their pillows allowed them to turn around and face backwards when the continually blowing wind became too much for their faces to bear.

The first night they camped in sight of the famous step pyramid at Saqqâra, which stretched for 8 kilometers across the desert. Tired and stiff, they arrived at camp to find their tents pitched and the tea poured. The Arab tents that had been erected were round, approximately 12 feet in diameter, their walls and ceiling decorated in a brightly colored patchwork of geometric designs, their borders embroidered with Arabic scripts invoking happiness and good fortune on the dweller. Their tent was outfitted with two iron beds, a washstand, and chairs. "A little comfort station," of similar style but simpler, had been placed within steps of the entrance.

Louise awoke the next morning in time to see the sun rise from behind the Great Pyramid. The desert stretched before them as a monotonous expanse of sand, the sparseness of the vegetation unlike anything she had ever experienced. The tents were soon dismantled and the camels loaded. A cold wind blew from the west all morning and at noon their guide made the camels lie down together to form a windbreak during lunch. The caravan traveled twenty-five miles more that afternoon before stopping for the evening on the edge of a cultivated field. The women were cold and sore but dinner proved to be a welcome surprise—soup, fish, two meat courses, asparagus, salad, fruit, and coffee. In the distance, Birkat Qârûn (Lake Morris) glittered in the moonlight. On the far side of the lake, the women could see their destination, a fossil formation that was marine at its eastern end—an almost unfathomable concept amidst the vastness of the desert sand. After dinner, Annie set out some traps on low hillocks covered with tamarisk. She caught one gerbil but no jerboas.

Despite an early start, the range of hills to the northwest proved farther away than it appeared. The women and their guides took four hours to reach the base of the escarpment, after which they spent a great deal of time trying to find a route suitable for the camels to climb. Before they were able to reach the fossil site, it was time to return to camp. Frustrated and disappointed, Alexander wrote to Chester Stock, "Our failure to reach the place was due to lack of time and knowledge of the country and also to the fact that the dragoman could not get it out of his head that we were not a tourist party bent on seeing temples instead of horizons [geologic formations]."[7]

Their last two days were spent camped in the Faiyûm amidst an oasis of fertile soil anchoring grove after grove of date palms. Birkat Qârûn, Egypt's only large natural lake, lay to the south and had irrigated the land for millennia. Annie made note of the birds that they saw and set traps for small mammals, but with minimal success.

One last camel ride took the women to the capital of the Faiyûm. The town boasted a hotel and taxis and its main streets gave quite the air of being a city. But shops in the narrow side streets with their overhanging wooden balconies were no more than holes in the wall, the alleys filled to capacity with noisy crowds of long-robed figures. With this image impressed upon them, they bid farewell to their guide and boarded the train for Cairo.

Joining their tour group in Cairo, Alexander and Kellogg spent several days sightseeing in the city. The party included twenty-three persons but Alexander did not feel that they were being "herded." They visited the mosque of ʿAmr in the Old City, the first place of Muslim worship built in Egypt. Nothing remained of the original structure erected in A.D. 642, but the mosque had been rebuilt circa A.D. 827 by the victorious general ʿAmr. Nearby, they stopped at the Coptic church where Joseph and Mary were believed to have taken refuge with their infant son. At the tombs of the Mamelukes, many of the sarcophagi had been constructed of marble painted in rich hues of blue and red, their chandeliers dripping with crystals, their floors covered in exquisitely embroidered rugs.

They visited the Cairo Museum, which now held the first of the relics to be removed from Tut's tomb. If Alexander and Kellogg were impressed with these treasures, they made no mention of them or their visit to the museum. Rather, Louise recorded that they returned to their hotel to find the "Arabs from the desert" waiting for them with another lot of snakes and a jerboa for sale.

As the hotel lobby was hardly the place to conduct business of this sort, they retired to a secluded corner in the adjacent gardens to examine the specimens. Annie purchased several snakes, larger than the ones she had previously paid for, and a small black-and-white mammal the Arabs referred to as "Pharaoh's Rat." As Louise noted, it looked suspiciously like a skunk and proved to be a zorilla (*Ictonyx libyca*), a striped polecat that is an Old World relative of North American skunks. "Skinning and the smell were the proff [sic]," she noted matter-of-factly in her journal. "It was not as oderiferous as our skunk or we never could have done it [prepared it] in the hotel."[8]

From Cairo, the group drove twenty-five miles south. Switching to camels, donkeys, and sand cars, they traveled an additional eight miles to their camp overlooking the Great Pyramids. For Annie and Louise it was a return to Saqqâra: large brightly colored circular tents in two rows stood ready, with a "dining salon" at one end and an evening of fortune-telling to follow the meal. And the following morning, for those who wished to climb the small pyramid and watch the sun rise, an Arab guide followed with a tray of tea. The pair were the only members of their tour to climb to the top of the Great Pyramid as well, and they were richly rewarded with a magnificent view of the desert and the surrounding Nile Valley.

On the morning of February 12, while the Temple Tour group reexamined the antiquities in the Cairo Museum, twenty-two other individuals stood in the heat of the desert sun at Luxor for what was anticipated to be an even more spectacular event—the opening of Tutankhamen's sarcophagus. For nearly a year, the careful removal of artifacts from the tomb's antechamber marked the end of daily work at the site. This event resulted in a chaotic media frenzy as dozens of correspondents immediately dashed from the site atop donkeys, horses, or camels and raced across six miles of desert sand to become the first to file the day's news from the local telegraph office. While those in the desert now awaited the opening of the sarcophagus, the Temple Tour members returned to their hotel in Cairo, packed their bags, and prepared to depart for Luxor by train at 6:30 that evening.

Aside from the extravagance of the treasures buried with the young king, people with an interest in arts or antiquities felt that this might well be the most significant find in the history of archaeology. Religious leaders became embroiled in the frenzy as well. Because Tutankhamen had reigned during the period of the biblical exodus, there were great hopes to find a contemporary Egyptian record of events amongst his treasures. Yet Annie and Louise seemed oblivious to the day's excitements. That evening, they merely gazed upon the green fields along the banks of the Nile from their hotel verandah, the escarpment rising in the distance on the far side of the river.

The loveliest view of the hills across the river occurred at approximately 7 A.M. when the first rays of sunlight brought out the pink hue of the cliffs. A few hours later, their group paid a short visit to the Temple of Luxor just steps from their hotel. Construction of the great temple had been started by Amenhotep III and finished by Ramses II, who had placed twenty statues of himself in his court; one of the statues remained, almost perfectly preserved. Little more than a mile and a half away, they toured the Temple of Karnak, an enormous complex constructed over two thousand years. By the time that the last pylon had been hammered into place during the Roman era, the first portion of the temple had fallen into ruins. It was an impossible site to view in just a brief visit, but the guides pulled their group away. The day's tour also called for a trip across the river and a visit to the famed Valley of the Kings and Tombs of the Queens.

In the Valley of the Kings, each participant mounted a donkey or was secured in a carriage for the two-mile ride to the Colossi of Mennon, two huge statues that stood in an open field along the road, seeming to serve as sentinels to the valley entrance. Each cut from a single block of stone, now badly weathered, they had flanked the entrance to a funerary temple long since destroyed by the ravages of time and earthquakes. The necropoli of the kings who had been buried in this valley were built upon sloping ground at the base of the steep escarpment that Annie and Louise had viewed from their hotel. While secret tombs carved out of the hillside protected their royal bodies and wealth, Louise mused, worship in temples on the plains below perpetuated the illusion of their immortality.

The group visited the three most important tombs in the valley, Tut's being one of the first that they came upon. All that Annie wrote to Edna of their visit to the famed site was that the tomb seemed to be "well guarded. We looked over a stone wall and could see the wooden door—all we were allowed to do. It is closed for the season and there are various stories going around as to the reason. There seems to be friction between Mr. Carter [Howard Carter, the English archaeologist who discovered the tomb] and the government."[9]

The visit merits no further mention of others at the site or, given the stirring events of just two days earlier (the revelation of a gold-covered wooden sarcophagus inside the first, stone coffin) rafts of reporters scurrying about. In fact, throughout their trip, detailed descriptions of the landscape and of anything to do with farming practices and irrigation devices, both modern and ancient, occupy the greatest number of lines in their letters and journal. Annie likened the Valley of the Kings to the "rocky canyons in Nevada, without the vegetation." In describing the wall paintings within Menna's tomb, Louise commented, "I thought the most interesting thing was the portrayal of the whole process of harvesting grain—first measuring the field, then cutting it with sickles and loading it in baskets to be carried to the threshing floor where oxen are treading it. Then it is winnowed in the wind and measured into heaps for the granary." Her "word about sakias and sadufs," goes on for half a page, vividly describing the elaborate systems involving oxen, pottery jars, buckets, and wooden wheels that the Egyptians used to convey water into the desert region near Aswân. "We wonder how many acres a day can be irrigated in this way," she concluded. "It must be effective for all their grain looks in first class condition and is now well headed out."[10]

From Luxor, the six-hour train ride south to Aswân was hot and dusty. The tracks followed the river past fields of ripening grain and provided a view of the desert and of the mountains off in the distance. Forty thousand acres had been reclaimed from the desert through irrigation. Annie marveled at the mud homes built among the rocks above the high-water mark of the Nile.

At Aswân, they boarded a small steamer for the round trip to the second cataract, passing through Nubia on their way to the border of the Anglo-Egyptian Sudan at Halfa. Annie regarded this leg of the journey as one of the finest excursions she had ever taken, and altogether different from anything that she had previously experienced. "We have been seeing the real African desert from a shady deck with comfortable arm chairs. There could be no more ideal way to travel for we can walk about, talk to the people we like or sit and watch the scenery as we glide along," she recounted to Edna.[11]

Their small steamer was to carry the group to the great temple of Abu Simbel, the finest structure built by Ramses II. It was a visit that would stand out in Alexander's memory from among the dozens of other splendors that they viewed during the tour. Four seated figures, each sixty-five feet high, guarded the entrance to the temple, two on each side. The hypostyle hall bore eight pillars, four on each side of its entrance and, against each of these, stood the figure of a colossus with folded arms and an expression of calm serenity that belied the passage of time. Alexander found the whole effect tremendously impressive.

Throughout the trip, the women set traps for small mammals wherever possible. At Dendur, where there was a small rock temple, they caught three frogs in a pond fronting it. Sightings of lizards at any stop, and the offer of five piastres for a good one, would send dozens of local boys scurrying after the creatures on their behalf. Intermingled with descriptions of temples and deities, Louise duly noted their trapping success: "We caught one gerbil mouse and one Spiny [mouse] in our traps. The latter we have nvere [sic] seen before. It has very stiff, spiny hairs on the back near the tail which is like that of a rat. A scarab had eaten one ear of the mouse and part of the face so we could not make a very good looking specimen out of him."[12] Overall their trapping success was only fair.

The tour reached Halfa in the Sudan at midday on February 22. This marked the starting point for their return trip to Cairo. With occasional stops at temples along the way, their boat reached the capital on March 4. The following evening, the tour members boarded a train and, at the Suez Canal, switched to a ferry. After clearing customs and crossing the canal, they boarded a second train bound for Jerusalem.

Their first look at Palestine in the daylight was as their train left the plain of Philistia on its approach to a rocky valley in the Judaean hills. The winter grain was well up and farmers were scratching at the ground with primitive wooden plows in preparation for winter planting. The bare limestone hills appeared a lush green. Wildflowers were abundant, with masses of red poppies and cyclamen interspersed among the rocks. Children in padded jackets offered these bunches of color for sale whenever their train stopped.

Jerusalem more than fulfilled any expectations that Annie and Louise might have had of the city beforehand. Their hotel was situated on the Jaffa Road, just outside the gate of the same name. Much to their surprise the city outside the wall looked new, the buildings constructed of the handsome native limestone quarried from the surrounding countryside and topped with red-tiled roofs. But once through the Jaffa Gate into the Old City the women found themselves seemingly back in time, plunged into cobblestone alleys lined with vegetable stands and small shops that made them feel that the place had not changed appreciably since the time of Christ. They strolled through bazaars covered with vaulted stone arches that permitted only thin streaks of light to filter through into the dark alleys. Shoemakers, silversmiths, weavers, and garment makers plied their wares in stalls dating from Roman times. In front of the wool merchant's shop, baskets of snow-white fluffy wool competed for space with decorated saddlebags and an assortment of leather straps on display by the harness maker next door. As millions of tourists have done, the women visited the Church of the Holy Sepulchre where the histories of Greek and Roman Catholics, Copts, Armenians, and Abyssinians intersect under a single roof.

The sun shone brightly the following morning as their caravan of cars began its journey toward the Dead Sea and Jericho. The most striking feature of this trip was the changing scenery and temperature that they experienced as the road passed from the cool Judaean hills through the stark splendor of the Judaean desert, before descending 3,800 feet as they approached the hot, lush oasis surrounding Jericho. They drove past the Garden of Gethsemane, circled the Mount of Olives, and traveled through Bethany. Another day they visited the Inn of the Good Samaritan where a "picturesque old shepherd" allowed them to take his picture before the tour went on to Bethlehem, Hebron, and Nazareth. Although they visited all the Holy Land sites most sacred to Christians, Jews, and Muslims, in no way did they seem touched by the religious fervor of other visitors or inhabitants. Rather, they remained fascinated by the agricultural practices that they observed and by the significance of the region's long and tortuous history.

A little over a week after their arrival in Palestine, their group traveled by boat to the southern end of the Sea of Galilee where they boarded a train for Damascus. There they toured the tomb of Fātimah, the mosques, and the bazaars. They crossed limestone formations and fertile valleys to visit the acropolis and temple of Bacchus at Baalbek. Crossing the Lebanon Range, they continued on to Beirut, where Annie and Louise visited the American University and examined the museum's collection of birds, mammals, fossil fish, and invertebrates.

From Beirut, the group traveled south by car, following the coast 110 miles to the port city of Haifa. The following morning they caught a train that crossed the plains of Sharon and Philistia to the sandy desert of the Sinai for their return trip to Cairo and the beginning of their trip home.

Annie had become enamored of the festive, circular Arab tents in which they had camped at the start of their trip, and she spent her last day in Cairo going around to the bazaars in search of rugs to carpet the floor of the one that she had ordered for herself. She also posted the specimens that she and Louise had acquired back to the museum in Berkeley. However, by lunchtime Louise was confined to bed with what Annie initially described as a "feverish cold." In actuality, she had contracted typhoid fever. Their imminent departure was canceled.

For the remainder of the day, Louise's temperature hovered above 102° F. Three days later she was transferred by ambulance to the Anglo-American hospital in Cairo where she remained for two months. At one point she nearly died. The hotel room that the women had occupied in Cairo was closed for fumigation and Annie was forced to move to another hotel. After several weeks, she asked to be given a room at the hospital in order to be closer to Louise. When not sitting by her partner's side, she spent her days consulting with doctors and nurses, running brief errands, and letter writing, her emotional escape and perennial catharsis. She wrote to Grinnell of the paucity of forests in the Middle East and expressed her concern about the implications that this would have on the remaining wildlife. And she remained abreast of museum business, adding in one letter, "It did not occur to me until the other day that you had not sent me the budget. I hope everything is all right. You are generally so prompt."[13]

Alexander herself developed intestinal trouble during this difficult period, attributing her own illness to the "pernicious effect" of the relentless Egyptian sun. She alerted Grinnell to their delay in returning home, expressing disaffection with the land and its people—"It seems rather dreadful to me to live in a country where every [?] of fertility is used to support human life. Land is too valuable to be grown to forests. Manure dried into cakes is the fuel for the people. And if you want to get away from this mass of humanity, where are you to go? The most appalling deserts confront you on either side. Think of living a life-time and not knowing what a native forest is!"[14]

By late May, Louise had recovered sufficiently to travel. Annie packed their bags and, with great relief, the women took the train from Cairo to Alexandria. Here they boarded a ship bound for Naples and began their long journey home.

21
The "Amoeba Treatment"

It took nearly a month for Annie and Louise to reach Oakland from Cairo. Following their return, both women began treatments for "endoamoebic dysenterine." When they saw no immediate improvement in their conditions, they checked themselves into a local sanatorium for several weeks. During this relative confinement and inactivity, Alexander battled the familiar feelings of alienation and restlessness that surfaced when she was in the city. Once again, she and Louise contemplated the desirability of maintaining an apartment in the Bay Area. Annie vacillated. It was not just the question of whether or not they should take an apartment, but also where it should be. In 1920, two years after her mother's death and the sale of the house in Oakland, she had written to Martha,

> It came over me the other day that I was living like a heathen with no home that I could call my own, where I could entertain and earn a place in the Community, and where I could have my own belongings about me. The farm is just a make-shift. I used to think that a tent with the roaming life it signifies would satisfy me to the full, but I'd give anything for a quiet little corner just now. If I did have a home I think I should like to live in Berkeley where I could keep in closer touch with my interests there and university life.[1]

The subject had come up again in 1922 after the death of Louise's mother. By 1923 Annie seemed to have changed her mind and was expressing a preference for living in San Francisco, while Louise favored the idea of renting an apartment in Oakland. Then, less than two weeks after their discharge from the sanatorium in 1924, they moved into the Hillcourt Apartments on Bellevue Avenue in Piedmont. The following summer they moved again, this time settling into a furnished apartment in the Regillus, a building at

19th and Jackson streets near Lake Merritt in Oakland where they remained for more than two decades. Alexander hinted at her previous misgivings when she wrote to Mary Beckwith a year later, "I decided to rent an apartment as an experiment. I have the curious feeling that if I tie myself to a place [by buying] that will be the end of me."[2]

The restlessness that Annie had been experiencing before her trips abroad continued after she and Louise were released from the sanatorium. At the same time, her intestinal ailments persisted. Her need to flee the feelings of confinement that plagued her at home, and dissatisfaction with her accomplishments, culminated in a new and intense period of collecting expeditions. Although her view of herself as an amateur naturalist never wavered, the expeditions on which she and Kellogg now embarked gradually became more directed and purposeful. With the exception of a trip to Kern County in 1924, the goal of these expeditions was to find small mammals, a shift that probably reflected Alexander's disappointment and frustration with the unstable situation in paleontology. In addition, she no longer seemed content simply to enlarge the MVZ's collections. Increasingly, the women focused on gathering specimens that would have direct bearing on the research questions being undertaken by the museum's staff and students.

The greatly anticipated Ford truck arrived from Herrick Bros. in September 1924. Eager to test the new six-wheeled vehicle with its detachable tractor treads, the women immediately set out on a six-week fossil-hunting expedition to Last Chance Gulch in Kern County, California, accompanied by Annie's cousin Mary Charlotte. The trio took an overnight train from Oakland to the small desert town of Mojave. There they were met by Eustace Furlong, who had driven the unusual vehicle down from the city. Not surprisingly, it attracted a great deal of attention. Mary Charlotte quickly christened it "I'll-away" after a song that Annie and Louise sang, "I'll away to the fossil land!"[3]

Because there was no way to accommodate four persons in the truck's cab, Louise and Mary Charlotte insisted that Alexander ride up front with Furlong while they sat perched atop the field gear and baggage, holding on to guide ropes for balance. Between the rough road and high winds, the experience seemed to Louise as precarious and exhilarating as a camel ride across the Egyptian desert. The Egyptian analogy seemed even more apt when the group stopped northwest of Red Rock Canyon near Ricardo. Here they set up the enormous Arab tent that Alexander had purchased in Cairo. In Egypt, it had required the combined strength of fourteen men to raise the great circular canopy. Now, with effort, the four of them pulled on the guide ropes and succeeded in hoisting the center pole 15 feet above the ground. Mary Charlotte was dazzled by its twelve brightly colored panels embroidered with Byzantine designs and Arabic good-luck phrases. When the Bedouin rugs were spread beneath it on the sand, the tent offered the weary collectors a cool refuge infused with subdued light, a welcome respite from the glaring heat of the midday desert sun.

Much to Alexander's surprise, the canyons in the vicinity of Ricardo offered little in the way of interesting fossil material. On their second day out the reason quickly became apparent—they learned that Frick and his party had already spent three weeks scouring the region.

Wasting no time, the group relocated their camp to Last Chance Canyon. The move gave Alexander the opportunity to try out the Ford's special device. Furlong eagerly deflated the four rear wheels of the truck and attached the specially constructed belt that had made the vehicle such a success traversing the deep snows of Maine. He and Kellogg then tested the vehicle, attempting to capsize it on all the steep hills and difficult terrain in the vicinity. Alexander then joined the pair on a drive to Cantil, during which Furlong took great delight in turning the truck off the road and into the deepest sand at the approach of each oncoming vehicle. Throughout, the Ford performed beautifully.

At Last Chance Canyon the party camped at 3,300 feet. The sedimentary formation in which they worked dated from the Lower Pliocene and contained unusual deerlike skulls with branching horns, bear and rhinoceros jaws, the ubiquitous assortment of camel bones, and a number of other miscellaneous fragments. The Lower Pliocene had been a period of great faunal interchange between Asia and America, and the women hoped to find representatives of species that were ancestral to many of the carnivores and ungulates that now flourish in North America.

Each evening the sunset transformed the buff-colored ridges and long lines of exposed, twisting strata on the slopes surrounding their camp into vivid patches of brilliant color. Mary Charlotte wrote to her family of the beauty and fascination that the scenery held and of the excitement and enthusiasm that Alexander brought to the work:

> I can understand now why Annie loves the desert so much, and why it is particularly alluring to her with her rare scientific knowledge and insight. She has never yet made such a trip without finding some rare fossils that made it worth while. She says life will become prosaic later when everything is discovered! The uncovered, weathered strata here are something like an open book to her, as she knows the periods so

well; she instantly gives the scientific name of any bone that we find, telling what animal it belonged to; and she recognizes the animal tracks and knows intimately all the birds we see.[4]

Near the end of their journey Mary Charlotte wrote again, reflecting on the expedition with words that Alexander herself might have used:

> I wonder how it will seem to go back to houses after having lived with all of out of doors for our beautiful room! The sunsets have been gorgeous, and we have been having marvelous moonlight nights again, and best of all we are having remarkable success. . . . The intense fascination of this work, I think lies in that fact that what we find has real scientific value, and as Annie says, "Throws light on a closed chapter of the world's history." She says that the place grows more and more interesting as we understand more of its significance.[5]

Not until late January 1925 did Alexander eat what she considered her first good meal since returning from Egypt. Although her doctor could no longer detect signs of dysentery, he professed that she was not producing bile and he placed her on bile capsules as a cure for her intestinal ailments. Standing a mere 5 feet 3 1/4 inches tall, in an undershirt and stockings she weighed just 99 3/4 pounds. She was no less cynical now about the medical profession than she had been in her youth. In an otherwise newsy letter to Grinnell written from Honolulu shortly after the New Year she commented, "I am sometimes a little skeptical myself that the amoeba was the cause of *my* trouble. Of course dynamite is a good explosive to use if you want to destroy anything but it is quite wholesale in its work—something like the amoeba treatment."[6]

Upon her return from Hawaii in March, Annie continued to be treated for intestinal problems. Her doctors subjected her to an unrelenting series of pokings, proddings, X-rays, and appointments, but with few positive results. With her patience for the treatments and her tolerance for Oakland both at a low, in early April she and Louise jumped in the Ford and once again headed for the desert to collect fossils. Alexander herself affectionately referred to the Ford as "Blundie," short for "Blunderbuss," and she seemed remarkably pragmatic when first one, then another device on the vehicle succumbed to harsh conditions in the Mojave—first the altimeter, then a drive rod, then the steering device required the attention of a trained mechanic. Problems in the field never seemed to dismay her. When a chain fell off twice en route to Mina, Nevada, in late June, the women left the vehicle there for service and returned to Oakland by train.

The desert provided Alexander with the quiet solitude she craved, far away from the barrage of physicians and from the oppression of life in Oakland. Grinnell's genuine empathy for her feelings in this regard further underscores the successful nature of their relationship. In reply to a letter from the university's president in September 1925, expressing concern for her health, Grinnell wrote, "Miss Alexander's health is somewhat improved; but she is still far from well. She shows marvelous grit in the way she keeps going. Incidentally, she obviously dislikes having reference made to her physical condition. Perhaps that is why she stays so little of the time within reach of people who continually make solicitous inquiries."[7]

For the first month that Alexander and Kellogg were in the field, they explored canyons and scoured strata in search of fossils that they hoped would delineate one or more branches of the mammalian evolutionary tree. Then, in late April, Grinnell wrote to Alexander with the casual suggestion that pocket gophers from the extreme eastern portion of California and up into Nevada would be a welcome addition to the museum's collection—"*If you find the opportunity for any mammal collecting at all,*" he added.[8] He contended that even one or two individuals per locality would suffice, although he acknowledged that half a dozen from each locality would be preferable.

Although pocket gophers were better known as agricultural pests than as fascinating subjects for evolutionary study, Grinnell was becoming increasingly interested in this group of subterranean rodents. He knew that the animals did not disperse widely and wished to measure the degree of differentiation among disparate populations that were now relegated to isolated water holes or seepages scattered throughout the desert. The formation of isolated populations coincided with the gradual transformation of much of southern California and western Nevada from wet to dry habitat. Because these populations had had the opportunity to evolve independently over long periods of evolutionary time, Grinnell reasoned, variation among them would offer insight into the fundamental mechanisms of speciation.

Alexander was certainly not averse to the idea of collecting specimens, in this case pocket gophers, for the museum. Quite to the contrary, she took great joy in being useful to its research mission. She also knew from experience that Grinnell's requests were never made casually, but always with the strong intention of following through to the ultimate goal of creditable publications. Approximately five weeks later she wrote to him of their success, "In spite of a storm of snow and sleet lasting three days we have taken in quite a number of things though represented in several cases by only a single individual."⁹ Belying this modesty, the list of small mammals she appended included twenty pocket gopher specimens.

Alexander and Kellogg returned to Nevada in early September for a three-month collecting trip through Nye, Lyon, and Mineral counties. In addition to their usual camp gear, on this trip they brought with them the replacement rods that Blundie required, having been forced to abandon the Ford in Mina in late June. Their first camp was on the slope of a hill overlooking Stewart Valley. They had purposely erected their tent a short distance from the only spring in the area as several bands of wild horses routinely came in to drink. Alexander noted that the animals were extremely shy, constantly on the alert for trouble, and that they had taken to visiting the site at night. She also commented on the abundance of bunch grass and white sage that had appeared as a result of unusually heavy summer rains, and she deplored the presence of cattle and sheep in the region, as they would soon browse the lush greenery into oblivion. One afternoon she watched intently as a coyote chased a rabbit out of the brush and into the wash where they were working. It grabbed the rabbit by the back of the neck in its powerful jaws, then just as swiftly dropped the frightened animal unharmed and quickly retreated when it realized that it was being observed.

The women divided their time between fossil hunting and mammal trapping. Each day they set three to four dozen traps apiece in the vicinity of camp, scouring the nearby strata for bone fragments between periods of checking and resetting their traps. For the first time, Alexander began keeping score of their success in her diary, for example, "L. gets 12 Micro. [*Microtus* = vole], 3 Perognathus [pocket mice], 1 dipo [*Dipodomys* = kangaroo rat] total 16. I get 13 Micro. 1 Peromyscus [deer mouse] 1 Dipo. 1 Perognathus total 16. L. 48 traps, I 42. Sandy flats covered with desert brush and bunch grass."¹⁰

It is difficult to guess why she began keeping track of such figures at this point. Among many field biologists, a certain machismo attends the number of specimens in a field catalog, because the time and effort needed to capture each specimen and prepare it as a study skin are substantial. Yet Alexander and Kellogg had never before felt the need to measure such effort or compete in this or any other activity. The prodigious number of specimens that they collected for the museum might lead a casual observer to overestimate the ease of the task, yet the number of animals captured relative to the number of traps and nights each trap line was run was often depressingly low. Thus, twenty-five years after she first began collecting small mammals, it is curious to think of Alexander being caught up in this good-natured professional competition.

Throughout their stay in Nevada, the weather varied considerably, hovering around 18° or 20° F at 6:00 some mornings, overcast and unusually warm other days. Nonetheless, the women continued trapping and hunting until early November, relocating their camp several times over the course of two months. They occasionally set traps for gophers, but with only moderate success. By the time they began making their way home, the chill of winter was unmistakably in the air. They drove west as far as Yerington before turning north to Wabuska. Here they parked Blundie in a garage for the winter and returned to Oakland by train.

Within three weeks of their return, Annie was on a ship bound for Hawaii. She was still not feeling fully recovered from her intestinal ailments and decided to consult a physician in Honolulu. The doctor recommended a hospital in London for further treatment. Returning to San Francisco in mid-March 1926, she and Louise left almost immediately for New York and, on April 17, sailed for Britain. Admitted to the Hospital for Tropical Diseases, she was diagnosed as suffering from sprue, a condition of unknown etiology characterized by abnormalities in the mucosa of the small intestine, malabsorption of essential nutrients, and the development of numerous nutritional deficiencies. Megaloblastic anemia is also a common symptom of the disease.

It is difficult to imagine how Alexander had been conducting intensive fieldwork for so long while suffering from sprue, but she wrote to Grinnell from her hospital bed, "My main symptoms would indicate it—intestinal trouble and low blood count (I have less than half the red blood corpuscles Miss Kellogg has for instance and what I have are misshapen). It is rather a relief to give up and become a bona-fide invalid—for a while at least. As long as one is up and around one has to pay the penalty by assuming to be a normal, sensible person."[11]

Despite a special diet, numerous medications, and two injections of arsenic plus iron per week, Annie did not respond positively to the treatment. She weighed just 101 pounds. Clearly dismayed, she wrote to Martha, "I wish I were headed to California as fast as my skinny legs could carry me! . . . The doctor is going to have me out soon, either to have me following up a strict diet for three months, keeping house somewhere near here, or discharged a failure."[12]

By early June, Annie had recovered sufficiently to be discharged, and she and Louise relocated to a flat in London. Still weak, she was unable to walk more than six blocks at a time before exhausting herself, and she was having trouble keeping down food. When her condition did not change markedly after a few weeks, she and Louise traveled to the far end of Lake Geneva where she checked into a sanatorium for additional treatment.

Alexander's state of health did little to distract her from the chaos that swirled around the situation in paleontology. Her stream of correspondence with university faculty and administrators continued unabated from Europe. While she rested and underwent treatment, she wrote letter after letter on the issue with her usual vitriol and frequency. In mid-July she was discharged, having completed the prescribed therapy (though her weight remained unchanged). From her New York–bound ship, she wrote to Camp of her intention to withhold support from UCMP until the university capitulated to her demands. But the sentiments expressed in that letter were no less adamant and heartfelt than those that she penned to Martha from their apartment in the Regillus upon their return, "I was so happy to get back, even despised Oakland looked good to me."[13]

One of Alexander's first acts on returning home was to purchase a new Franklin. The day after the women arrived at their apartment, they had headed north to check on their interests at the farm. They had gotten no farther than Cordelia when the Franklin threw a rod, forcing them to be towed the remaining distance. This was only one in a long series of problems that Birdie had been experiencing in recent years and Alexander felt that the car could no longer be relied upon either for fieldwork or for local commuting. While Blundie served its purpose in extremely sandy habitats, the Franklin had its own unique strengths. With its air-cooled engine, it too was an ideal vehicle for desert travel. And the unusual placement of the rear bumper directly behind the back wheels meant that it could clear unexpected ditches with relative ease.

Confident that their ranch manager had operations on the farm well in hand, in October the women set out for a three-month collecting trip to Lander, Churchill, and Nye counties, Nevada, in search of additional pocket gopher specimens for Grinnell. Alexander's often wry style of letter writing had not changed over the years, as when she began a letter to Grinnell by stating, "I hope we are not getting too many gophers for I know that tray room space [for specimens] is something to take into consideration." Grinnell fired back a week later, his own serious style similarly unchanged over the years, "Don't have the least hesitation about taking gophers and other animals in quantity, on the grounds of Museum space. We have just gotten in 15 new cases; and I figure that each case will hold *800* gophers (skins-with-skulls)!"[14] Although Alexander's letter did not list the number of specimens the women had captured to date, it did close with a postscript, "By the way, we shall need more [specimen] tags." These were dispatched immediately.

As the women collected, they paid special attention to the vegetation in each of the habitats in which they trapped, carefully noting differences in plant associations from one locale to the next, and frequently photographing each. This kind of detail would prove invaluable to Grinnell in his analysis.

While Alexander wrote wryly and confidently to Grinnell, she continued to express deep-seated intellectual insecurities in her letters to Beckwith, feelings of dissatisfaction with her accomplishments that seem to have remained unaltered over the course of two-and-a-half decades. In early December she wrote from Austin, Nevada, "I feel destined to become some day a recluse. The world is too much for me, too masterful. I fly from it to the lonely places that make no demand upon my [mental] strength." Her affinity for Oakland proved similarly short-lived and in late December she wrote almost predictably from Millis, Nevada, "I'd like to stay on indefinitely in this country.... Long as I have lived in Oakland and thereabouts I've always felt an alien to those parts."[15]

The pleasures of winter camping declined somewhat when the temperature dropped to −16° F one morning. Alexander and Kellogg decided it was time to break camp and move farther south. Up to this point they had endured driving rains and the occasional brief blizzard, availing themselves of the opportunity that the weather presented to collect, among other species, muskrats in thick winter pelage. By the time that they reached Oakland in mid-January, they had collected 220 gophers from 15 different localities in Nevada's Great Basin. In all, they had shipped almost twice that many specimens back to the museum during the course of their three-month foray.

As was his custom, Grinnell wrote to the president of the university informing him of the size and research value of the museum's most recent accession. His reason for doing so seems to have been twofold. First, although Alexander's endowment had ensured the MVZ's stability of purpose and had guaranteed the security of its collections for many years to come, Grinnell wisely realized that the overall health and well-being of the museum was dependent upon its relationship with the university. As the institution grew, with each passing year the endowment contributed proportionately less to the museum's operating expenses and he wished to earn for it a secure and prestigious place in the eyes of the administration.

Second, Grinnell remained cognizant of Alexander's belief and chagrin that the administration viewed her merely as a funding source. He wished to impress upon each of the university's presidents that her contributions to the museum were far more than just monetary. As the number and value of her specimen donations increased over time, these letters ensured that Wheeler, Barrows, Campbell, and eventually Sproul remained intimately familiar with all aspects of Alexander's benefaction. Grinnell also frequently suggested that Alexander might be pleased to receive a personal letter of thanks from the president for her specimen donations. Whether or not this was true, these small exchanges between presidents and patron seem to have been a positive link between the administration and an often-demanding benefactor throughout the course of their long and frequently contentious relationship.

22

Fieldwork–The Later Years

Alexander and Kellogg's trip to central Nevada in the winter of 1926–27 was the first in a series of extended collecting expeditions that continued until Grinnell's death in 1939. The trips varied in length from one to six months and in the types of specimens that the women sought to collect. At Grinnell's suggestion Alexander and Kellogg now focused on building up the museum's collection of topotypes—specimens from localities from which new taxa had been previously described—based on the belief that such material contributed to the MVZ's independence and resourcefulness as a research institution. Those trips were of inestimable value in strengthening the museum's collections. Succeeding years saw ever-increasing habitat destruction throughout the western United States and, with it, loss of access to historical patterns of species' distributions and natural history. Coupled with the women's field notes and photographs, each specimen collected documented the occurrence of a given taxon and added insight into its habits.[1]

At the outset, neither Alexander nor Grinnell quite anticipated the difficulty involved in such a quest. Much of the land in the western United States had come under cultivation in the twenty to thirty years since the animals they sought had first been described. How were they to find suitable habitats in which to place their traps? Alexander was both disturbed and astounded at the almost complete lack of natural vegetation in some regions they visited, particularly in northern Utah. Other areas exhibited perturbation as a result of grazing by sheep and cattle, or by humans. In 1931 she and Kellogg wrote to Edna from El Paso, Texas: "Types [type specimens] of a number of the specimens we have collected were named from here so we thought it quite important to get the topotypes for comparison, although we hate collecting in a big city where the original territory is covered with houses and auto camps and one has to go miles out of town to set traps in suitable ground."[2]

But the extended field trips that the women undertook from the late 1920s through the 1930s represented more than an altruistic mission for Alexander. Her letters during this period continually point to the field as a haven from life in Oakland. In the middle of the depression, she commented pragmatically to Edna, "We are glad to have our minds occupied with other things while the stock market remains so depressing." Away from the social obligations of city life, and the hay fever that she said plagued her at home, she wrote to Grinnell in the summer of 1931, "I wish you might get off for a little run into the mountains. A visit once a year, however brief, I believe is essential to one's well being."[3]

Alexander strictly followed her own prescription. In July 1930 she and Kellogg took a break from the rigors of collecting topotypes in northern Utah so that Louise could fish in the clear mountain streams of Pine Canyon, at a much higher and cooler elevation than where they had been trapping. From this secluded spot the women wrote to Mary Charlotte,

> Every time we see a mountain, especially if we see a forest on it, we want to get up into it and camp—the only place [to be]. Life otherwise is humdrum, a confusion of dodging autos, unholy noises and processions of people whose business you don't know nor ever will. They are just humans on two legs going somewhere—you meet them all the time and never give them a thought. Louise says this doesn't make any sense but Anne says it is just the impression she gets when she descends like Elijah from the chariot of fire.[4]

While the mountains seemed to provide a haven for Alexander, she was never really able to relax, even there. She could not be idle. Even though the fishing trip was to provide a break from collecting, Annie discovered that Pine Canyon was inhabited by a species of pocket mouse that had eluded their capture thus far, as well as a number of "other interesting animals," which she and Louise proceeded to trap. Whether this compulsion to collect, to be constantly productive, was a result of her upbringing or a manifestation of her insecurities is unclear, but her need to be continually and constructively engaged is unmistakable. Martha Beckwith once sent Alexander a poem from Goethe's *Wilhelm Meister's Apprenticeship*, which the latter frequently quoted in letters to others. Her German translation varied slightly each time she set pen to paper, but it began, "Keep not standing fixed and rooted, Bravely venture, briskly roam." By her own admission, Alexander's philosophy of life was best embodied by the maxim—It Is Better to Wear Out than to Rust Out.[5]

Shortly after their stay at Pine Canyon, Alexander wrote to Grinnell that she and "Miss Kellogg" planned to meet her sister Martha at Pennask Lake south of Kamloops, British Columbia, in the latter part of September for another brief fishing trip. Her letter noted that they had not brought a checklist of type localities for Washington and Oregon with them but, if Grinnell was willing to send one, they would attempt to collect specimens along the way. Alexander added, "Is there anything in particular you would like us to get? You once mentioned needing Perognathus from this region."[6]

Grinnell's own drive and devotion to work were much like Alexander's, a fact that provides an additional key to their successful collaboration. Pleased at Alexander's request, he immediately obliged by sending the women a list of type localities and the species he most desired from each. When specimens were slow to accumulate, however, Alexander went out and purchased an additional twelve dozen traps in the hope of improving their success. Between them, she and Kellogg began setting four hundred a night. Still displeased with their rate of capture, she wrote to Grinnell, "The situation of The Dalles on the Columbia River is very scenic but what charm has a place if topotypes are not forthcoming."[7]

Many examples of Alexander's driven personality exist. On a large scale her creation of the museums and simultaneous management of the farm were massive undertakings and few individuals would have attempted both. On a much smaller scale, the volume of correspondence that Alexander maintained, without benefit of a secretary, is astounding. From the farm or in the field, she might write as many as a dozen letters each day to university staff and administrators, to her ranch manager, and to family and friends. These ran the gamut from conducting business, to looking after her financial interests, to making social inquiries.

Each spring, as soon as the last of the asparagus spears had been cut and shipped, Alexander and Kellogg packed the car with their field equipment and set out. Armed with a list of type localities, they most frequently headed east. In late May 1928 they spent a month collecting in Mineral County, Nevada, just across the California border. Alexander was now sixty. One of their most pleasant camps, if not their most productive, was in the Excelsior Mountains. At approximately 6,500 feet they stumbled upon an abandoned miner's cabin replete with tables, chairs, and what Annie considered "all the comforts." The scarcity of small mammals in the area afforded them ample opportunity to enjoy long, leisurely hikes. Piñon pines dominated the northern slopes of the Excelsiors and in them large flocks of piñon jays kept up a perpetual banter. A small clearing in the trees or a sudden turn in the trail often provided magnificent views of the surrounding countryside. To the west, the Sierra Nevada rose majestically and, in the clarity and freshness of the desert air the two women were often able to see Mount Whitney to the south, and even Lake Tahoe to the north.

On this particular trip the women "collected the usual run of desert mammals, including a large Kangaroo rat that we found as delicate and as good eating as chicken."[8] The rodent proved to be their only source of fresh meat, although Louise did noose two large gopher snakes with her fishing line, and two rattlesnakes. These, however, along with a variety of lizards, were preserved in formalin.

In addition to vertebrates, the women collected half a dozen species of plants and shrubs that were new to them. Although their plant collections were still limited in scope, Alexander's growing interest in botany and in the botanical communities that she observed is evident. She easily recognized that the evolutionary questions Grinnell and his staff were addressing with regard to vertebrates could be applied equally well to plants. She explained in a letter to Beckwith, "[L]ike the small mammals they harbor[,] one form [of plant] will predominate in a given locality and another in another according to the character of the soil and altitude."[9]

Although their primary mission was to collect topotypes, Alexander and Kellogg often collected additional specimens that they knew would be of interest to particular researchers in the museum. On this trip, after purchasing supplies in the small town of Mina, Nevada, just east of the Excelsiors, the women decided to travel a route that was new to them. They had not driven more than a mile or two when they noticed fresh gopher mounds by the side of the road. Ordinarily, pocket gophers dig in moist soils, so their presence in this dry rocky habitat covered with upland desert shrubs was unusual. Stopping to investigate, the women noted that the mounds housed a peculiar form of pocket gopher that they had previously collected only at two widely disparate, but equally dry, localities. Here was an opportunity to collect additional specimens at a third, intermediate site.

Nothing seemed to discourage the women in the field. To be out in the open was Alexander's supreme joy. In July she wrote a newsy letter to several of her friends from Arlemont, highlighting the details of their trip thus far. It began, "We have just come from one of the lonesomest valleys that Miss Kellogg and I have ever visited in Nevada." But the loneliness of the valley was not what made this particular leg of their journey worth recounting. Rather, it was the approach they had taken in getting there. The women had been warned that the only road into Huntoon Valley was a steep climb, so steep in fact that an old miner had given them a block of wood to shove under a rear tire if the Franklin did not prove equal to the challenge. Because they had been following a dry streambed out on to a plateau, the sudden—almost vertical—ascent they confronted after a sharp turn caught them by surprise. Neither gears nor brakes were able to keep the car from slipping backwards. Grabbing the wood, Alexander shoved the block under the tire but "before I could snatch my hand away, the wheel had run over one of my fingers, crushing it but not breaking any bones."[10]

Fortunately, an embankment had been built at the turn and it halted the Franklin's reverse descent. While Annie soaked her finger in a jar of disinfectant, Louise unloaded more than half their gear onto the side of the road. Only then did she succeed in coaxing the reluctant vehicle up and over the almost vertical slope. But their troubles did not end there. Alexander's narrative continued:

> At the mouth of the canyon there was a bunch of willows but no water or buildings and the road became so sandy we decided to turn around which proved no easy matter. It was then just about noon, not a breath of air stirring and no shade. We got caught on a sage brush and in a gully and were hours getting out. Louise wore her fingers to shreds pulling the chain of the jack. I was of very little assistance except to kick the sand out of the way with my heel and my cup of disinfectant soon became full of sand and flies. On top of it all we had a flat tire.[11]

Alexander suffered no lasting effects from her misadventure with the Franklin. The following year she and Kellogg packed up the beloved vehicle and headed for southwestern Utah. The topography of this region pleasantly surprised them. Alexander had pictured Utah as one large desert. She had not anticipated its high plateaus cut by deep canyons, its forests of pine, fir, and spruce, or its towering mountain ranges. On the downside, the weather was hot and, in some localities, the collecting poor. They also had difficulty finding several of the type localities that they were after.

As was their custom, wherever the women set their traps they sought out old settlers, hoping to glean information about the exact localities described on their list. On this particular trip they met with little success. On one occasion they queried numerous individuals for directions to "Briggs Meadow, Beaver Range Mts.," the type locality of a marmot, a chipmunk, and a pika, but each person swore up and down there was no such place as

Briggs Meadow—although there *was* a Briggs Creek *and* a Briggs Corral on a nearby rocky plateau. After Alexander tracked down an old woman who claimed to have boarded the two gentlemen who described the locality twenty-five years earlier, she became convinced that the collectors must have meant "Britts Meadow" at 8,500 feet on the road to Puffer Lake, rather than "Briggs Meadow," which is how the locality had been published.

The women went out of their way to climb several of the highest peaks in the Pine Valley range. These were over 11,000 feet and offered breathtaking views in almost every direction. One day Annie rented horses and she and Louise embarked on a twenty-mile ride along the crest of the mountains though forests of fir, spruce, and giant aspens. Near the summit they encountered white pine and a few foxtail pine, but on the lower slopes Alexander observed that the original yellow pine had been logged long ago. She and Louise then camped for two days in a grove of fir trees on the slope of Brian Head Mountain. With only seven traps, Kellogg succeeded in capturing four pika on the talus slopes of their type locality. By the time that the women relocated their camp to Britts Meadow, however, they had equipped themselves with a "baby shotgun and a .22 rifle," Alexander's only regret being that "it was not a 30–30 when we came to get the big marmots."[12]

In early July the women drove north to Ogden to meet Annie's sister Martha, who was traveling west from New York. The three visited Yellowstone National Park for several days before Martha reboarded the train bound for San Francisco and from there caught the steamer to Hawaii. Annie and Louise continued trapping. More than a week later they stopped in Kamas, Utah, to retrieve their mail. In addition to the general accumulation of letters and newspapers that awaited them at the post office, Annie received a telegram from Hawaii. Wallace, Martha and John's twenty-one-year-old son, had been killed in a freak polo accident after his horse tripped and fell on top of him. The women immediately loaded the Franklin and returned home. After repacking her bags, Alexander sailed for Honolulu on the next available ship.

By mid-September 1929 Annie was back in Oakland. As the month drew to a close, she and Louise readied the Franklin and set out for Utah to continue the trapping that had been interrupted earlier in the summer. This time their destination was Box Elder, Weber, and Utah counties in the north, and San Juan County in the southeastern corner of the state. They still planned to collect topotypes, but the onset of winter weather ultimately limited their access to many of the localities at higher elevations.

From the beginning, luck seemed to be operating against the women on this particular trip. While climbing the Sierra Nevada near Gold Run, their car started to buck, advancing only a few yards and then stopping. Louise tried blowing out the gas line several times, to no avail. By fits and starts they hobbled to a garage where the mechanic was able to clean out the screen over the carburetor that had become clogged with dirt.

Approximately thirty miles beyond Wells, Nevada, they left the highway. After making careful inquiry about a shortcut to Kelton, Utah, and having been warned that the first ten miles or so of road were in extremely poor condition, they set off on a back road toward Rosebud Ranch. They had just been congratulating themselves on successfully negotiating the worst of its numerous gullies with their exhaust pipe still intact, when the Franklin slid crosswise into a sandy wash and would not budge. Because they had slid into the wash at an angle, all four doors of the car were jammed shut and the women were forced to crawl out of the vehicle through its windows. The mishap occurred at midday and, in the heat of the desert sun, Alexander realized that they had failed to fill their storage containers with water. After unloading the car, they managed to jack up the two low wheels and jam rocks under them, but it was nightfall before they were able to extricate the car. At that point, Annie was relieved simply to throw her bedding on the ground and stretch out.

When the women eventually reached their destination near Kelton, the area seemed anything but promising for trapping small mammals. The ground was covered with alkali and a scanty growth of saltbush. Nonetheless, they each set out four dozen traps. They hoped to catch pocket mice and chipmunks and then head for higher ground—that is, until the gas line under the vacuum tank of the Franklin cracked. In attempting to fix the line, they broke off the elbow at the carburetor. The nearest garage was in Brigham, seventy miles away. Alexander had the car towed the distance and the mechanic there soldered the line. However, the line broke again several miles from town and the car had to be towed back to the garage a second time. While the mechanic wired to Salt Lake City for new parts, Annie listened to the falling rain. Certainly, the farm in Suisun desperately needed the moisture but, as she watched, the low-lying areas around them began to flood, washing out the surrounding roads. It was clear that further fieldwork would be impossible, at least in the immediate future.

After a week of rain that forced the women to remain idle, the weather abated sufficiently for them to set out. Although the mechanic in Brigham had fixed the broken part on the gas line by replacing the threads, he could not completely repair the gas line itself, which continued to suck air. This required Louise to stop every four miles or so to fill the vacuum. By the time they managed to get the line wrapped with friction tape, Annie had taken to carrying a 5-gallon container of gasoline with them as a precaution.

Even though the rain had now let up, the nights had become cold and their trapping success remained uninspiring. One day they set out six dozen traps for kangaroo rats. This was an area of alkali flats amidst swarms of fierce mosquitoes, which bit mercilessly. After driving to town to purchase ammonia for the itching, they were nearly hit head-on by a drunken driver who suddenly slammed on his brakes and spun around 360° in front of them. Fortunately, no one was seriously injured in the crash.

Not all their misadventures on this trip were so serious. Alexander's interaction with the skunk was one example. When the women trapped in the vicinity of urban areas, or when the weather looked precipitous, they increasingly took to staying in auto camps instead of hotels. By 1929 such facilities dotted the highways from coast to coast, a natural outgrowth of the increasing number of cars on the road and the general public's new interest in the outdoors. Auto camps varied somewhat in the amenities they offered, but most provided tent spaces adjacent to one's car, as well as cabins equipped with bed frames, mattresses, stoves, and electric lighting, a welcome comfort on those evenings when the women spent their time putting up specimens late into the night. In Ogden, for instance, the auto cabin that Annie and Louise occupied was 12 × 18 feet and furnished with a coal stove for warmth as well as for cooking, a double bed, a folding couch, a table, and three chairs. Annie wrote jokingly to Martha, "[W]e are becoming so addicted to auto cabins that we may give up our apartment in the Regillus to live in one in Oakland."

When she and Louise spent a night near Provo, the cabin that they rented there was even more elaborate, and cheaper. The floors were tiled with linoleum and the room was equipped with running water and an electric hot plate for cooking, in addition to the heating stove—all for $1.25 per day. With their belongings flung helter-skelter throughout, Annie wrote to Edna, "It answers all our needs perfectly and we see no reason why we could not entertain in one of them. A bunch of flowers and a few pictures on the wall would make it as homelike as any place."[13]

One of the specimens that Alexander and Kellogg hoped to catch near Ogden was a small spotted skunk, referred to locally as a civet cat. Almost prophetically, Annie's letter to Martha continued, "If we succeed in getting any specimens we promise to be very discreet, as we would not like our fellow auto campers to detect any significant odors about us. The man in charge of the camp might not be so affable."[14]

To catch the skunk, the women purchased an old hen and baited their traps with its feathers, guts, heart, liver, head, wings, and legs. To make the bait more enticing, they smoked it. While some of the bait was taken, they did not catch their prey. No matter, much to Alexander's delight a boy in Provo Canyon succeeded in trapping a young individual for them.

Louise was in town running an errand when Annie began to skin the immature animal. Perhaps because of its small size, she almost immediately punctured the skunk's scent gland. The smell quickly filled their tiny room. In a matter of moments the odor became so overwhelming that she could barely walk a straight line. As Louise reached the entrance to the camp, the unmistakable scent greeted her nostrils, the smell intensifying as she approached their cabin door. Trying to appear nonchalant, she casually walked passed a man with two children who was sniffing the air suspiciously while looking under each of the buildings in the immediate vicinity.

From Provo, the women planned to drive to Bluff in southeastern Utah, an area of natural bridges and cliff dwellers. Annie was looking forward to this leg of the trip but, once again, they experienced "a piece of ill luck." She described the accident to her insurance agent this way:

> [W]e were driving down the Provo Canyon, when, at a rounding curve, we encountered another car, traveling at high speed. We put on our brakes and turned to the right as far as possible, knocking down the guard rail in our effort to avoid the car. It hit us, damaging both fenders, stripping off and crushing the wooden box we had on the running board, ruining the front tire and shattering the glass wind wing.[15]

Her description to Edna was more colorful and contained the following additional details:

> We ran into the guard rail, breaking one of the posts, in an effort to get out of their way but they side swiped us, smashing both fenders, ruining the front tire, breaking the glass wing and tearing off the box from the running board, scattering all the contents over the highway. Seven gophers we had collected that morning, were drenched in kerosene and formaldehyde and the contents of the glass jars of pears, coffee, traps etc. were thrown broadcast. The car had gone some distance without stopping and Louise started storming after them, her chin dripping blood onto her shirt and the boulevard.[16]

The driver's arm was badly cut, and Louise sustained a cut on her chin from the broken windshield. An employee of the Utah Light and Power

Company was fishing nearby and witnessed the accident. He offered to take Louise to the power station for first aid and, should the need arise, testify that the driver was traveling at a high rate of speed. Annie was sure that the driver had been intoxicated. After his friends helped the women change their damaged tire, she and Louise returned to Provo. Their car went to a garage; Louise went to a doctor. Neither emerged looking better for the incident but, fortunately, neither incurred serious injury.

Their string of bad luck had not ended. When the women were finally able to leave Provo and continue on to Bluff, Alexander discovered that she had left her camera in their auto cabin. She wired the manager of the camp asking him to retrieve the item and send it to her. He replied several days later that the cabin in which the women had stayed had been occupied by someone else the night of their departure and that when he went to look for the camera it was no longer there.

Bluff, Utah, proved to be a desolate spot. Beauty could be found in its red cliffs clear cut against a blue sky and in the cottonwood trees following the windings of the river. But the pursuit of topotypes had brought the women here, and in her elliptical manner Alexander wrote to Grinnell, "We started out to find the site of Noland's Ranch—went to three remote trading posts for information and had to go out of Utah and into Colorado, around a mountain and back into Utah again before we finally reached the spot.... We felt like [we were] looking into the Promised Land!" To Edna, Alexander wrote simply, "[I]t seems like the end of the earth."[17]

By late November winter was well on its way with temperatures hovering around 14° F. The women had collected 177 mammals, 4 birds, and 1 amphibian. As for the species they had gathered, Grinnell's first thought was that 3 were new to the MVZ collections and 2 were probably new to science. After further examination of the material, he remarked that the women had collected 14 species that were new to the museum, and the specimens of *Clethrionomys* (red-backed voles) and of *Phenacomys* (heather voles) that they collected became the first records of those genera from Utah. Despite the low number of specimens that they had captured, the women also secured 8 specimens of the Colorado wood rat and 1 Arizona striped skunk, the first recorded findings of these species in Utah.

The spring of 1930 was unusually hot, and the asparagus on Grizzly Island ripened quickly. Annie and Louise worked night and day to keep up with the crop. Finally, in late May the harvest ended and the women found themselves at liberty to escape. This year they planned a much broader circuit than last and their primary objective was topotypes of the long-tailed pocket mouse, *Perognathus formosus,* for Seth Benson's dissertation. The promising young mammalogist was examining color variation with respect to habitat differences in this species, the kind of study for which large series of animals from a variety of localities were essential.

From the end of May through mid-October the women collected in Nevada, Utah, Idaho, Oregon, Washington, and British Columbia. The hot weather that they had experienced during the spring harvest became oppressive once they crossed to the eastern side of the Sierra. Even Annie, who loved the warmth of the desert sun, now found it difficult to work under its scorching gaze. She wrote to Edna from St. George, Utah, "We are sweltering in our little auto camp cabin, with the thermometer at 122 in the shade outside our door. It is one o'clock and two hours from now we must go out on the blazing hot hillside, pick up our traps and re-set them again."[18] The temperature in their auto cabin was not much more pleasant, hovering between 95° F and 100° F for most of the day and night, never dipping below about 68° F in the early morning when the women set out to check their traps. Their intake of liquids increased markedly and Louise discovered Becco, a nonalcoholic cereal beverage made in Ogden that served as a substitute for beer. When stopping for drinks, she and Alexander also purchased large blocks of ice to keep their specimens from spoiling until they could stop and put them up.

In addition to pocket mice, the women were also interested in obtaining specimens of chipmunks, prairie dogs, and a particular species of shrew, only a single specimen of which had ever been captured. Alexander felt that their chances of trapping the tiny mammal were slim and did not plan to devote much time to it. However, they found the chipmunks they needed in great abundance in the foothills of the Wasatch Mountains but, they wrote to Edna, "Annie was disgusted one morning when she found at least a dozen traps in a line sprung by the squirrels and they not only deliberately sprung them, but dropped their dung and kicked dirt on them."[19]

Approximately a week later, Annie was about to set a trap under a rock when she realized that what she took at first glance for a coil of rope was, in reality, a rattlesnake, its head emerging from the center of the coil, its eyes shining brightly. From this point on she vowed to be more vigilant, but such hazards never deterred either her or Louise from sleeping out under the stars whenever possible.

In August Alexander wrote to Grinnell asking that he send her more specimen preservative and she warned that she and Kellogg might also need additional specimen tags. Delighted at these signs of success, he immediately shipped the materials that she requested. Alerting her to their arrival, he wrote that the preservative he was shipping was "13 parts alum, one part saltpeter, one part borax, one part talc—by weight. By the way, I myself use an alum and arsenic mixture. The alum-saltpeter mixture is preferred by those who are susceptible to arsenic poisoning—which apparently I am not." Acknowledging receipt of the tags and the recipe, Alexander thanked Grinnell and, in her businesslike manner replied, "We have been using pure arsenic lately on the small skins but perhaps your formula of alum and arsenic mixture would be better."[20] Although arsenic is no longer used in the preparation of natural history specimens, Alexander never reported any ill effects as a result of using the chemical to prepare skins.

The women enjoyed collecting in areas of large sand dunes and Alexander frequently commented on the beauty of the ripple marks, of the white sand sparkling in the moonlight, the smooth feeling of the fine grains under her feet, and of her amazement at the innumerable small mammal tracks that appeared in the early dawn on the great expanses of white sand. Grinnell commented on more than one occasion that Alexander's letters served as a valuable adjunct to her field notes, usually relating meaningful features of the landscape through which the women had passed while detailing significant natural history observations about its animal life. In 1930 the mammals that the women collected in Utah contained three previously undescribed subspecies and increased the list of known taxa from that state by eleven. Many of their specimens also represented significant range extensions and several exhibited evidence of intergradation between species.

Alexander routinely mixed her unembellished anecdotes with historical details about an area of interest, elaborate habitat descriptions, and replies to information in letters received. As Grinnell frequently noted, a single sentence, or several sentences, might contain a wealth of natural history information. Tucked between such lines is a glimpse or two of Alexander's emotion, the overriding passion that obviously drew her to this way of life. From Fort Grant, Arizona, she wrote to Grinnell in 1931,

> Our next objective was the Graham Mts. to secure topotypes of Microtus a. leucophaeus [a vole] and Sciurus f. grahamensis [a squirrel]. A road goes up to what is called Turkey Flat and is being continued along the south slope of the range but there were no meadows within walking distance so we packed in six miles to the head of Marijilda canyon, 8850 feet, and camped in a beautiful grove of fir and spruce and by a stream whose mossy and flowering banks had been unmolested by either cattle or sheep. It showered afternoons and nights

and the dampness and song of the thrush put us in mind of Alaska. The forest ranger told us there were no grouse but we were sure we saw one by the stream.[21]

Grinnell was not an effusive individual himself, another area in which his personality matched Alexander's. Nonetheless, he never failed to acknowledge each and every letter that Alexander wrote, frequently paragraph by paragraph. Often he annotated the margins of her letters for future reference, or with notes that he wished to include in his reply. Letters acknowledging specimens that Alexander shipped from the field usually included a list of identifications if these differed from what the women had assigned, and frequently they detailed the importance of each taxon for the museum collections as a whole, or for a particular research project. In this instance he replied:

> Your letter of July 9, from Fort Grant, and the box of seven skulls mailed from Bonita, Arizona, have arrived. The three [mountain] lion skulls are of especially high value; for previously we had just one skull from Arizona—that from the Colorado River a little above Yuma and representing, I think, a different race from that represented by yours. Your letters, describing so vividly the country through which you pass, are full of just the right kind of information needed for permanent record in connection with the material you are gathering. Especially important are your findings with regard to the present-day identity of the type localities you visit. Place names are changing right along, and the older names, like your "Priesto Plateau," are forgotten as the old timers drop out. You are playing a "Sherlock Holmes" role to very good purpose![22]

Alexander and Kellogg continued their quest for mammal topotypes in the summer of 1931. From mid-June until early November they collected in the White Mountains of Arizona, the Chiricahuas, and the Huachucas. In New Mexico they set up camp in Animas Valley, at Hachita, at White Sands, at Escondido, and in Dona Ana County. They set out traps near El Paso, Texas, and they returned, yet again, to southern Utah in the vicinity of Kanab.

In Arizona, the women camped along the western fork of the Black River amidst a forest of yellow pine. At 8,000 or 9,000 feet the trees were interspersed with acre upon acre of wild iris in full bloom, a beautiful sight and unlike any that Annie had ever experienced in the Sierra Nevada. She and Louise climbed Mount Baldy, camping for two nights along Reservation Creek amidst a cluster of dense firs three miles from the mountain's summit. At approximately 11,500 feet, the peak offered them a magnificent view. An almost unbroken stretch of forest lay to the east, south, and west, a surprise to Alexander who had always envisioned Arizona as mostly desert.

From Mount Baldy the women headed south over the Coronado Trail. They climbed to the Forest Lookout Station on Rose Peak (8,700 ft.), after which they recounted unemotionally to Edna, "Two days were spent trapping near the peak for a chipmunk which we did not get but instead dorsalis [*Eutamias dorsalis*, the cliff chipmunk], which is a much rarer form, one of the largest of the chipmunks. Anne made a specimen of one she took away from a rattlesnake that was about to swallow it whole."[23]

Alexander's letters from the field recount her actions in the nonchalant manner of someone crossing the street to look in a dress shop window, and her narratives favor the history or beauty of a spot, but never the real or potential danger at hand—stalking a bear in Alaska or snatching a chipmunk from the jaws of a rattlesnake. Throughout their careers as field biologists, she and Kellogg slept out under the stars whenever possible, both in bear country and in areas where they had observed rattlesnakes in and about their camp, yet there is never a mention of fear, anxiety, or even an adrenaline rush.

In their amusing details, the letters describe activities in which the two women engaged, and the daily routine of their lives, as quite ordinary. Their letter to Edna following the snake and chipmunk encounter continued:

> For once in our lives we were routed out of our bed in the open by insect pest—innumerable ants that had crawled over us and bit us. We had to spend the balance of the night in the car with our feet sticking out of the windows. About midnight Anne heard a lapping of water and looked out to see what is called a hog-nosed skunk, a handsome creature with a broad single white band on its back drinking water out of a saucepan. We were supposed to get one of these—the type locality is about sixty miles north—but Anne did not want to wake Louise to get the [shotgun] shells that were under her.[24]

The women obviously had a high tolerance for the minor discomforts associated with fieldwork, for rarely is there any mention of the desire for clean sheets or a hot shower. If they did complain about conditions in the field, it was generally about the heat, which adversely affected Louise more than it did Annie. Such expressions were usually couched in a snippet of dry humor, as when they wrote to Edna in late June from Safford, Arizona, "This is an awful hot spot but we have to endure until we get the rat which no one in town has ever heard of or seen." Three weeks later they replied to news that it had been over 100° F at the ranch for several days in a row by noting, "We are getting rather used to the heat here as we wear next to nothing and Louise sheds that whenever she can."[25]

In the early part of 1938 Louise became sick. The doctor related her illness to the typhoid fever that she had suffered in Cairo more than a decade earlier and recommended the removal of her gall bladder. Alexander believed firmly in the restorative powers of the mountains and desert, and by summer the women were off again. On this occasion it was Annie who suffered from the heat to the extent that her clothes became oppressive. From their camp along Haypress Creek, she wrote to Martha, "I paraded about an hour yesterday in the warm sunshine of the open places with nothing on but shoes and stockings but don't be alarmed—there are no campers here. The arrival of the forest ranger was of course unexpected but then that was this morning when I was busy writing a letter."[26]

If the women were somewhat unconventional in the conduct of their lives, they were no less so in their choice of field attire. Although a photograph of Alexander taken in 1901 shows her wearing a full-length skirt while poised against a boulder aiming a shotgun, she soon abandoned such dress in favor of more practical, and presumably more comfortable, garb, namely shirts and knickers. In August 1930 she wrote to Abercrombie and Fitch in New York City from the field,

> Enclosed please find check for $25 for a pair of knickers, tan wool gabardine, with buttoned cuff below knee, size 28 which I think sell for $16 or $18, and one ladies viyella flannel shirt, tan size, 36. Not having a catalogue I do not know the exact prices so if there is a balance send me the bill. Will you kindly ship these, parcel post, to me at Pasco, Washington, General Delivery.[27]

The women's field attire did not ignore all sartorial conventions. A proper dress or a skirt and shirtwaist remained a standard part of their baggage, and one or the other was called into service whenever they found themselves in an urban area and wished to take in a concert or movie.

In July 1931 Alexander bought a .22 repeater rifle. Though the women had not carried a rifle as part of their regular field equipment for a number of years, its utility became obvious now, given some of the larger squirrels and small carnivores that they were seeking. Prairie dogs were their immediate objective but they managed to secure a large kangaroo rat first, "the hind legs of which furnished us meat for a meal and were as large as those of a cottontail." Alexander described the rodent as measuring over 14 inches in length, with a beautiful bushy black and white tail. She also mentioned the huge dirt mounds that it formed out on the open plain. This was not the first time that the women had dined on kangaroo rats (*Dipodomys*). In 1929 Alexander wrote in her diary, "Large dipo eaten—put up two others and a Microtus [vole]."[28] This and other diary entries point up the fact that the women did not prepare all the specimens that they trapped.

Alexander's letters to Grinnell in 1931 are filled with details about habitat perturbation and the disappearance of black-tailed prairie dogs from localities they had historically inhabited. Although most persons interviewed swore that the animals no longer existed, by persistent questioning the women found a colony of some fifteen individuals six miles west of Willcox, Arizona. Alexander hoped to collect several of the rare mammals, particularly when she learned that the colony was slated for extermination by a federal agent. Although she and Kellogg failed to capture any themselves, on their second visit to Willcox, Alexander hired a trapper who secured the specimens for them. He collected three, almost all that remained after the poisoning campaign had been initiated.

Grinnell considered the prairie dog specimens "prizes" and wrote, "I confess I cannot see why any kind of wild animal on *un*cultivated ground should be pursued to outright extinction. I think such an act is a *crime*. I am all wrought up over the biological injustice of it! And the case which you have been witness to is only one of what promises to be a long series of wilful extinctions."[29]

Alexander shared Grinnell's thinking on the issue of wildlife extermination, and the extended field trips that she and Kellogg were making coincided with increased advocacy on his part for greater understanding and appreciation of wildlife at the national level.[30] Destructive actions toward mammals that were regarded as "pests" and "vermin" paralleled the sweeping transformation of natural habitats into rangeland and agricultural farms at this time. The difficulty that the women frequently experienced in securing topotypes merely illustrated the rapidity with which the landscape in the West was undergoing change.

In addition to topotypes, the women went out of their way to secure specimens of special scientific interest to museum faculty and students. Although Alexander disdained flattery and remained averse to platitudes, sincere expressions of gratitude and genuine calls for assistance did not fall on deaf ears. When Seth Benson felt that he was on the verge of discovering a new species of *Perognathus* (pocket mouse) in Arizona—if only he had a few more specimens from designated localities to substantiate his theory—the women readily complied, considering his need more pressing than any other at that particular moment. Then, in late September Alexander wrote to Grinnell suggesting that Benson join them, at her expense, to study firsthand the two color phases of pocket mice that they had been collecting for the young mammalogist, a light one on the white sands of Alamogordo and a dark one on the black lava beds to the north. He did. Benson also wished to discover whether the white phase was found anywhere other than in white sands. Helping answer that question, the women subsequently trapped them on practically white quartz sands west of Tularosa.[31]

The onset of winter rains in the delta signaled to Alexander and Kellogg that it was time to return home. In summarizing the results of their 1931 field season, Grinnell wrote, "The 720 specimens you collected represented 83 kinds [of mammals]. There were 181 topotypes, representing 25 species or subspecies; and 71 of the specimens pertain to forms which are being newly named."[32]

Such statistics were not unusual for the women to achieve. But the quality and type of specimens, as much as the quantity of material that they collected, are worthy of note. Perhaps the most remarkable specimen that they gathered on this particular trip was the skull of an Arizona grizzly bear that had been sitting in the window of a gas station in Willcox. A Mexican sheepherder had shot the grizzly in the fall of 1928, and Alexander had a hard time getting the owner to part with the specimen. She went back to the garage three times before he relinquished the trophy. In 1931, however, only two grizzlies were known to exist in Arizona and Grinnell predicted that these, like their forebears, would be killed in the near future by cattle ranchers. Alexander recognized that the specimen's scientific value far outweighed its role as a curio.

If increasing age had begun to affect Alexander, she rarely acknowledged it. Less than a month after returning from Utah in 1929, she did comment to Martha, "I'd like to know how it would feel to feel young again. When I drink a good cup of coffee I feel exhilarated and enterprising but it isn't feeling young!" Alexander was now sixty-one. A year later, she again alluded to age in a letter to Martha, although she had not necessarily conceded to it in any way. She and Louise were fishing on Lake Pennask in British Columbia. It was late September and the mornings were cold. Often the temperature at their camp varied as much as forty-five degrees between dawn and dusk. She acknowledged, "This is the first break in our collecting work and it does seem good not to get up at 4:30 every morning, and to take life a little more leisurely."[33]

Despite such admissions, 1932 brought no reduction in the women's schedule of fieldwork. Immediately after the New Year, they left for a brief visit to the islands. They returned in time for the spring harvest, with plans to set off again for the field as soon as the asparagus had been cut and packed. During their absence, the museum received a series of more than eight hundred mammal specimens from an amateur naturalist on the East Coast, Morris Green. As always, Grinnell immediately informed Alexander of this valuable addition to the museum's collection. With the issue of age apparently still on her mind she replied:

> Your friend Mr. Morris Green must be a rare and exceptionally nice gentleman to offer in such a generous fashion to donate to our M.V.Z. a portion of his valuable collection of Eastern mammals. His letter quite thrilled me. He must have gathered the specimens himself judging from your remarks on the fine make-up of the skins and it is inspiring to know he is still keen on collecting—at the age of 62! Tell him for me, being 64, that I consider the sixties a very appropriate time in one's life to do field work—an out-of-door quest that always will have, I believe, a certain charm and excitement about it.[34]

From May 22 through November 20, 1932, the women collected across northcentral Arizona, southeast as far as Roswell, New Mexico, and finally north into Utah. From various points along their route, Alexander shipped back boxes of specimens to the museum. In mid-November she wrote to Grinnell that she and Kellogg did not feel like returning to Oakland until they had visited most of the type localities in Arizona north of the Colorado River. By the time that they turned the Franklin toward home, the women had collected 1,647 specimens.

Alexander and Kellogg's search for topotypes that year took them to the town of Oraibi, Arizona, where they spent a week trapping. The Oraibi pueblo occupied a commanding position on the edge of a high mesa and from their camp a quarter-mile away, the village appeared to crown the heights like a medieval town partially in ruins. The onset of heavy rains during their stay was cause for great celebration. The Hopis collected water for drinking, then bathed and washed their clothes. One man showed them the Hopi method of trapping animals with a stone and they watched in awe as a Hopi woman baked paper-thin tortillas on hot flat stones. The bread was made from finely ground corn mixed with water until it formed a thin paste. Alexander pulled out her scale and weighed one—twenty grams!

While they were camped in the area, the women attended the famous Indian Snake Dance performed on the Second Mesa at Shipaulovi (Chapalivy), one of the oldest Hopi pueblos, which was perched high on the rocky outcrops. They watched as visitors slowly toiled up the long stone stairway to the citadel while Indian women in brightly colored shawls looked down on them from the high walls. The ceremony concluded when four priests, each with an armful of snakes, ran away from the circle—in the direction of the four winds—to release the serpents on the rocks below. Annie thought it a truly spectacular sight to see the priests race down the steep slopes, their bodies painted in hieroglyphics, feathers waving on their heads and fox skins flapping behind them.

From May 5 through October 2, 1933, the women roamed through Yucca Grove, California; Boulder City, Nevada; Hoover Dam Ferry, Arizona; and numerous sites in Colorado, ending their five-month excursion in St. George, Utah. Increasingly, they collected vegetation samples as well as mammals, and these they sent to Herbert Mason at the University Herbarium for identification. On this trip, they pressed 591 sheets of plants, most of which they gathered in association with the mammals they collected.

In mid-June the women were camped on a flat at the base of Navajo Mountain in Utah when they learned that the mammalogist E. A. Goldman was in the vicinity. They quickly relocated their camp to War God Spring at 8,400 feet. Alexander wrote to Grinnell,

> It was somewhat of a surprise and shock to him to hear we were already entrenched here on the self-same mission but [he] warmed up with a cup of coffee and was quite chatty, telling us we must be sure to look for shrews along the edges of the spring. Miss Kellogg and I were especially interested in meeting the man whose type localities we have been pursuing for so long.[35]

The biologist from the Smithsonian was familiar with Alexander by reputation and apparently enjoyed his visit with the women as much as they enjoyed theirs with him. He had just collected what he believed to be an undescribed species of pocket gopher and that fall he named the new taxon in Alexander's honor (see appendix). In turn, the women wasted little time in acting upon Goldman's suggestion to trap shrews in the vicinity of the spring and Louise's subsequent capture of four individuals belonging to the "leucogenys" group earned the women a commendation from Grinnell who wrote, "That was a great find of Miss Kellogg's. . . . The employment of plant indicators to find mammals of narrow habitat requirements is good field tactics."[36]

From late December 1933 through March 1934 the women embarked on a three-month collecting trip and, between late January and early March 1935, they covered more than 2,700 miles through San Bernardino, Imperial, and Riverside counties in southern California. After they expressed their preference for working in desert areas, Grinnell recommended a triangle in the Mojave Desert that extended across its base from the Colorado River to the Salton Sea and from which specimens would be useful for his own research. He sent Alexander maps of the region marked with localities from which the museum had specimens, as well as those areas from which the mammalogist Vernon Bailey had collected pocket gophers for the Biological Survey but for which the MVZ lacked representatives. He cautioned the women to plot their route carefully, look closely throughout, and mark on their map instances where they might have expected to find gophers but did not. Grinnell's belief in their ability to discern such subtleties, and his willingness to accept their conclusions, speaks volumes for the respect that he accorded them as field biologists. However, his attitude seems to have had no influence on Alexander's self-perception of her own capabilities as a naturalist and field biologist.

Alexander wrote to Grinnell that the whole area in the vicinity of Twentynine Palms was now much more accessible in the wake of extensive mining efforts and the work of the metropolitan water district than it had been even a decade earlier. Along with their gophers, she and Kellogg also began to collect soil samples in areas from which they had difficulty describing its composition.

The women returned from the desert in time to help cut their asparagus crop, but in 1935 the harvest was late and Alexander did not wish to leave for the field again until the cannery had opened. Recognizing the importance of the specimens that Grinnell desired, she sent him a personal check for $500, directing him to send "two energetic young men" to continue collecting down south while the weather was still relatively cool.

With the harvest behind them, Alexander and Kellogg made a brief collecting trip to Arizona and New Mexico in May, followed by a two-month trip to Siskiyou, Modoc, Plumas, Amador, and Shasta counties in northern California to collect gophers between mid-August and mid-October. Between early February and March 1936 they explored the lowest elevations within Death Valley, an area that Grinnell had assured them was workable in winter. It was an area from which the museum had relatively few specimens, and the promise of discovering new or novel taxa lured the women south through Greenwater Valley and into the Kingston Range in what would prove to be a most memorable trip.

23

Saline Valley

The trip through San Bernardino and Inyo counties in February and March 1935 had been the women's first to the eastern side of the Sierra in California. John Muir had awakened a national consciousness to the beauty and natural diversity of the Sierra Nevada, yet this equally remarkable biotic region had received significantly less attention. The author and activist Mary Austin was one of the few who had written feelingly of the eastern Sierra and its native inhabitants, who struggled to eke out an existence from its parched soil. Alexander once mentioned a desire to visit Austin but nothing further came of this wish.

The stark beauty of the eastern Sierra landscape, coupled with a fondness for winter camping, enticed Annie and Louise to return to the region the following year, this time to explore the valleys and mountain ranges to the north of San Bernardino County. Beginning in late November, the women camped and trapped pocket gophers from the eastern Mojave north. Alexander wrote to Grinnell from their camp at Deep Spring Lake,

> This is a cold place— . . . my first experience for setting traps for microtines [voles] on ice! . . . Expect to move to Eureka Valley Sunday. May camp at Willow Springs. They say there is no road down the valley— big sand dunes there, only the road that crosses the northern end is shown on your topographic map. . . . We shall try for Saline Valley from Big Pine—any suggestions you might have to offer write me at Big Pine.[1]

Grinnell's reply must have lent an air of excitement to their impending adventure. He responded, "Your next prospective stop (beyond Eureka Valley), in Saline Valley is a perfect unknown in MVZ's history—not a specimen here of any kind from that 'dead' basin. *So*—whatever you find will be good. I don't know enough to 'advise.'"[2]

The women spent December 20 and 21 in the town of Big Pine checking their mail and purchasing supplies for the trip. With the Franklin fully loaded, they set out along Waucoba Wash the next morning. The car now towed a small trailer carrying extra gasoline and the majority of their camp gear. In addition to their tent, cookware, foodstuffs, water, traps, pinning boards, cotton, wire, scissors, forceps, and the little sheepherder's stove that perpetually accompanied them, the women carried a plum pudding to be enjoyed on Christmas day. As their vehicle crossed over the 7,200-foot summit at Westgard Pass, the sky was clear, the sun shone brightly, and snow lay only on the highest peaks outlined against a distant horizon.

From the summit of the pass, the road into Saline Valley descended into a narrow canyon which paralleled a long wash before rising and passing through a forest of piñon pine where the women stopped for lunch. Following sandwiches and a cup of tea, they went on down the grade toward what they expected was the valley floor. When they finally crossed the wash, however, they were surprised and perplexed to discover that the road did not continue south toward Willow Creek as indicated on their map. Instead, it started to climb. Having no alternative, they pressed on. Often traveling less than five miles an hour, the Franklin labored at times with the trailer in tow.

After three tortuous miles they arrived at the mouth of Lead Canyon, marked by an old cabin and a piece of flat ground large enough to allow them to turn their vehicle around. In the distance, a mine dump sat perched on the side of the mountain while opposite the mine, Saline Valley opened up before them. The dry white sand and alkali of the valley floor gleamed in the fading sunlight, likening the landscape to that of an enormous crater guarded by towering peaks—the 9,000-foot New York Buttes rising abruptly to the south and the bare and eroded flanks of the Last Chance Range looming majestically to the northeast.

Morning brought bright sunshine, and the lure of the mountains became irresistible. After a stiff climb up the canyon, the women reached a spring tucked beneath a thicket of willows. Louise had carried a dozen gopher traps up the slope and these they placed in the vicinity of fresh mounds. Annie noticed at least four species of grasses in the area, including one that resembled the tule commonly found in the marshes near their farm. After lunch, the women set out more traps for kangaroo rats and canyon mice.

By midafternoon the sun was no longer striking their tent and the temperature in the canyon had begun to drop. The night was cold. After checking their traps the following morning, they decided to break camp. Annie had captured three deer mice and one wood rat; Louise just one gopher. They hoped that their next site would prove more productive.

Because the main road through the canyon had been washed out and abandoned, the women chose to follow a secondary road to the old Borax works near Salt Lake, eighteen miles to the south. For three hours they bumped along in low gear over the rocky substrate, the road at times seeming to disappear amidst drifting sand that held little more than a few scattered and scraggly creosote bushes.

Their second camp was made near the old mill at the junction of the road leading northeast to Warm Springs. Although the ground here was alkali as well, they set out three dozen mouse traps and several rat traps under bushes, along the edge of a mesquite thicket and along an old stone wall near the mill where Annie had spotted antelope ground squirrel droppings. They captured a squirrel, two kangaroo rats, and several deer mice, but no pocket gophers or kangaroo mice. Annie watched as a lone roadrunner devoured two mouse carcasses that she and Louise had discarded. A single gulp for each was all it took.

The women awoke on December 27 to the sight of snow on the peaks of Last Chance Range. On the valley floor it had rained, making the dirt road they were traveling muddy and slick. Undeterred, the following day they drove eight miles to Lower Warm Springs on the eastern side of the valley. Here hot water bubbled up from three small pools and a variety of grasses nurtured by the springs offered hope for trapping new species of mammals.

But this pleasant warmth did not last. Snow fell on their camp in the canyon and then retreated. The morning of December 30 dawned clear, but by afternoon the sky was increasingly overcast. That night it rained in the valley, leaving the surrounding peaks draped in an ever-thickening blanket of snow. By sunrise, a heavy mist and a sense of concern hung over their camp.

Annie wished to escape the canyon for what she hoped would be better weather and increased trapping success in neighboring Owens Valley, but to do so she knew that she and Louise would need help. Their only exit to the south was already blocked by several tons of fallen rock and a 6,800-foot pass, an impenetrable barrier this late in the season. A crew of workers at the nearby Bunker Hill mine had broken the snow-covered road to the north as far as Waucoba Springs but had been forced to turn back when their vehicle broke an axle.

While collecting in the vicinity of the hot springs, the women had noticed two cabins inhabited by an old pair of miners. Much to Annie's disgust, the men were of little help. The chains on their truck tore to shreds almost immediately. She and Louise then unhitched the trailer from the Franklin and tossed out two sacks of split wood, lightening the car's load enough to allow it to travel four miles through the snow up the narrow canyon, a mile farther than the truck from Bunker Hill had gone but nowhere near far enough to get through the ever-deepening drifts that blocked their exit.

By now it was late afternoon and the group had no choice but to make camp in the snow in the middle of the road next to their abandoned vehicles. The women slept in their tent; the men slept in the trucks, complaining bitterly about the cold. Morning brought sunshine and the decision to drive back to the valley floor. This proved no easy task. First came the placement of a frying pan of hot coals under the fly wheels of both vehicles, then the shoveling away of snow and maneuvers to turn the cars around, another hour and a half. The tracks that they had made up the canyon road the previous day proved slippery. Time and again, one car or the other slid into the drifted snow and had to be extricated. And they suffered a punctured tire. A rung that extended over the rim of the wheel had broken and the stem was so badly bent that they were left without a spare.

Back on level ground, the women parted company with their companions and drove back along Saline Valley Road to the point at which it left the wash and turned up into Lead Canyon, roughly four miles from the Bunker Hill mine. Determined to find out what plans the mining company had for clearing the road as far as Big Pine, the women hiked to the mine to talk with the foreman. Annie was disappointed to learn that not only had the axle broken on the Bunker Hill truck, but its engine block had cracked as well. The manager of the mine assured them that as soon as they fixed it, the truck would again start up the road.

Though their campsite on the valley floor was exposed, the women chose to keep their camp near the big wash where driftwood was readily available, rather than relocate closer to the mine. Two days later they watched from a distance as the Bunker Hill truck with four men aboard made a second attempt to climb out of the valley. The following morning two miners, visibly exhausted, arrived in the Alexander and Kellogg camp. Their truck had gotten within half a mile of the summit when the axle broke for a second time. The other two miners had decided to walk out of the canyon to Big Pine. There they would buy a replacement for the broken axle, repair the truck, load it with firewood, and return to the mine, after which they would assist the women in getting out.

Toward evening it began to storm. For three days a fierce north wind blew through the valley. The women spent their waking hours splitting driftwood with a cold chisel and stoking their camp stove to keep warm. At night they slept in the tent, having pounded every available peg into its perimeter to keep it from blowing away. The traps that Annie had set before the storm had to be pried from the ground with screwdrivers; the bait in them remained untouched.

For two weeks the snow continued to present an impenetrable barrier to any thought of escape. It was almost a week before the icy wind slacked off enough to allow the women to resume setting traps. This activity distracted them from their primary concerns—the persistent need for firewood and a dwindling food supply. The absence of snow on the valley floor made water hard to obtain. They stopped bathing and washing dishes. Wanting to keep their dwindling supply of gasoline for what they hoped was an imminent departure, they began to walk a mile and a half each way up the canyon to fill buckets with snow that could be melted for water.

The miners at Bunker Hill invited the women to stay in a little shack near their camp. Alexander declined but asked to borrow a few provisions if the men could spare them—some bacon, a can of apricots, three cans of milk, and a half a loaf of bread became a veritable feast.

One afternoon an airplane flew in low over their camp and then up the canyon toward the mine. There it circled several times before heading south. The following day the women hiked to the mine and learned that the pilot had dropped a message: every effort was being made to have the road cleared, but could the miners hold out for another week? The men's reply had been affirmative.

Given Alexander's penchant for letter writing, this lull in an otherwise continuous stream of correspondence did not go unnoticed by her family and friends. Even members of the museum staff were beginning to worry. The 1936–37 winter was proving to be one of the coldest on record. At night the temperatures in Big Pine hovered around −10° F and two to three feet of snow lay on the ground around town. Finally Grinnell decided to break his silence. He wrote to Alexander in care of the post office in Big Pine. His letter traced a fine line—at once unwilling to suggest that the women were incapable of dealing with whatever circumstances might prevail in the field, at the same time determined to express the pervasive concern for their welfare.[3]

Four days later, with no reply or acknowledgment forthcoming, he wrote again, this time deftly balancing his understanding of Alexander's independent spirit against his very real concern for both women. He began,

I have decided to launch a "field trip" to the Inyo county—this very quietly, to get first-hand information that will reassure friends of you [sic] and Miss Kellogg, who are, I am convinced, truly suffering from *worry*. Then there is a chance that you, yourself, would find use for assistance in retrieving your equipment—in case you decided to come over the "divide" without it. . . . I know, and Mrs. McDonald [Edna Wemple] knows (she is a *brick*!) that you and Miss Kellogg are thoroly [sic] well qualified field people—to care for yourselves under any sort of conditions imposed by Nature. But worry (on the part of others) is, more or less, a human attribute. It *might* be that your Franklin is out of commission.[4]

As soon as he mailed the letter, Grinnell dispatched Ward Russell, the museum preparator, and his assistant, Bill Richardson, to Big Pine to secure the women's rescue. Grinnell made sure that Russell understood Alexander's disdain for publicity and cautioned the young man in no uncertain terms against attracting unnecessary attention. Given this prelude, Grinnell must have been greatly relieved a week later when he received a letter from Alexander that began,

We are no end grateful to you for coming to our rescue in sending Russell and Richardson to Big Pine to look us up. You can imagine our thrill coming back to camp near noon Wednesday—we had been after wood we had split in the washes—to find a man sitting by our tent, backpack and skis by his side who proved to be Norman Clyde, trapper commissioned by Russell to find us. I must confess this enforced stay is not contributing anything to vertebrate zoology. Keeping warm is a prime necessity and mornings are spent splitting wood in the wash with our invaluable cold chisel and axe. . . . In the afternoons we chop the various pieces we have brought and I add to the various journal letters I am keeping to my friends. The spring season ought to be wonderful this year in these mountains with so much moisture in the ground.[5]

The man whom Alexander and Kellogg found sitting outside their tent that cold winter day was far more than just a trapper. Norman Clyde was perhaps the most famous winter explorer in Sierra Nevada history. It had taken him two days to reach the mine at Bunker Hill on cross-country skis and he had been astonished to find that the women were not there. He had spent the night with the miners and the following morning made his way down the canyon to their camp. While Kellogg prepared lunch and Clyde waxed his skis, Alexander wrote a letter to Russell. With it she enclosed a traveler's check for $100—"to use at your discretion—hire a caterpillar tractor and bulldozer if possible. We would like to get our car out and [camp] outfit.... I would be willing to spend much more in order to get out."[6] She also asked Russell to wire "Mrs. McDonald" that they were well, and to bring with him ten gallons of gasoline.

Clyde's backpack weighed fifty pounds and included an extra supply of food in case of emergency. This he left with the women—eight eggs, bacon, butter, sugar, and cookies. After subsisting for almost two weeks on "collector's cornmeal," as Alexander referred to it, pink beans, and the few cans that the miners had given them, the women felt enriched. In gratitude, they used what little gas remained in the Franklin to drive Clyde six miles up the road to where the snow still blocked their exit from the valley. From there he planned to ski to Marble Canyon by nightfall, arriving in Big Pine the following day. By his own calculations, they might have to wait as much as another week before the road into the valley would be cleared.

Much to their pleasure and surprise, just four days later the women were awakened at 2 A.M. by the sound of heavy equipment approaching their camp. It was a caterpillar tractor, sent at Governor Merriam's behest. Against Grinnell's advice, the preparator had found a way to mention Alexander's name to someone close to the governor. Knowing that Alexander represented votes, rich votes, and that Merriam was a governor to whom such things were important, Russell reasoned that only a caterpillar tractor with an enormous blade attached to the front would be able to clear a path to the pass, and only the governor could commandeer such equipment on short notice. The county snowplow abandoned by the side of the road, its engine still running, was testament to the severity of the current conditions. Clearly, the old Reo truck that Russell and Richardson had driven to Big Pine was no match for the 15-foot drifts that clogged the valley road. It was barely able to get around town.

While the tractor detoured and cleared a path to the Bunker Hill mine, Russell and Richardson helped the women pack. Given Alexander's comfortable financial circumstances, Russell was more than a little surprised when the museum's benefactor insisted that he lie down on the cold ground and retrieve a quart jar containing a small quantity of motor oil from beneath the drip pan of the Franklin. This he then carefully returned to the crankcase. He was further surprised, and relieved, when the party finally reached Big Pine and Alexander took great pleasure in poring over the newspaper accounts that had been published about the rescue. The *Daily Californian* (the Cal students' newspaper), the *Los Angeles Times*, and the *Sacramento Bee* all reported the story with varying degrees of accuracy, bearing titles such as "Rescue Crews Fail to Find Lost Graduates," "Snowplows Used to Reach Eleven Trapped in Inyo County," and "Rescue Crews Buck Snow to Marooned Party." Perhaps it was the inaccuracies that Alexander found amusing. In one article she was even misnamed. Yet another surprise occurred after Russell and Richardson returned to Berkeley. Alexander bought each man a pair of cross-country skis as a token of her appreciation; both had spoken admiringly, and with a twinge of envy, about Clyde's prowess on them.[7]

The Saline Valley expedition almost instantly became part of museum lore, but the experience did nothing to dampen the women's zeal for winter camping and collecting. The following year, which marked Alexander's seventieth birthday, she and Kellogg returned once again to the Mojave Desert in search of pocket gophers for Grinnell. Before their months of exploration, the animals were thought to be scarce in that region. Following their capture of 95 specimens from 23 localities, however, Grinnell noted in a personal memorandum:

> It was geographic representation of the genus *Thomomys* that Miss Alexander was especially seeking, over an area whence practically nothing previously was known. Indeed, it was supposed that no pocket gophers at all were likely to exist over most of that area—that such as *were* there, persisted only as isolated colonies on mountain tops high enough to provide moisture from rain and snow to support adequate perennial plant-life for food, or in desert depressions around permanent springs or seepages, with like permanency of food. Miss Alexander's latest findings, supported by 95 specimens of *Thomomys*, from 23 localities, prove, rather, that this burrowing type of mammal is widely, though in ordinary (driest) periods of years, sparsely, distributed over the Mohave Desert.[8]

In the winter of 1938 the women returned to the southern California desert in search of pocket gophers one last time. Neither inclement weather nor age seemed to deter them from seeking out the most remote sites where these elusive and highly variable rodents might occur. Throughout November and December they trapped in Riverside and Imperial counties, collecting an additional 128 specimens.

The women's ability to divine rich or unusual collecting locales perpetually surprised brash young field biologists. During the 1940–41 winter, Charles Sibley, John Chattin, and John Davis set out on an MVZ expedition to collect small mammals in the Anza Borrego Desert. Much to their pleasure, they found the area almost deserted. A park ranger recommended a region known as "the Badlands" for good trapping but almost immediately retracted the recommendation, remarking that the road into the area was in poor condition and that the trio might seriously damage their vehicle en route. Then he paused for a moment and reflected. An elderly pair of ladies driving an old Franklin had driven out into the Badlands to trap mammals several weeks earlier, and when he had gone to check on them a few days later he had found that they were doing just fine. They had knocked a hole in their drip pan but had placed a bucket under it to catch the oil and all was well.[9]

24

The End of an Era

Grinnell's sudden death in the spring of 1939 from a heart attack was perhaps the biggest blow in Alexander's life since the death of her father. After suffering a first attack the previous fall, Grinnell had gradually been regaining his strength. Ironically, at the time of the attack he was on sabbatical, his first since assuming the directorship of the museum. When Alexander stopped by his home for a brief visit before leaving for the field "he seemed like his old self so full of interest and enthusiasm in the subject so dear to him."[1] The pocket gopher specimens that she and Kellogg had recently collected lay spread across the dining room table and he was eager to share with her his analysis of their variation.

The women were conducting fieldwork in Nevada when news of Grinnell's death reached them by wire. They left their car and camp gear in place and returned immediately to Oakland by train. Funeral services were private, but Alexander accompanied the museum's acting director, E. R. Hall, and his wife to the cemetery to pay her respects to the man who had a profound influence on her life. Acknowledging the loss, Annie wrote to Martha, "We have been so closely united in our ambition to make the Museum a center for the study of vertebrate zoology with a reputation for high quality output that his sudden death leaves me stranded in a way. So bright, so ardent a spirit, it is hard to conceive that he has really left us."[2]

The condolence letter that Alexander wrote to Grinnell's wife, Hilda, offers the key to the success of their relationship:

> I have had occasion again and again to congratulate myself at my good fortune in the choice of such an able man. Throughout the years he never seemed to lose sight of the fact that I was his backer in this splendid and inspiring enterprise, was ever solicitous to keep me in touch with what was transpiring at the Museum, to shield me from

worry and that I should share in all the triumphs that came our way. It is my conviction that the influence of his leadership which aimed at the highest quality of output will long survive.³

In a more general context, Alexander viewed the director of any museum as "the life of an Institution. In hundreds of ways he can further its causes if he be single-minded."⁴ Thus, her hope of replacing Grinnell with a man of similar vision and purpose now rested on the reputation that the MVZ had built for itself and the safeguards that she and Grinnell had tried to put into place several years before his death.

Since the museum's founding in 1908, Grinnell had reiterated the institution's objectives in each of the biennial reports that he prepared for the president and, subsequently, for the chancellor and the Board of Regents. In 1935 he had taken that process one step further by drafting a formal statement of the museum's functions and its role within the context of the university and the community at large. He intended that the document be signed jointly by himself, Alexander, and the university's president, anticipating that the museum would eventually come under pressure from a changing administrative structure that would be less responsive to Alexander's influence and the original goals set forth. He hoped that such a statement would ensure the perpetuation of the vision and the program that had governed the museum's founding and growth for almost three decades.⁵

Grinnell sent the two-page document to Alexander, soliciting both her opinion as to the advisability of making such a statement and her comments on the draft itself. From Kelly Spring in Modoc County, California, where she and Kellogg were trapping small mammals, Alexander replied, "I do not see that I can either add or subtract from the principles enunciated and as a precaution against uncertainties of the future I am signing one of the copies of the draft as outlined by you. It was this same reason that prompted me to increase the endowment of the Museum to insure its continued activity."⁶

Whatever the exact cause or causes of Grinnell's uneasiness, his action shows his foresight. He was fifty-eight, ten years Alexander's junior. In his letter accompanying the final draft of the "Analysis of Functions" he wrote to President Sproul, "I may say that, through the years, I have become deeply impressed with the wisdom of a principle Miss Alexander originally voiced and which she has followed so consistently in her own benefactions; namely, slowly and carefully to formulate a plan of major activity, and then to adhere to it on long-time schedule."⁷ In 1935 such an approach seemed to fly in the face of rapid technological and social changes in a world impatient with the old order of things, a world on the brink of a second global conflagration.

The document that Grinnell drew up spelled out the museum program's six foci. Briefly stated they were, first, to preserve the collections and their associated documentation; second, to increase the value of the collections through continued fieldwork; third, to publish research based on the collections and associated data; fourth, to disseminate the results of research through publications, teaching, participation in professional societies, and service to other organizations; fifth, to train naturalists; and sixth, to promote biologically sound principles of wildlife conservation and management.

Indicating his full support of this program, Sproul signed four copies of the "Analysis" without changes. One was for the university's files, one was for Alexander, one was placed in the museum's safe for posterity, and the fourth was kept on hand by Grinnell for "current reference."

The document proved useful on several occasions. In 1946 Sproul wrote to the chairs of all departments and research units on campus (at that time they were all chair*men*) requesting from each a five-year plan. Then MVZ director, Alden Miller, responded that he and his staff had found Grinnell's "Analysis of Functions" still so accurate and broadly conceived that it remained a substantially correct statement of the museum's current and future goals and objectives.[8]

During Grinnell's sabbatical leave, the administrative responsibilities of the directorship had been shared among several of the museum's staff. Upon his death, no small amount of jockeying ensued between Hall and Miller, each of whom viewed himself as the rightful and logical successor to Grinnell's legacy.

The difference in personality and temperament between Alden H. Miller, curator of birds, and E. Raymond Hall, curator of mammals, could not have been more striking. Miller was a gentleman and his manner was both proper and formal. Raised by a Victorian mother, he was a quiet individual, not demonstrative in any way, and his level of reserve contrasted sharply with that of his father, Loye.[9] He disliked conflict and would avoid open confrontations at almost any cost. He had been a faculty member in the zoology department and only months before Grinnell's death had become curator of birds in the MVZ, a split appointment that reflected changes within the university and a more intimate and complicated relationship between the museum and the department.

In contrast to Miller, Hall lacked polish. He had grown up in the Midwest and had a stubborn, rough-edged persona and doggedly persistent nature. He was relatively short in stature, a robust, pipe-smoking, extroverted individual who adopted a much more paternal attitude toward his graduate students than did Miller. As one of the first and brightest of Grinnell's academic progeny, Hall owed much of his eventual success to Alexander's generous financial support of his research. In 1926, while he was still a graduate student, Alexander had funded a two-year fellowship that permitted him to conduct research in lieu of holding down a teaching assistantship, and she continued to send him money for fieldwork even after his departure from Berkeley. Of equal importance, and little known, is that Hall's classic monograph, *Mammals of Nevada*, would not have been published without her financial backing. He expressed his personal gratitude for her generosity and demonstrated the lessons that he had learned from Grinnell when he wrote to her from Lawrence, Kansas, "Back at the Museum [of Natural History, the University of Kansas, Lawrence] I find a copy of 'Mammals of Nevada.' I hope you will be pleased with it. I like it except for one feature. That my name is stuck prominently on the front wrapper, where really, if truth were told, you had better been mentioned as having had far more to do with the undertaking than I."[10]

Alexander's financial support of this 1946 publication was all the more poignant in light of the events that immediately followed Grinnell's death. A tremendous cloud of uncertainty hung over the museum. Alexander did not wait to see what would happen. Before she and Kellogg returned to Nevada, even before she had written a condolence letter to Hilda Grinnell, Alexander wrote to Sproul, attaching a copy of the 1935 "Analysis of Functions" to her letter. In the letter she noted, "I am hopeful that the general principles and policy of the Museum will be continued . . . and, therefore, I am hopeful that someone fully informed of his [Grinnell's] aims and especially someone who knew first-hand of his methods in later years be selected to carry on in his place."[11] Hall was serving as the museum's acting director and Alexander proposed that he be permanently appointed to the post.

Sproul was generally noncommittal in response to Alexander's proposal.[12] He indicated his intention to abide by the general principles in the "Analysis of Functions" but added that it was not within his power to appoint a new museum director. Under the university's regulations he was required to consult with a committee of the Academic Senate. He did, however, extend Hall's appointment as acting director while the final appointment remained under consideration.

In late October Sproul wrote again, this time indicating his intention to consult with Alexander about the directorship at some level—but only after the faculty committee looking into the matter had made sufficient progress in their deliberations to "render such a consultation fruitful."[13] At this point it became clear that she would have little input into the final decision. As Sproul was leaving for a month on a trip to the East Coast, he advised Alexander that there would be a further delay, with no decision forthcoming before the end of November.

The New Year came and went and the situation remained unresolved. In mid-February, Sproul informed Alexander that the faculty committee had recommended Alden Miller for the directorship, although Hall had also been regarded as qualified.[14] He solicited any further comments that she might wish to make—while acknowledging that the final decision rested in his hands.

Alexander did not fight the verdict. She was now seventy-three. She had approved Miller's appointment as a museum curator the previous year and had even offered to supplement her monthly stipend to the museum in order to offset the half of his salary that the MVZ was required to assume under the joint appointment. However, she did not know Miller well and their relationship lacked the warmth and mutual understanding that had marked her association with Grinnell. While Miller respected Alexander and was deferential to her, their correspondence over the years proved to be more or less perfunctory, much lighter in volume and content than the one she had carried on with Grinnell. Ultimately, though, it was Hall's unprofessional behavior in conjunction with Miller's appointment that caused Alexander to support the ornithologist and made her financial contributions to the publication of *Mammals of Nevada* in 1946 a truly magnanimous gift.[15]

Hall was stunned and incredulous at Miller's appointment, so much so that at the end of 1940 he resigned from the museum to take a position with the Department of Agriculture in Nevada. He returned to the MVZ within a year, but his hostility toward Miller remained undiminished. He wrote to Alexander asking her to intercede with the administration on his behalf. He argued both to her and to President Sproul that the museum had been much more productive under his leadership than it now was under Miller's.[16]

Alexander found Hall's conduct in this matter increasingly offensive. She wrote sympathetically to Miller,

> Referring to our talk yesterday, you are aware that I espoused Dr. Hall's cause at the start but his continued hostile attitude toward you and the conduct of Museum affairs have thoroughly disgusted me. Understand, I have not criticized him directly, only by letter, as I have

not wished to get into any controversy. He knows that, yet he often cannot resist the chance to take advantage of my courtesy in visiting him at his office to attempt to involve me in an argument, mostly by implication. One would have thought the unrivaled opportunity for research at his command would have more than offset his fancied grievances, but such has not seemed to be the case.[17]

Miller was deeply appreciative of Alexander's political and emotional support during this difficult period and, in turn, gained her respect for his demeanor throughout the protracted negotiations that ensued.

Hall remained outraged at the injustice that he felt had been perpetrated against him on the part of the university. Unable to remain in the museum given his hostility toward Miller, he successfully applied for a Guggenheim Fellowship. This solved the problem temporarily by providing him with one year's salary and the money to travel and visit other museum collections. But the situation in Berkeley did not change during his absence and he returned to the museum to find Miller still firmly ensconced in the directorship. Hall then negotiated a year's leave of absence from the university and promptly accepted an offer to become professor and chair of the zoology department and director of the Museum of Natural History at the University of Kansas in Lawrence, his alma mater.[18] If he believed that such action would alter the situation in Berkeley, he was sorely mistaken.

Alexander's ability to recognize Hall's potential as a researcher, and to place that consideration above his distasteful behavior, was testament to the steadfastness with which she maintained her vision for the museum and her more general commitment to basic research in evolutionary biology. Shortly after he had assumed this new post she wrote kindly, "Even in this great change in your affairs, you still have knowledge and experience in your line of studies far above any mammalogist in this country. I believe it should continue to be put to service for the benefit of science. I want to keep in touch with you and to know what your prospects are along the lines you have long had in mind."[19]

The move to Kansas did not diminish Hall's desire to assume the directorship of the MVZ. As his year's leave of absence neared its end, he clearly indicated his willingness to return to Berkeley in that capacity. The decision was not Alexander's to make. For its part, the university had no intention of altering its position. A financial settlement was eventually reached between the parties, and Hall remained in Lawrence.[20]

Hall had learned much from Grinnell, not only about natural history but also about a cordial relationship with Alexander, and he applied those lessons from his new home. He and Alexander corresponded regularly. He faithfully recounted to her his program of fieldwork, the progress of his students and, most important, the nature of their resulting publications. He boldly communicated to her his desire to do research that would ultimately result in a publication on the mammals of North America.[21] This was exactly the kind of research that excited Alexander, contributions that would enlarge the general pool of scientific knowledge in the area of vertebrate natural history. In her mind there was no reason not to support his work financially, at least on a modest scale. He was certainly not the first bright, productive, but difficult biologist with whom she had had to deal.

While obviously desirous of her continued support, both financially and politically, Hall's respect for Alexander seemed to go beyond the perfunctory. In a Christmas letter that reflected a personal and sentimental side of him that was probably unknown to his colleagues, Hall wrote that he considered Alexander to be one of the "world's three finest women" that it had ever been his good fortune to know, the other two being his wife and his mother. The following year, when the University of Kansas wished to approach the paleontologist and KU graduate Ruben Stirton about his interest in leaving Berkeley for a faculty position in Lawrence, Hall wrote to Alexander, "Before the matter goes any further, I wanted to inform you so that if you care to do so you could express your own wishes. I would not want to make any offer from here that would upset any part of the program that you have in mind."[22] Alexander's reply is not available.

From the outset, Hall intended that the Museum of Natural History at KU would compete successfully with the MVZ and immediately set out to build up its collections and its research productivity. One example was his acquisition of the 65,000-volume natural history library belonging to Ralph Ellis, an obsessive bibliophile whom Hall had befriended while still in Berkeley. Ellis had also collected mammals and several thousand specimens of birds and eggs that came to reside in the MVZ during Hall's tenure there. To everyone else's surprise, Ellis's mammal collection moved with Hall when he went to Kansas. On a rare visit to Berkeley in 1946, Hall intimated to Alexander that the bird specimens belonging to Ellis also were to be transferred to KU at some future date. Whether or not this was wishful thinking on Hall's part is unclear. However, shortly thereafter Ellis died, bequeathing his bird collection to the MVZ. Alexander had seen no reason to question Hall's remark in 1946, but when Miller subsequently informed her about the terms of Ellis's will, she expressed her surprise at the former curator's impetuousness.[23]

Hall remained at Kansas for the duration of his career and built up that museum's vertebrate collections and its reputation for graduate education and research. Grinnell's death and Miller's appointment as his replacement coincided with a reconfiguration of Alexander's relationship to the MVZ, and she shifted her involvement from one of actively building its program through specimen and monetary donations to more passively protecting her investment. The university's continued growth as a multicampus system, Sproul's appointment as president, the joint appointment of MVZ staff as faculty members in the Department of Zoology, and a decrease in the proportion of the museum's operating budget that the Alexander endowment now represented combined to precipitate this change. Yet by the time of Grinnell's death, Alexander's hope of enticing the university to support research in vertebrate natural history had largely been realized. The MVZ continued to prosper under Miller's direction. Thus, for the remainder of her life, Alexander tended to observe the university's actions with respect to the museum from a distance and intervened only when called upon to do so. This left her free to direct her interests and seemingly undiminished energy elsewhere.

25

Hawaii—"My Only Real Home"

Grinnell's death was not the only factor affecting the museum's research program during 1939. Even before the United States' official involvement in World War II, the war impinged on activities at Berkeley and elsewhere across the country in a variety of ways. With the inauguration of conscription and draft registration, student enrollment and class sizes decreased. When the university demanded that a 10-percent increase in lower-end salaries come from within each department, Miller took the money from his supply and equipment budget, as the war prevented him from purchasing many of the usual items involved in the day-to-day curatorial operation. Most notably, the zinc cases to house the museum's collection of birds and small mammals were no longer available. Zinc was an essential component in shipbuilding and, upon U.S. entry into the war, it immediately disappeared from the domestic market. Gasoline rationing also curtailed the MVZ's field activities.

Before the war, Alexander and Kellogg had fallen more or less into an annual cycle of activity. Spring found them working at the ranch, summer and fall found them out in the field, and winter found Alexander visiting family and friends in Hawaii while Kellogg generally remained on the farm.

Immediately after the bombing of Pearl Harbor, travel between Hawaii and the mainland was halted for all but military purposes. Both women perforce spent the majority of the war years tending to farm business. Although the constant demands of the enterprise kept them physically active, the farm was no substitute for the field. Alexander now "counted each day lost . . . that she spent away from the mountains."[1] On rare occasions when she and Kellogg were able to accumulate enough gas coupons to make a short trip, they had trouble finding a guide who could pack them into the more remote mountain regions of Trinity County or the Sierra Nevada. Alexander was now in her seventies and the luxury of a guide seemed to be her single concession to age. During the war, even older gentlemen were unavailable as guides, having to care for families and businesses that their sons, grandsons, nephews, or cousins had left behind.

As soon as Japan surrendered, Alexander anxiously awaited word from the 12th Naval District in San Francisco regarding the availability of passage to Honolulu. She had been given priority for a sailing date, one she hoped would be as early as February 1945. After a hiatus of almost three years, she was eager to visit Martha and John. The absence of their youngest, Montague, who had been killed in the Solomon Islands in January 1944, would mar the joy of that reunion but his death only increased the urgency that Alexander felt to be reunited with her sister and family.

When Martha and John Waterhouse married in 1900, they settled in Honolulu. Within a few years, they began construction of a house near the top of Mount Tantalus. Nostalgically, they named it Olindita, after the family's much-beloved summer home at Olinda. On the cool, damp slopes of Haleakala, the Alexander children had spent countless hours playing on the mountain in its wet tangle of ferns and trees. As a young girl, Annie recalled building bonfires on the mountain. Standing in the smoke until her clothes reeked, she would imagine herself to be a genie out of the *Arabian Nights*.

Mount Tantalus was equally wild and undeveloped when Martha and John began building a home there. There was no road to the top of the mountain in those early days. The area was a virtually uninhabited landscape across which their children rode fearlessly on horseback. At 1,400 feet the wind on the mountaintop was fierce, the house literally anchored to the earth by enormous ropes to prevent its blowing away. But the temperature at that elevation was always a good ten degrees cooler than in the city below, and the house served as a retreat from the oppressive summer heat and humidity that characterized the lowland.[2]

Alexander loved the islands and returned there almost annually. On most of her visits she stayed at Olindita, often taking long hikes in the woods surrounding the property. The family would come down from the mountain to swim in the ocean, attend the children's sporting events, or have tea with family and friends. Periodically, Alexander visited the family homestead on Maui. In 1910 she wrote to Mary Beckwith from Haiku, "This is such a beautiful place! I have been so happy here these five short days—tomorrow will see me adrift from what will always be my only real home."

By 1934 much of the island landscape had changed but Alexander was no less enthralled. She wrote to Grinnell, "I have been spending one of the happiest weeks of my life here on Maui revisiting with two old friends from Chicago the scenes of our early childhood. . . . My old home at Haiku is changed of course—many of the trees are gone but it is still beautiful and the views of the ocean and mountains that used to delight us are still the same."[3]

Alexander loved the cool, sweet air atop the mountains in Hawaii and continued to climb them with remarkable stamina and regularity throughout her life. It did not matter that she might find the summit too clouded for a panoramic view of her island paradise, or that she might be soaked to the bone by intermittent showers. On such occasions she was content to watch the mist sweep down the ravines and be enveloped by the lushness of the surrounding vegetation and the palpable feel of the moisture-laden atmosphere. Following her father's death in 1904, she had climbed several of the peaks on western Maui with her uncle James, just as James and Samuel did as young boys. Even the volcanoes in Hawaii thrilled her. In October 1919 when Mauna Loa was erupting and a tremendous river of lava flowed down its side, Alexander and her sister's family had traveled to the Big Island to witness the dramatic spectacle for themselves, drawn by the reflection of the fire on the clouds which had been visible to them in Honolulu each night. Annie revisited the volcano in early 1920, this time with Louise. She wrote to Martha Beckwith, "When I breathed the delicious air . . . and looked at the wide sweep of the slopes of Moana Loa [sic] I said to myself—This is where I belong!—There is no spot in the world that thrills me more—not even Maui."[4]

Annie and Louise also hiked in the nearby bird forest, a veritable oasis amidst the destruction caused by the flowing lava. The forest still housed grand old trees found nowhere else on the islands and filled Alexander with delight as she listened to the birds sing from their boughs. "The impression on the mind takes hold of one's very soul," she exclaimed to Martha. "The world of sensation is by far the bigger world. I had the feeling as I was walking to the bird forest that I was breathing a beneficial atmosphere. We are a part of nature and when we get away from her—into town and among too many people—we are unhappy, at least I am."

Alexander was enthralled with the physical changes in the landscape being wrought by the flow of molten lava. At the end of 1920 she, her sister, and her nieces and nephews spent an additional two weeks at Mauna Loa. From the porch of the cottage where she and Kellogg had stayed the previous winter, she wrote in detail of the path of the lava's flow, of its up-welling to engulf large sections of previously hardened crust, and of her dismay that only a single geologist was studying the eruption, a phenomenon whose implications she felt sure must be important to an accurate understanding of the evolution of the universe. She herself spent hours at a time watching the boiling morass and marveling at the changes taking place minute by minute.[5]

Ever since 1901—when Alexander first collected marine fossils near Pearl Harbor—she had been learning about the history of the islands' flora and fauna and acquiring specimens of Hawaiian taxa for the collections at Cal. When residing in Honolulu, she frequently visited the Bernice Pauahi Bishop Museum, whose collections are dedicated to the documentation and preservation of Hawaiian natural history and culture. It was there that Alexander met William Bryan, the ornithologist who wished to undertake a comprehensive survey of islands in the South Pacific and make the Bishop a center for research on all scientific subjects relating to Polynesia. In 1907 Bryan gave Alexander an enlightening tour of the museum's bird collection, pointing out the remarkable radiation and diversity of honey creepers that existed throughout the island chain. Fascinated, Alexander described all that she learned from Bryan in a letter to Beckwith, recounting the evolutionary problems that the group presented to researchers and embroidering her pages with drawings of the most divergent forms.[6]

Alexander had known that terrestrial vertebrates with limited geographic ranges offered interesting questions to the student of evolution. Yet it was not until her visit with Bryan that she recognized that birds, with their tremendous power of flight and their ability to migrate long distances, could offer similar problems for study. Once awakened to the unique position of Hawaii's avian fauna, Alexander donated to the Bishop those specimens that she felt would be of special value to its researchers. Of these, her most outstanding contribution was the Munro collection of Hawaiian birds.

George C. Munro had come to Hawaii as a member of a collecting expedition for the Rothchild Museum at Tring. At the conclusion of that assignment he simply stayed on, in time becoming internationally known as a naturalist and leader in chronicling and preserving Hawaiian wildlife, both zoological and botanical. But his primary passion was birds. Best known for his popular volume, *Birds of Hawaii*, Munro also wrote articles alerting the public to the dangers of introducing exotic birds to the islands, a concern that Alexander shared and expressed to Grinnell on more than one occasion.[7]

Alexander and Kellogg had the "good fortune" to be Munro's guests on a brief visit to Lanai in 1926. The three hiked in mountains still covered with native vegetation and Alexander took great delight in seeing and hearing several of the native bird species. When Munro showed the women his personal collection of avian specimens, Alexander was astounded. She immediately realized the value that they would have as part of a museum collection. Sorely tempted to purchase the specimens for the MVZ, she refrained.

Upon her return to Honolulu, Alexander spoke with the director of the Bishop Museum about the Munro collection. Aware of how anxious the museum was to secure the specimens, she offered to purchase the collection anonymously on their behalf if Munro was willing to sell at a reasonable price.[8]

The ornithologist was caught completely off-guard by the Bishop Museum's sudden inquiry and set the price at $5,000. This figure included his archeological specimens, his land shells, and his duplicate sheets of botanical specimens, as well as the birds. Even so, the price was far in excess of the Bishop's modest budget. Alexander thought it reasonable, however, after estimating what Munro might be able to get for the specimens on the open market. She offered to provide half the needed amount if a donor could be found to cover the remaining sum. At the next meeting of the museum's trustees, George Carter stepped forward to offer the matching $2,500.[9]

Following publication of the *Birds of Hawaii* eighteen years later, Munro wrote to thank Alexander for her kind words concerning the volume and her order for copies to give to friends. For the first time he also acknowledged her philanthropy, adding, "It was extremely generous of you to buy my collection for the Bishop Museum when so sorely tempted to acquire it for the Museum of Vertebrate Zoology at Berkeley in which you take so much interest."[10]

The number of Hawaiian specimens that Alexander presented to the MVZ over the years was relatively modest, owing in large measure to Grinnell's avowed policy to restrict the museum's focus to the vertebrate fauna of the Pacific Coast.[11] Thus, Alexander had been pleasantly surprised in 1909 when Grinnell sent her a list of Hawaiian bird taxa and expressed interest in having representatives of as many as possible in the MVZ's collection.

Alexander was never granted a permit to collect protected species of native Hawaiian birds, despite accompanying letters of support from island biologists. Species that were not protected, however, included house (English) sparrows, house finches, rice birds, mynah birds, crows, and any imported species of perching birds. In 1910 Grinnell wrote excitedly to her about the prospect of obtaining representatives of this latter group: "The impression you received, that the English Sparrows in Honolulu are *paler* than those around here, is *very* significant. *Please* don't fail to gather in at least a few." To Grinnell's delight, Alexander responded three days later, "I have secured so far specimens of the following birds—English sparrow, linnet, Golden Plover, Wandering Tatler and Sanderling, Chinese dove & mynah." A weeklong trip to Molokai followed and, after camping by herself on the beach about a mile from Palaau, she added quail, skylarks, curlews, night herons, and spoonbills to her list.[12]

Throughout the years, Alexander supplemented the collections that she herself made by purchasing a number of valuable specimens for the museum. In 1911 she acquired a collection of 117 native Hawaiian birds, representing 22 species, all of which had been gathered between 1900 and 1903. In 1934 she purchased a lei for the museum that included a large number of clear yellow feathers of the extinct O-o bird. And in the 1930s when Alden Miller undertook his study of anatid anatomy and systematics—the anatomy and evolutionary relationships of ducks, geese, and their relatives—Alexander contacted her nephew Jack Waterhouse about securing even a single specimen of Ne-ne, the native Hawaiian goose, to send to Miller. Although the Ne-ne was a protected species, Waterhouse arranged to have several specimens exported to the mainland (the birds had died by accident or natural causes).[13]

The beauty and diversity of bird life in Hawaii did little to diminish Alexander's enthusiasm for capturing the much more mundane-looking small mammals as well, in particular the Hawaiian rat, described as approximately half the size of an ordinary rat and presumed extinct until its presence was reported in 1915 on Popoi'a Islet in Kailua Bay. Often referred to as Flat Island, this insular piece of real estate had been home to many bird species but most of its other wildlife was gone in 1919 when Alexander set out in search of the Hawaiian rat. The island was little more than a three-acre plot of coral overgrown with pickleweed.

Alexander had sailed to Honolulu in 1919 without camp gear, traps, tags, or field notebook paper, but the possibility of acquiring a series of specimens for the MVZ now convinced her to purchase the needed equipment. On a Thursday morning in early January her sister Martha drove her to Kailua Bay, where the women found a boatman willing to row them out to the island. "I must confess my heart sank as we approached the island and I realized its smallness—only three acres!," she wrote to Grinnell in a ten-page letter that detailed her adventures and the rat's behaviors.[14]

After landing on a small stretch of sandy beach, the only opening in an otherwise continuous ring of coral surrounding the island, the women carried Alexander's gear to the dense stand of "milo" where she set up her camp. "As I was looking through it and remarking to my sister that my chances looked dim of finding the rat, an individual of the species emerged from behind a rock, walked slowly along and disappeared about ten feet beyond so I knew it was up to me to secure a specimen."[15]

After lunch, her sister went back to the mainland while Alexander set to work. The rat that she had glimpsed that morning was just one of many that she observed over the next two days. She baited her traps with bacon and rolled oats and looked for animal signs. The majority of the traps she placed in the stand of "milo," under and along the edges of rocks. A few she placed near burrow entrances closer to the water's edge. While the sun shone she surveyed the island, rebaiting her traps several times and setting the treadle mechanism as lightly as possible. More than once she came upon a rat sitting calmly next to a trap, having managed to eat all the bait without tripping the treadle. By sunrise the next morning, however, five of the traps had produced six specimens.

On the Mokapu Peninsula, a stretch of land jutting north from Oahu's eastern shore and bounded by Kaneohe Bay on the west and Kailua Bay to the east, Alexander also looked for rats but caught none. Not easily deterred, two weeks after her return from Popoi'a Islet, she hired a boatman to row her out to Moku o Loe, a volcanic island approximately seven acres in extent located in Kaneohe Bay. Alexander's field notes are filled with elaborate descriptions of the island's vegetation, its bird life, and the innumerable rat runways that she discovered there. After surveying the terrain, she set out a dozen rat traps and several mouse traps. The result the next morning was eight rats and three house mice. Pleased with her catch, she wrote to Beckwith a month later, "I took my little Hawaiian Rats to the Museum and exhibited them with great pride to an admiring circle. Really, they did a lot for my morale, and I'm ready for another quest."[16]

As a young woman, Alexander could not imagine herself living in Hawaii. She viewed its culture as revolving around families. In such a tight community, without a husband and children, how would she fit in? Yet with the passage of years Alexander increasingly professed that Hawaii was where she truly belonged. In the early 1930s she wrote to Grinnell, "I have been staying at the beach with my sister and family since coming here. The booming of the surf is constantly in our ears. Through a fringe of coco palms we can catch the waves crashing upon the reef. The balmy air and tropical foliage make this a different world from California. In spite of all the years I have lived there, I still feel an alien and that this is my rightful home." Less than two months after Grinnell's death, Alexander mused thoughtfully on the same subject in a letter to his widow, Hilda, " I sometimes wonder why I don't settle here permanently—perhaps I shall when I can't climb mountains anymore."[17]

Despite such declarations, Alexander struggled to reconcile her passionate feelings about the beauty and natural diversity to be found in Hawaii with an inability to feel truly at home there. Her childhood friends and family—and particularly close ties to her sister Martha and her nieces and nephews—belonged to a world apart from her life with Kellogg. For almost twenty years when she traveled to the islands she went alone. Perhaps until she and Kellogg were able to hire a ranch manager whom they trusted, they could not both be away from the farm for extended periods. Yet their farm interests did not preclude collecting expeditions. On rare occasions when Kellogg did visit the islands with Alexander in those early years, the two of them apparently took a room in a hotel in downtown Honolulu rather than stay with Martha and her family, as was Alexander's practice when she visited Hawaii alone.

For a number of years, Alexander owned property in Kahala, a section of Honolulu just east of Diamond Head. Then in 1927, while visiting a friend in the remote and undeveloped area surrounding Kailua Bay, she noticed that the adjoining beachfront lot was for sale. In those early days, Kailua was to Honolulu as Cape Cod was to Boston, a summer ocean retreat removed from the noise, the heat, and the throngs of the big city. On this far eastern shore of Oahu, Alexander envisioned life in a beach house in the cool air of the countryside, away from the rapidly developing areas around Honolulu.

Within a week of seeing the empty lot, Alexander placed an offer on the property, put her place in Kahala up for sale, and hired an architect to design a single-story house and garage on the new site.[18] Several years later she added a second floor when she realized that it would keep the ground floor cooler.

Once the house was complete, Alexander's letters and diaries make regular mention of Kellogg traveling with her to the islands for at least part of the time she spent there each year. This fact and Alexander's failure to mention Kellogg in family correspondence suggest the family's coolness toward her partner. She and Louise occasionally met Martha and John at a mountain lake on the mainland where the four would relax and where Louise and John would fish. However, Alexander's brother, Wallace, was apparently less tolerant of the women's relationship and refused to travel with them.[19]

Among her nieces and nephews, Alexander seemed closest to Martha's firstborn, Pattie, and to Jack, one of Pattie's younger brothers, who himself maintained a long-term relationship with a same-sex partner.[20] While Pattie was a student at Wellesley College in the 1920s, Annie and Louise made a point of visiting her over the Christmas holidays. In addition to a day or two in Boston, Annie generally invited Pattie and one or two friends on vacation during the winter break. One year they went skiing in Maine. They spent another holiday at Lake Placid, and a third year Alexander took her niece and two of Pattie's cousins on vacation to Florida.[21]

Most of Annie and Louise's travels with Jack began a decade or more after Pattie's graduation from college. Jack did accompany the women on a relaxing trip to the Sequoias in 1919, but it was not until March 1934 that Alexander wrote to Grinnell about a trip to Samoa and the South Pacific that the three planned to make the following summer. Jack had visited those islands on many previous occasions and had a number of friends there. Grinnell responded in predictable fashion—by sending Alexander a book on Samoan birds. He realized that this was not to be a collecting expedition but, knowing Alexander's tendencies, he readily added that the museum had little Samoan material and *"Anything* else would be well worth saving."[22]

The trio made Leone, the second largest settlement on Tutuila, their headquarters. After relaxing for several days and touring the nearby villages, Alexander chartered a forty-foot launch to take them to the islands of Upolu and Savaii. Although Savaii is the largest of the islands in what is now Western Samoa, in those days anything other than small freighters carrying oranges and copra seldom visited it. A stay there required a special government permit obtained in advance. Most of the island was uninhabited, its appeal to Alexander being an almost impenetrable jungle that awaited those with the audacity to explore it.

Despite the heat and extreme humidity, for the next two weeks the women enjoyed long walks along steep hill trails that led deep into lush forests. At each new village they were warmly welcomed, feted with kava-drinking ceremonies, elaborate feasts, siva dances (similar to the Hawaiian hula but uniquely Samoan), and singing late into the night. Alexander loved the festivities. After their first evening on Savaii she wrote,

> We had some siva last night!—the most formal yet. A bugle was sounded throughout the village to announce that a siva was to take place. It was the signal for those who were to take part to prepare. A dozen young men finally came in, decorated with banana leaves and vines and sat down. It was nine o'clock by then and it looked as if the whole village had gathered, sitting on the ground outside, to witness

the ceremony, mostly young children and they sat until the last gun was fired—after eleven o'clock! The leader of the chorus had a white and green grass skirt on over his red lavalava and a necklace of small red and white flowers. He was the most agile, snappy dancer we have seen yet, leaping into the air like a rooster, with his legs wide apart, suddenly whirling around and flinging his arms about, all in perfect time, impossible to describe his antics. He was smeared all over with cocoanut oil [sic], so were some worthy dames, led by the high chief's sister, their heads crowned high with flowers. The chief's sister was as dignified looking as any member of a Ladies Aid Society but I should like to see anyone of them going through the contortions she did when the spirit of the dance was well upon her, prancing around on one foot while the other was pointed in the air. Too bad we self-contained Americans can't express our heathenish feelings sometimes in this natural way.[23]

After each such evening, Annie would present the men of the village with a barrel of salt beef and the women with a large tin of corned beef, while Jack handed out chewing gum to the children. The honored guests would then retire to a *fale*, a native guest house, falling gently asleep atop small stacks of mats that had been piled on the floor and draped with mosquito netting, discreetly screened from view by tapa curtains.

On Upolu, the trio went by car from Apia to Vailima, the property Robert Louis Stevenson bought, and made the steep fifty-minute climb up Mount Vaea to Stevenson's tomb. At every opportunity, they would swim in one of the many clear pools that dotted the island. If not alone, the women took to wearing lavalavas while they swam, long bolts of brightly decorated cloth wrapped around the body like a bath towel, with Louise expressing "some difficulty in making them stay on."[24] They also visited the tiny islands of Manono and Apolima that occupy the 18-kilometer strait between Upolu and Savaii. Apolima was most unusual, the remnant of a volcanic crater whose walls rise abruptly from the sea. A village of two hundred lay nestled in its hollow center. On approach, the narrow entrance to the island seemed impenetrable in the crashing surf, but a long boat manned by four natives skillfully maneuvered through an almost invisible channel in the coral reef. The experience seemed even more adventuresome upon their return at low tide when Alexander was able to see the reef jutting menacingly from the water. Manono was a flatter and much larger island, its houses almost buried amongst luxuriant breadfruit trees and coco palms.

Their return to Tutuila at the end of two weeks proved long and harrowing. Driving wind and rain, coupled with lightning and rough seas, left five of them crowded on the ship's deck for twenty-four hours, huddled beneath a single bedsheet on three thin mattresses. Unable to reach Leone, the captain finally changed course and put into Fagasa Bay on the island's northern shore. Exhausted and hungry, but grateful to be on dry land, the group shared coffee and hardtack before spending the night at the captain's home. Rather than reboard ship the following morning, Annie, Louise, and Jack walked across the mountains to Pago Pago where they caught a bus back to Leone.

The women spent the final week of their Samoan visit in a forest near Leone accumulating a representative collection of specimens for the museum. These included doves, fruit pigeons, honey suckers, kingfishers, and fruit bats. Ants were a persistent problem and Louise worked diligently to position the study skins out of their reach, as earlier in the trip ants had eaten large holes in the necks of several pigeons that she had carefully prepared.

Two years after their trip to Samoa, Annie, Louise, and Jack again traveled together, this time to Tahiti, Moorea, Raiatea, Huahine, and Bora-Bora.[25] By Alexander's reckoning, their most interesting stop was Rapa Iti, almost eight hundred miles to the south, a cold, rainy island rich in a variety of land shells and plants.

The impetus for this trip is unclear, but a year earlier Alexander had written to Grinnell that her cousin, "Dr. Cooke," had just returned from Tahiti after a six-month cruise in a sampan with a party of scientists collecting land shells, plants, and insects. Alexander seemed intrigued that the group had remained on good terms despite sleeping, eating, and looking after their specimens in a room only 11 × 14 feet. Or, because Jack loved the islands and the peoples of the South Pacific, he may simply have convinced his aunt to make the voyage.

Upon word of her impending adventure, Grinnell immediately sent Alexander directions for salting down skins of seabirds, along with his personal copy of W. B. Alexander's *Birds of the Oceans*, a volume he thought she might wish to take with her. In Tahiti, she and Louise settled themselves into a small but comfortable bungalow on the Blue Lagoon. From the verandah, Alexander wrote letters and watched as the surf broke upon a distant reef. With the exception of herons, native birds were scarce. She wrote to Grinnell about those that she and Kellogg saw and included detailed descriptions of those that they could not positively identify. The women also visited the small collection of birds that Rollo Beck had deposited in the museum at Papeete, but they shuddered upon seeing that the case in which they were housed was swarming with wasps.[26] Many of the specimen tags were faded to the point of being barely legible. Alexander copied down what data she could and sent the information on to Grinnell.

Two Hawaiian excursions that the women made with Jack in the latter years of Alexander's life also deserve mention for the affectionate manner in which she noted them in her diary and letters. The first was in 1947 when Jack met the women at Kokee on the island of Kauai. A friend had loaned them a cottage on the edge of a lush, semitropical forest at 3,600 feet. Three miles from the house the road ended at Kalalau Lookout. In one direction the embankment dropped precipitously into a valley refreshed by never-ending cascades of water. The opposite direction offered stunning views of the Na Pali cliffs backed by an immense ocean. Other trails from the house led them on hikes through deep, fragrant woods with an undergrowth of ferns, various native trees, and the songs of tropical birds. It was the type of habitat from which Alexander wished never to emerge. Following a week at the cottage, Jack drove the women to the valley of Hanalei. Annie visited her grandfather's homestead at Waioli, her father's birthplace, and was pleased to see that it was well maintained, still occupied by descendants of the missionaries who succeeded him.

The following year Jack joined the women for a tour of the Big Island. He borrowed a jeep and drove them around Mauna Kea, a distance of 85 miles. At 7,000 feet they found themselves in an extensive forest situated above the clouds covering the rain forest. Here, Annie savored the taste of akala berries, plump red berries about five times larger than raspberries, sweet reminders of her childhood days on Maui.

Following World War II, Alexander began to wonder whether many of the native plant species that she had known as a child still existed on the islands. Her uncle David Baldwin had been intensely interested in ferns and, as a young man, had made a creditable collection of the native Hawaiian varieties. Drawn to the ferns' special beauty, Annie now purchased a Ford truck in Honolulu and she and Louise set off to explore the remote reaches of the island in search of this vanishing flora. Unfortunately, on their third day out, the ground along a cliff trail gave way and Louise slipped, breaking her leg in the fall.

Kellogg's accident forced Alexander to face her own physical limitations. Now seventy-nine, she gradually conceded her inability to search on the highest peaks and in the most remote locales for many of the plant species that she desired. The terrain was just too steep. On several occasions, she had hired local hunters or ranchers in California to pack her and Kellogg into the more remote mountain and canyon areas they wished to explore as "the fact is we can't carry more than 6 to 8 pounds on our backs on our climb, especially on the sliding sands of the Sierra canyons," she confessed to Herbert Mason, director of the University Herbarium.[27] Now she also began to hire others to search out and collect selected specimens for her.

During the war, a young naval air officer by the name of Warren Herb Wagner had become familiar with much of the Hawaiian flora while on layovers there. After being discharged from the service, he had accepted a fellowship to study botany at the University of California in Berkeley. There, he made determinations on many of the specimens that Alexander and Kellogg had brought back from their trip to Kauai in 1947. When Alexander approached Mason about the availability of an individual to study and analyze the status of native Hawaiian ferns, Wagner seemed the obvious choice. A well-spoken and animated individual, he was single, adventurous, and already somewhat familiar with the islands' flora. After a meeting at the herbarium, Alexander agreed to pay all Wagner's expenses and arranged to have a jeep placed at his disposal. Despite his subsequent assertion that discovery of a particularly elusive fern species was purely fortuitous, his benefactor was delighted and urged the young man to press on.[28] While Wagner continued his quest in Hawaii, Alexander and Kellogg simultaneously focused their own efforts on gathering representative floral samples from regions in California that were unexplored or unfamiliar to professional botanists.

26

The Switch to Botany

By 1939 Alexander and Kellogg had collected birds and small mammals for more than thirty years when—with characteristic zeal—they redirected their energies toward collecting plants, an activity they had already begun to pursue to a limited extent. Alexander did not abandon her museums; she merely stopped collecting large series of specimens for them. Financially the museums were now secure, their reputations established, and she shifted from actively building their programs and collections to more passively managing their growth and continuing to structure their relationship with the university.

Whereas Alexander's unusual collaboration with Grinnell had clearly motivated her and Kellogg to collect specimens that would prove valuable to his research, their shift from trapping birds and small mammals to collecting plants was not solely related to his death in the summer of 1939. After thirty years, it had become increasingly rare for the women to be surprised by the types of vertebrates they captured, a factor that gradually diminished their enthusiasm for the task. Plant collecting provided an opportunity to wander through familiar landscapes with a fresh eye. Perhaps most appealing, it held the promise of discovering new and different taxa and increasing their knowledge of the natural world.

And animal trapping was physically strenuous work, even more so now that Alexander was over seventy. The women had to carry dozens of traps, a sufficient quantity of bait, plus a shotgun and shells while they hiked for hours along deep canyons or scaled intriguing slopes. With plant collecting, they could leave the presses and blotter paper that comprised the necessary equipment in camp while they hiked. All they had to carry in the field was a notebook for recording specimen data and a satchel into which they could place the plants.

Like her father, Alexander loved trees and remained captivated by the beauty of the natural vegetation that surrounded her. The vibrant colors and fragrant aromas of tropical plants—the rich purples of bougainvillea in full bloom, the showy red-leafed flowers of poinsettia, the unmistakable scent of ginger—had filled the world of her childhood. The fruits and flowers that she and Kellogg planted on their farm in Suisun paralleled the trees and the garden that Samuel had planted so lovingly at his own home in Haiku.

Alexander had been collecting plants to some extent for as long as she had been collecting any natural history specimens. Following the trip to Crater Lake in 1899, she arranged to have the specimens that she and Beckwith had collected identified by a trained botanist. Among the sheets of dried stems and flowers were a number that aroused interest and inspired her to continue making collections. It was perhaps this formative experience that caused her and Kellogg to prepare at least a few sheets of plants for the University Herbarium on many of their early field trips. While trapping small mammals or scouring the landscape for fossils, they often collected vegetation samples from the habitats as well, a logical adjunct to the photographs they took and the field notes they so meticulously compiled.[1]

As Alexander had been attuned to the speed with which the native fauna in western North America and in Hawaii was disappearing, she was equally aware of the regions' loss of native flora. By the time that William Patterson Alexander and his fellow missionaries arrived in Honolulu, the Sandwich Islands had already become established as a major port of call in the Pacific. In the burgeoning trade, native forests rapidly disappeared, especially from the higher plains and grass-bearing slopes of the mountains. Missionaries and exporters harvested sandalwood at an unsustainable rate. Less desirable tree species they burned for fuel, and free-ranging cattle browsed on young shoots. The native forests also fell prey to introduced grubs and pests brought in on foreign ships. Soil aridity began to increase as loss of this vegetative cover enhanced evaporative water loss. In some instances, the volume of water flowing through streams was noticeably less than it had been just several decades earlier. Eucalyptus trees introduced from Australia as a replacement for the depleted forests merely produced a sterile monoculture; they did not replace the capability of the native forest as an effective watershed.[2]

Another problem the islanders faced related to nonnative animals, most of which had been gifts to Hawaii's kings and queens throughout the period of the monarchy. Such animals were referred to as *kapoo*, taboo. Crown property literally could not be touched. Allowed to roam freely for generations, these nonnative species had multiplied unchecked. By 1893 when the monarchy ended, trees and shrubs that had not already been harvested for export or fuel had largely been destroyed by grazing; aggressive weed species began to appear in these perturbed areas. By the late 1800s, forests and valleys that centuries before had comfortably supported one to two hundred native Hawaiians now yielded food for less than half that many.

Alexander frequently wrote to Grinnell of her impressions of the changes taking place on the islands, vividly describing the parallel loss of native vegetation and introduction of exotics:

> The C.C.C. men have cleared an old trail along the precipitous cliffs of Manoa Valley, just a short distance from here. I was on it yesterday and felt sad to see how many foreign grasses are invading the native undergrowth. We encountered flocks of small birds—the little white eye from China which have increased greatly the last years. There seems to be a tendency to try out foreign birds promiscuously here—as the Chinese thrush, the Japanese robin, the Kentucky and Brazilian cardinals—all thriving here. It is a pity the native birds are confined to the almost inaccessible mountain ridges. No one ever sees them so the inclination is to introduce birds from the outside.[3]

Similarly, in California, Alexander recognized that many plant communities were threatened, either through economic exploitation or as a result of agricultural development, logging, and spreading urbanization. The overwhelming need for water, both for agriculture and for the state's growing population, was reshaping the environment. One of the last letters that Alexander wrote to Grinnell before his death spoke of the changes being wrought along the Colorado River. Construction of Parker Dam had created an expansive lake in the midst of a desert, with a system of tunnels and canals more than three hundred miles long to carry water to a rapidly growing and thirsty Los Angeles County. Almost overnight, every adventurer with a car and the desire to explore or exploit the desert had perturbed relatively pristine habitats. By the mid-1940s, grazing over much of the state had also seriously impacted the native California flora. On a trip to the Sweetwater Mountains in northern Mono County in 1945, Alexander was "horrified" to find sheep everywhere; not a vestige remained of the plants that she and Kellogg had collected there many years earlier. By the late 1940s, she was "appalled" at the way that southern California had been built up "and the tremendous volume of traffic everywhere."[4]

The transition from collecting animals to collecting plants required a marked shift in the women's annual cycle of activity. While winter is the period when vertebrate specimens are at their finest—birds in bright plumage, mammal skins sleek and luxuriant—plants blossom and display their finery in late spring and early summer. In part, this shift was facilitated by the hiring of Don Wilson as ranch manager. In addition, asparagus proved to be an ideal crop to complement their new activity as the stalks were ready for harvest by early spring. And, unlike Hawaii, California's highly varied topography supported a tremendous diversity of vegetative types—from the moist coastal belt in the north to the Mojave Desert in the south, from sea level in the west to the highest peak in the contiguous United States in the east— and opportunities for botanical fieldwork seemed limitless.

In the summer of 1939 Carlotta Hall, a former botany student affectionately referred to as "the fern lady," and the widow of Harvey Monroe Hall, professor of botany and director of the University Herbarium, joined Alexander and Kellogg in the field. Hall's friendship with the women blossomed after the death of her husband in 1932. To save money for her daughter's college education, she shared a home in Berkeley with Annie's cousin Mary Charlotte. Hall and Hilda Grinnell were also acquainted, presumably because both were the wives of university professors and museum directors. Carlotta and Mary Charlotte belonged to the Fifteen Club, but it is unclear whether Hilda belonged to this book club or another. When Martha Beckwith moved to Berkeley in 1938 for a brief period of time, she joined the club to which Hilda belonged. The women all played bridge together and letters written to Alexander and Kellogg in the field faithfully reported on books read and various comings and goings. Both Hall and her husband, Harvey, had been interested in conservation and in California's changing natural history in the wake of westward expansion. During the first decade of the twentieth century, the Halls spent a great deal of time collecting botanical specimens in Yosemite National Park, work that culminated in publication by both husband and wife of *A Yosemite Flora*.[5]

Even as late as 1939 it was still unusual for a woman to travel alone or conduct fieldwork on her own. Like any woman who had experienced the joy and freedom of being in the field and of pursuing a scientific interest, Carlotta Hall was delighted to take up these activities again with Alexander and Kellogg. Whereas Hilda Grinnell would forever remain "Mrs. Grinnell" in their letters to her and they, in turn, "Miss Alexander and Miss Kellogg" in hers to them, Hall joined the ranks of Edna Wemple and Martha Beckwith in referring to Annie and Louise by their first names.

The women began their 1939 expedition at Parker Dam by following the newly created water district roads through the Whipple Mountains. In the town of Cross Roads they rented a little cabin, replete with a "homemade air cooler"—a fan that sucked air through a water-soaked burlap curtain into their room. Although it was only mid-May, the temperature in southern Nevada was already climbing. Hall met Alexander and Kellogg in Needles, California, and the trio started north across the desert, Needles to Goffs, then to Lakeview. As they climbed in elevation, the Joshua trees became more prevalent. As they reached the summit of the grade, they were greeted by barrel cacti. Farther along, cholla cactus and creosote bush predominated. They crossed the New York Mountains to Clark Mountain Station and spent a week collecting wildflowers and photographing the countryside. Alexander experimented with color film. "The blooms were unusually abundant due to recent rains—the most striking was Penstemon Palmeri, tall stalks with pink flowers and a fragrance like orange blossoms. Besides these there were three shrubs of the rose family, beds of blue salvia, a brick red mallow, paint brush and larkspur. In all we collected about 80 varieties," she wrote in one of her many letters.[6]

From Clark Mountain, the women crossed into Nevada to explore the mountains around Charleston Peak, stopping first to visit Boulder Dam. The area proved delightfully cool and they soon found themselves comfortably ensconced in a housekeeping cabin amidst white fir and yellow pine, the surrounding cliffs still wearing a good deal of snow. In a nod to the past, Louise set a few gopher traps while Annie and Carlotta collected plants. Over the course of the following week, she managed to capture a golden-mantled ground squirrel, a few gophers, and a chipmunk while helping the others collect botanical specimens from a variety of elevations, atop mesas and along deep canyons cut through the mountains.

Moving north to Indian Springs, the women headed off the main road, following up the side of a rocky, gravel-covered mesa formed by the disintegration of the surrounding mountains. As they climbed toward 3,000 feet, they passed through a landscape of yucca plants and cacti, stopping to lunch under a juniper tree. Hall recorded in her field notes, "The view of the desert was wonderful. I was scanning the mountains in front of me and gradually turned around toward the desert. When it came into view I was startled. It was like another world. From that distance it was shades of blue and seemed without vegetation. It seemed so soft and later as we came home in the late afternoon—six o'clock—the country looked like velvet, so soft and without sharp lines."[7] Two days later, camped among pines and firs in a cabin at 7,500 feet in the mountains of southern Nevada, Alexander received the telegram notifying her that Grinnell had died. The women left their gear in the cabin, their car in Las Vegas, and returned immediately to Oakland by train.

The following spring, Alexander and Kellogg spent a week at Twentynine Palms in San Bernardino County, returning to southern California several weeks later to spend almost three months revisiting much of the area that they had traversed a year earlier with Hall. Alexander loved the southern California deserts and the eastern Sierra and, for the remainder of the decade, focused her attention on little known or poorly collected sites in these two regions. By repeatedly visiting locales at different times during the year, or even different times of the day, she and Kellogg were able to collect the same plants under different conditions and note the peculiarity of their habits, for example, on the same steep, rocky slope one species of plant had its petals open, displaying white blossoms, while another species with yellow flowers had its buds tightly closed. On their return down the trail that afternoon, Alexander observed that the condition of these two plants was now just the reverse, white closed and yellow open.

As they had done when planning field expeditions in search of small mammals, Alexander requested lists of type localities from the herbarium staff. Throughout the spring and early summer of 1944, she and Kellogg collected in the eastern Sierra Nevada and the Sweetwater Mountains. Camping above 6,000 feet, they found the weather unseasonably cold, but the days were usually sunny and the air crisp. The presence of a warm, spring-fed pool delighted them. Surrounded by white pine and fir in Sweetwater Canyon, they chose to sleep in an aspen grove but used a nearby cabin, complete with stove, for cooking and drying specimens. The steep old mining road that they had used to make their ascent ended at 8,000 feet. From that point they continued on by foot to the summit of Mount Patterson (11,654 feet, according to Alexander's calculation). On the western side of the mountains, they packed in to over 10,000 feet, camping at Deep Creek for two weeks and exploring its basin to the summit of the divide. Alexander was seventy-six. While hiking, they heard pikas calling and, one day, Kellogg came across a snowshoe hare. An old miner told them about marmots in the mountains but, regrettably, the women never saw any.

Alexander concluded that the great diversity of plant life that she and Kellogg had observed in the Sweetwaters, including a surprising mix of both desert and Sierran forms, was sufficient to warrant a more detailed analysis of the collections that they had made, and she suggested as much to the herbarium director, Herbert Mason. To E. R. Hall she suggested that a faunal study of the region might be equally interesting. This is because biogeographically the Sweetwaters are considered part of the Great Basin, yet many of the plants that they collected in the region had previously been reported as occurring only in the Sierra Nevada. After receiving and examining a portion of their specimens, John Thomas Howell, then curator of botany at the California Academy of Sciences, concurred:

> Your 1944 Carex [sedge] collection is quite remarkable—not only does it include a large number of unusually interesting and critical things but it contains a large number of species and in no instance have species been mixed in the same number. You couldn't help the fact that you got interesting things, especially in the Sweetwater Mts., but the fact that you distinguished so many kinds is somewhat of a record for a non-caricologist! I have put an X before those of special interest from a botanical point of view and many of the others are equally interesting from a phytogeographic point of view.[8]

Mason similarly praised their collection, noting that it included several rarely collected species.

As a result of Alexander's suggestion, in 1946 Seth Benson surveyed much of the region for small mammals. The steep, rugged topography, the weight of his traps, and the gun he carried hampered Benson's work, but his results were equally revealing—many of the mammals captured resembled the Sierran fauna as opposed to that from the Great Basin. Alexander and Kellogg were delighted with his observations and visited the MVZ to examine his specimens personally.

Unlike the MVZ or the UCMP, the University Herbarium did not owe its existence to the vision or patronage of any single individual. The plant collections of the California State Geological Survey date from the 1860s.[9] Owing to the economic importance of agriculture and the pharmacological role that plants played in medicine, botanical science figured in the state university's curriculum from its inception. By 1890, the new College of Natural Resources at the University of California, and the incipient Department of Botany within it, claimed a small botanical garden and an associated herbarium. Edward L. Greene had been appointed as the first instructor of botany in 1885, but his departure a decade later left the department in a precarious position and its collections severely reduced in both quantity and quality. As was the custom of the day, when Greene left Berkeley, he took his collections with him.

Following Greene's departure, William Albert Setchell assumed the positions of professor of botany and chairman of the department, joint appointments that he held until his retirement in 1934. Setchell officially established the University Herbarium and the Botanical Gardens on campus and his pioneering systematic studies on marine algae and his interests in plant evolution and phytogeography raised the visibility of the department both nationally and internationally.

Herbert Mason assumed control of the herbarium upon Setchell's retirement and became its director four years later. Alexander and Kellogg began their botanical fieldwork in earnest during Mason's tenure, and his comments in letters to Alexander parallel Grinnell's (though his relationship with Alexander seems to have been more like the one Miller maintained with her). Mason personally thanked Alexander for each stack of pressed plants that the women donated to the herbarium and he acknowledged their importance to the collection and to research. Often he praised the data contained in their field catalogues as well.

If Mason hoped that Alexander would supplement her specimen donations with financial gifts to the herbarium, he kept such thoughts to himself. He seemed content with the number and diversity of plants that she and Kellogg proffered. Infrequently, Alexander sent Mason a check, usually in the amount of several hundred dollars, to be used as he saw fit "in furthering the work of the Herbarium." The money seems to have been sent spontaneously although, on occasion, such contributions were intended to subsidize field expeditions. Following the three-month trip to Baja California that Alexander and Kellogg undertook in 1947, Alexander funded one or two botanical field expeditions to the peninsula in which she was not a participant.[10]

Alexander and Kellogg genuinely desired to know the names of the plant species that they were collecting, and Alexander became increasingly focused on receiving such determinations in a timely fashion. Her impatience with the identification process occasionally lapsed into sarcasm, as when she wrote to Mason, "We were glad to find your letter of July 31 and the fourth list of determinations—Fine to have them before we forgot what the plants look like!"[11]

Alexander understood that determinations were not always easy to make. Floral descriptions in the literature were limited and often based on examination of only a few specimens. Variations in plant height, leafiness, amount of pubescence, or dentation of leaves were just some of the characteristics left to herbarium botanists to reconcile. And the women intentionally chose to collect in relatively remote and inaccessible areas, regions from which the flora was unlikely to be well known. Periodically, Alexander would ship from two to four boxes of dried plants back to the herbarium from the field. Botanists there would unpack and sort the plants by family, tentatively identifying them if necessary, before mounting each specimen on a herbarium sheet for permanent storage. Duplicate specimens routinely went to specialists at other institutions who were familiar with a particular group and could confirm or correct the determinations.[12]

Often Alexander and Kellogg themselves kept duplicate sheets of the specimens to refresh their memories of determinations or to use in the field as supplements to the limited plant keys and guides that were available at the time. If they were completely stumped in reaching even a tentative determination in the field, they would generally make a special trip to the herbarium and ask Mason or Annetta Carter to identify the specimen for them, paying close attention so that they might learn how the determination was made.

The speed with which Alexander and Kellogg became intimately familiar with the California flora after switching from zoology to botany is remarkable, but not surprising, given the extent to which the women had spent their lives in nature. While readily acknowledging their amateur status, Alexander was not loath to question Mason if the final determination that they received from him seemed erroneous or inconsistent with their identification, for example, "We take exception to the composite found in Mazourka Canyon no. 3003 being called Hulsea algida. It certainly doesn't resemble the description in Jepson's book." Mason conceded that Lincoln Constance's determination of no. 3003 indeed did not match Jepson's "Manual" but, he argued, Jepson had based his description solely on the original collections of the plant from Mount Dana and had not taken into account the extreme variability of that species.[13]

Some of the principal players in the history of California botany were as contentious and egotistical as those with whom Alexander had had to deal in paleontology and in vertebrate zoology. John C. Merriam and E. R. Hall had their botanical counterpart in Willis Linn Jepson, the first individual to receive a Ph.D. in botany from the University of California. After receiving his doctorate, Jepson was promoted from instructor of botany to assistant professor, publishing numerous books and papers, including the first truly informative guide to all the flowering plants in California and several volumes of the definitive *Flora of California*. Jepson had begun collecting plants as a teenager and had donated his specimens to the University Herbarium until he established his own herbarium. Over time, however, he became increasingly compulsive and guarded about his collections, more and more unwilling to allow others even to examine his voluminous material. Hoping to keep all plants collected in California for his exclusive use pending the publication of his California flora, he refused to permit duplicate sheets of specimens contributed to his herbarium to be sent elsewhere. He was once believed to have stolen specimens from the University Herbarium and to have falsified data by transferring specimens and data from one collection sheet to another.[14] From Alexander's perspective, such action was unacceptable, and his general behavior stood in direct opposition to her belief that collections should be available to the scientific community at large. She and Grinnell had fashioned a museum philosophy based on the premise that the number of biological questions to be answered approached infinity, imposing what amounted to a moral obligation on researchers to make specimen data readily available to all interested parties.

Jepson tried relentlessly to convince Alexander to donate her collections to him rather than to the University Herbarium and became almost obsessed with her resistance. Mixing data and pilfering specimens reflected the degree of his exasperation, as well as his somewhat unbalanced state of mind. He clearly recognized the quality and value of the specimens that she and Kellogg collected. Alexander was not unsympathetic to Jepson's requests but, after weighing his proprietary stance against her admiration for his knowledge of the California flora and his record of publication, did not capitulate. Yet on one occasion, frustration and impatience at not receiving determinations of her specimens from the University Herbarium in a timely fashion led her to resort to a form of academic blackmail, using Jepson as the implied threat. She wrote to Mason,

> A very eloquent and urgent appeal has come from Dr. Jepson to turn over our collections to him for the benefit of his Flora. Many of his arguments appear quite sound and probably the determinations of the plants would be available sooner. One loses interest when months pass before knowing what one has collected.[15]

Mason responded honestly and earnestly to her letter with a lengthy explanation for the delay, citing a woeful shortage of staff, resulting from the recent death of one individual and the temporary relocation of a second to Washington, D.C., and an unusually heavy teaching load. Answering immediately, Alexander acknowledged his chagrin and discomfort, and she reassured him that her comments were not meant to reflect on his character in any way. She understood his constraints. From her own vantage point, however, the determinations and an occasional comment were the only rewards that she and Kellogg received for their efforts. Several years later she wrote to Annetta Carter, wondering if Carter could be relieved of some of her office work so that she could devote a greater portion of her time to making determinations. "This work instead of a pleasure becomes an oppressive burden and the collector in turn is led to feel that his efforts were perhaps not worthwhile."[16]

In one instance, the specialist who made a specimen determination was Alice Eastwood, curator of botany at the California Academy of Sciences. Eastwood was not only famed for her botanical expertise, she also singlehandedly saved the academy's invaluable collection of plant type specimens by tossing them out an upper-story window as fire bore down on the building in the wake of the 1906 earthquake. She and Alexander knew of one another at least as early as 1903, when Alexander participated in several long walks with the "across country club of San Francisco" (led by Eastwood and Katherine Hittell, the young woman who introduced Alexander to C. Hart Merriam several years later). Eastwood closed her letter to Alexander by noting, "The specimens are the finest ever and we are very grateful to you and Miss Kellogg."[17]

Alexander and Kellogg's plant collections often elicited a level of admiration from botanists comparable to that from paleontologists and vertebrate biologists at the university. Their preparations were not only beautiful and detailed, but equally often they contained unusual, and occasionally undescribed, species of plants. Whereas Alexander disdained publicity of any sort, be it formal accolades or species named in her honor, she seemed to value the unforced and enthusiastic appreciation of scientists who had received nothing more than the specimens she proffered.

Since the specimens that Alexander and Kellogg collected were customarily sent by the herbarium director to other researchers, compliments and acknowledgments from these researchers were generally sent to him and then forwarded to Alexander. In this instance the writer was Ivan M. Johnston:

> The lot of borages is of unusual interest, not only for its beautiful preparation and labeling but also for the species and localities represented. The material from the desert, collected by Alexander and Kellogg [,] is an answer to my prayers. I have long hoped that someday I might get such a collection of borages from the mountains of extreme eastern San Bernardino County. I had imagined that I would have to make the collection myself and now, lo and behold, the job has been done in a manner which I admit I couldn't better myself.[18]

The taxonomist David D. Keck also wrote to Mason about a collection of Alexander and Kellogg plants that he had received:

> The Kingston Range plants, numbers 2344 and 2344a are very unlike any on record, although they have no truly distinctive characters of their own, but rather recombine the characters of existing closely related species in a new pattern. As I am not prepared to combine all the close relatives of this group under one large inclusive species, I must agree with Miss Alexander's opinion that this is a new species.[19]

These accolades never changed Alexander's perception of herself as an amateur botanist who lacked the intellectual ability to "do science." Nonetheless, she seemed intuitively to grasp relationships among organisms and see a larger picture, as when she wrote to Lincoln Constance, assistant director of the herbarium, "You are right,—there is far more to the Science of Botany than describing and classifying plants. Dr. Grinnell in his studies of birds and mammals, always made a great point of geographical distribution and adaptation as necessary to an understanding of the causes and methods of evolution.... They [the problems of evolution] require a certain tenacity of mind and purpose far beyond that of the ordinary student but the rewards are just so much the greater."[20] She then went on to discuss a group of plants on which he was working and their probable geologic history.

Alexander's genuine interest in, and enthusiasm for, all things related to natural history became obvious to almost anyone who met her. In the summer of 1942 she and Kellogg were collecting specimens in the vicinity of Mount Whitney, California, when they encountered the collector P. J. Bole, Jr., of the Cleveland Museum of Natural History. Following the trip, Bole wrote to E. R. Hall, "Miss Annie M. Alexander and Miss Kellogg are certainly charming old ladies, and mighty robust ones, too, I might add. We had a truly delightful visit with them in Cottonwood Valley, where we were getting topotypes of coneys [pikas], chipmunks, pocket gophers and meadow voles. They were very much interested in our mammalogical activities, and it was a real pleasure to find someone who knew something about them."[21]

While decidedly opposed to poaching and the wanton destruction of wildlife, Alexander bitterly resented the ever-increasing government regulation imposed upon scientific collectors, the increasing number of permits required, and the limits on access to many parts of the West that had once been open to exploration. In the summer of 1945 she and Kellogg wished to collect in the vicinity of Cat Creek in Mineral County, Nevada, a desert region bearing few trees above 9,000 feet. However, the site's only access was a road across property under the jurisdiction of the U.S. military. To reach the site they needed to arrange for government approval and a military jeep to escort their vehicle twelve miles up Cottonwood Creek, where two U.S. marines were required to usher them through a locked gate that marked the entrance into the Wassuk Range.

With the help of Kellogg's nephew, Major Jack Howard, the women secured the necessary permit. A fairly good road took them to the summit of Mount Grant, where an old cabin provided welcome storage for their plants and presses. Predictably, they chose to spread their sleeping bags on the floor of the aspen grove that stood nearby. Although they noted only a few distinctly alpine species, they managed to collect almost one hundred varieties of plants at the site. Only two or three seemed to differ from those they had found across the state line in Mono County, itself an interesting observation. Of greater scientific interest, however, was their discovery that the floral life zones on Mount Grant were inverted, a conclusion that E. R. Hall had also reached with regard to the fauna that he had previously collected on the mountain.[22]

At the end of the week, the women packed their belongings and loaded the Franklin. Unable to find the individual authorized to unlock the gate at the end of the road and unwilling to wait in the hot sun for his return, Louise simply took out her toolbox and proceeded to remove the heavy door from its hinges. After driving through onto government property, she reattached the gate and the women drove off.

Elated at their cleverness and happy to be on their way, Alexander wrote to E. R. Hall, among others, about the incident. He replied, "I had a good laugh at your account of the gate episode. For young ladies past 18 you and Miss Kellogg are uncommonly daring at times. I, a mere man, probably would be in the military guardhouse yet had I flaunted U.S. Navy regulations by removing one of its gates by means of its hinges."[23] Ironically, the incident did force them to contact the military. Only hours later and many miles down the road, they suddenly realized that they had left two presses full of plants on a high shelf in the abandoned cabin. Kellogg wrote to the base commander and, without comment on their behavior, he dispatched a young lieutenant to the cabin to retrieve the specimens for them.

In the summer of 1946 the women collected plants in the Warner Mountains, a range running vertically along the northeastern border of California. With the end of the war and gas rationing, more extended field trips were once again possible. Part of their incentive in heading away from the warm deserts of the south and into the cool mountains of the north was to indulge Louise's passion for trout fishing. Roads in this region were not for the faint of heart. Driving required nerves of steel and kidneys to match. Flat tires were common and the women often experienced several a day. On their first ascent of Eagle Peak on foot, they encountered a snowstorm that extended from the summit far down into the basin. Their camp, nestled comfortably under pines and alongside a murmuring stream, was blanketed upon their return. Because they had left their car on the road and had packed into the area, they had only a small tent, some bedding, and their pajamas with them. Louise melted snow for tea and, as soon as the storm subsided, they continued to collect plants. Two days later when they climbed the peak a second time, the sun shone brightly.

Compensating them for the capricious summer weather and rough terrain were frequent sightings of deer, including one standing near the summit of Eagle Peak as they approached. One night they heard a coyote in the distance, no doubt attracted by the many sheep being herded about the mountain. From the snowy peaks stretching both north and south they looked down into Surprise Valley, its three great lakes spread on a floor thickly carpeted with grassy fields. Although the unexpected snowstorm they encountered had limited the size and scope of their botanical collections, the women also accumulated two small boxes of fossil leaves that Alexander described as "mostly rocks!"

The following summer, the women again camped along the edge of a stand of piñon pine at 6,100 feet on the southwestern flank of the Wassuk Range, intending to devote a significant portion of their collecting effort to uncovering additional samples of fossil leaves. A good spring ran a short distance below their camp and they delighted in the early morning sun as it first touched the distant hills. Don Wilson, their ranch manager, had given the Franklin a fresh coat of paint at Louise's behest, but this did nothing to hide the fairly dilapidated state of its underlying machinery. His personal opinion was that the car was liable to fall apart at any minute. Such conjecture was insufficient as a deterrent. Annie was fond of the Franklin, even without the fresh paint, and the women were realistic and knowledgeable enough about its machinery not to trust it on any out-of-the-way mountain roads. Among other failings, it had no hand brake.

Because their primary objective had been to collect samples of fossil plants, the women prepared only twenty-six herbarium sheets on this particular trip. Even so, among them there were several unusual items of interest. Most of their fossil leaf collection came from the shale formation at

Coal Valley. Specimens there were not abundant and they broke open many more blocks of shale than they had leaves to show for it. Ruben Stirton and his party from the UCMP joined them for two days and asked them to seek out a locality known as Pine Grove on their return to the Bay Area. This they did. An old miner told them that what they wanted was Pine Grove Hills farther north, reachable via an unused, but passable, road. Their reward was an area not unlike the Badlands of South Dakota. For several days they explored the outcrops, finding fossil bits from two species of camel, including a jaw from one, a beaver jaw with three teeth, two good horse teeth, four peccary teeth, and a sloth tooth. They rejoined Stirton's party but nearly wrecked the Franklin on a terrible road approaching High Rock Canyon from Quinn River Crossing. After limping back to Fernley for repairs, they decided to head home. But Louise, not to be cheated out of a fishing trip, immediately repacked the Oldsmobile and the women headed for the mountains shortly after their return.

By mid-September they were off again to the desert, this time accompanied by Mary Charlotte. Annie once likened collecting to a "bad habit," one with which she was sorely afflicted, but she was quick to note that it was also "lots of fun."[24] The professed excuse for this trip was a need to break in the new Dodge Power Wagon that she had purchased for her intended exploration of Baja California. In reality, the trip was as much an excuse to go collecting, escape Oakland and Suisun for the desert, as it was to test out their new vehicle.

As usual, Louise took the wheel. If they had been after speed, the Power Wagon would not have been a good buy, but it proved its worthiness by slogging through the deep desert sand and over rocky terrain. But it was noisy, so noisy in fact that the women had to shout to make themselves heard above the din. That, coupled with unusually high temperatures, made their first two days on the road more than a bit uncomfortable (San Francisco recorded a high of 95° F the day they left town and, of course, the truck lacked air conditioning). Annie had had a "funny little wooden seat" custom-made and attached to the top of the truck. There she perched for twenty miles, the distance from Sonora to Long Barn, until the wind almost got the better of her and she capitulated.[25]

It took over two hours for the truck to lumber across Sonora Pass, 9,624 feet at the summit and sporting occasional sheer drops of thousands of feet on either side. The roads deteriorated as they crossed the state line into Nevada and, after turning off the main road, they passed the abandoned remains of Aldrich's Station, originally a stop on the old stagecoach route.

Whether for their own comfort (Alexander was now seventy-nine) or that of their guest, the women had brought inflatable rubber mattresses. Along with a gas stove, Louise had packed a cooked turkey, a leg of lamb, and fresh vegetables. They also had a wood stove in their tent and they dug a coal pit for roasting the lamb. Louise baked corn spoon bread and hot biscuits. Lunch was accompanied by maté, a bitter tea brewed from the leaves of the South American yerba tree (and a particular favorite with Annie), but with dinner they served coffee.[26]

In the same way that Louise always did the driving, she also always did the cooking. On this trip, Mary Charlotte was permitted to do the dishes while Annie chopped wood for the fire. In the evening, they read aloud about Arthur North's experiences traveling the length of Baja California.

This particular campsite was where the women had met Stirton's party earlier in the year, an area of shale outcroppings where they hoped to uncover additional specimens of fossil leaves. Although desert now, a great lake edged by trees had covered the area tens of millions of years ago, watered by clouds blown inland from the Pacific that now dropped their moisture on the western side of the more recently erupted Sierra Nevada. Each woman unearthed two leaves apiece on their first day out, a satisfactory showing by Alexander's reckoning as she considered the much sought-after specimens "scarce as hen's teeth."[27]

Mary Charlotte was struck by the almost incomprehensible beauty of the desert, the magnificent sunsets in which the Sweetwater Mountains became the deep color of amethysts, and the spell of darkness when the stars appeared like thousands of diamonds carelessly strewn across a velvet carpet. It was too late in the season for her to witness the wildflowers in full bloom, but a profusion of yellow rabbit brush blossoms hinted at the spectacle she had missed. In the evenings an owl might call; in the mornings the linnets sang. Annie and Louise pointed out deer, coyote, and sheep tracks to their guest and reported on a rattlesnake they had seen on one of their hikes.

The performance of the Power Wagon on this particular trip convinced the women that the vehicle would serve admirably on an extended desert excursion. Since the 1920s, Alexander had extensively funded MVZ fieldwork and exploration in Baja California.[28] Now, almost twenty-five years later, the prospects for a botanical expedition to the Baja peninsula began to intrigue her. As the winter of 1947 approached, she and Kellogg prepared to embark on their last great adventure together.

27

Baja California—*Tres mujeres sin miedo*

Stretching south from Tijuana, Baja (Lower) California extends like a long, contorted finger into the Pacific. For many species of California and Arizona desert plants, the peninsula encompasses their southernmost distribution. For many Latin American species, Baja California represents the northernmost extent of their range, and a great many desert plants are found nowhere else in the world. Having spent forty years collecting almost exclusively in California and western Nevada, Alexander would have been the first to admit that a great deal of unfinished fieldwork remained in those states. However, at the age of seventy-nine, she decided that the time had come to make a foray of her own into this southern region. Lincoln Constance's expedition with Carl Sauer of the UC geography department and Edward W. Nelson's published account of his travels through the peninsula had inspired her. Her primary purpose in making the trip was to collect plants, to contribute something to the knowledge of the flora of Baja California. Secondarily, it was simply to fulfill her perpetual longing for adventure. Alexander had also read *Enchanted Vagabonds* by Dana Lamb, an account of how the author and his wife traveled by canoe from San Diego to Panama, living off the land.[1] The book described in detail the game birds, deer, and other wildlife that the couple encountered en route and Alexander lamented the Lambs's lost opportunity to collect specimens as well.

The proposed expedition would not be Alexander's first excursion across the border into Mexico, simply her first visit to the state of Baja California. From December 1934 through January 1935 she and Kellogg had visited Mexico City and its environs with one of Louise's relatives. The women had flown south from Los Angeles to the Mexican capital, a distance that required them to stop five times either to change planes or refuel. International flights were still in their infancy at the time and, in a postcard to a friend and former Lasell classmate, Alexander described the beauty that she had witnessed from the air, the panoply of light and color that sunrise brought to an emerging landscape.[2]

Annetta Carter was forty years old in the fall of 1947 when Alexander and Kellogg invited her to join them on a collecting expedition to Baja California. Alexander offered to pay all Carter's expenses during the three months that the women would be gone, save her salary, an expense that Alexander insisted should continue to be borne by the university.[3] Carter had received her bachelor and master's degrees from the university in the late 1920s and early 1930s, but her plans to find a position teaching botany were stymied by the depression. Instead, she went on working in the University Herbarium, a job that she had held since her senior year in college. By 1947 she was in charge of all its day-to-day operations and had made determinations on many of the specimens that Alexander and Kellogg had collected on earlier expeditions.

Because of her sex, Carter had never been invited to participate in university-sponsored field trips, a reality that left her feeling increasingly bitter and disheartened. And so the day in 1947 when "Miss Alexander and Miss Kellogg" walked through the door of the herbarium and invited her to join them on their upcoming expedition changed both her life and her career.[4]

In 1947 Baja California was arid and sparsely populated. Self-sufficiency was mandatory for any expedition that hoped to be successful there. Because several of the MVZ staff had had experience conducting fieldwork on the peninsula, they gave Alexander a great deal of advice beforehand, including a five-page list that itemized essential general equipment.[5] For the truck, a siphon, fuel pump, fan belt, two long-handled shovels, bailing wire, gunny sacks, distilled water for the battery, tire irons, a patching kit, and a "useful assortment" of tools were routine items. Camping and cooking equipment included a machete. Under personal gear (clothes, toiletries, and stationery), Alexander added "dresses: 1 wash, 1 semi-good" (although there is no mention of wearing them). Their first-aid equipment included a "snake bite pocket kit," while the list of botanical collecting equipment was more standard: plant presses, paper, blotters, pruning shears, pruning saw, notebook paper, and pens. The list of foodstuffs began with 42 lbs white flour, 8 lbs other flours, 3 lbs cornmeal, and 25 lbs rolled oats. Also detailed were a variety of canned goods, for example, fruits, vegetables, meat, fish, soups, and a single 1-lb jar of peanut butter. It is unclear how much extra gasoline the women actually carried with them, but scrawled calculations showed that they estimated 8 miles to the gallon and 800 miles to Santa Rosalía, the first locale where "good American gasoline can be gotten," for a total of 96 gallons. It was also suggested that they carry 20 or 30 gallons of water.

For Alexander, the Power Wagon was the kind of car she had always dreamed of owning, much better than a half-ton Dodge truck—an order she had placed and then canceled. It was orange, the civilian model of a weapons carrier that had been used during the war. It had good clearance, four-wheel drive, eight forward speeds, and a bench seat that easily accommodated the three women. Before their date of departure, Alexander had a sturdy winch affixed to the front bumper and a gas tank with greater capacity installed. She had a steel mesh shell constructed over the bed of the truck and heavy canvas "curtains" hung from it to minimize the amount of dust that would get blown onto their equipment as the Power Wagon traveled over the peninsula's dirt roads. Lighter gear, like blotting paper and presses, was secured to a carrying rack that had been mounted on top of the shell and this was covered with a waterproof tarpaulin.[6]

Annetta was a strong woman. Like Louise, she was handy with tools and able to fix what machinery she owned. Before setting out, she spent quite a few hours learning basic truck mechanics, for example, how to identify a leaking fuel pump from something else that might be wrong with the engine. Their original plan called for taking two vehicles, Annie and Louise in the Power Wagon, Annetta in a red truck she had recently purchased. Perhaps the volume of gasoline or number of drivers needed for two trucks deterred them. In the end, they left in a single vehicle.

On the afternoon of November 3, 1947, with the Power Wagon fully loaded, the women crossed the border into Mexico. At the customs station in Tijuana, they declared their shotgun but were told by the officer in charge that they did not need a permit for it. They were advised, however, to register their camera with the American officials so that they would not have to pay duty on it when they reentered the United States.

Their first night in Mexico was spent in Ensenada, "and then we got into jeans and felt a little more ready to cope with the world."[7] The following day they continued down along the western coast of the peninsula to Rosario where the road turned inland toward the southeast. Mexican Highway 1 was the only road that ran through the peninsula in 1947 but today's new highway does not parallel it.

The women soon settled into a daily routine of rising early and enjoying a hearty breakfast before breaking camp, packing the truck, and heading off with Louise at the wheel. They stopped frequently to collect plants and Annie served as the expedition's photographer. From Tijuana to the tip of the peninsula at Cabo San Lucas and back again, the Power Wagon remained in first or second gear, never exceeding 15 mph. Any faster and they risked breaking a spring—or worse. On their second day out, south of Santo Tomás, the "good" gravel-surfaced road gradually metamorphosed into a dirt track and, from this point on, fifty miles a day was considered excellent progress. Louise took a good deal of comfort in knowing that they had packed tire irons and a patching kit although, miraculously, they never needed either.

At around noon each day they would stop for lunch, slipping moist blotters from between the folds of their plants and spreading them out in the midday sun to dry. If they had awakened to a heavy dew, as was often the case, they would also set out their sleeping bags to warm. Lunch included a cup of maté (which Annetta considered very much an acquired taste). After lunch, the women would pick up the warm blotters and replace them between the specimens, a task that sometimes required all six hands when the wind was blowing. Once everything had been repacked in the truck, they would start on their way, this time with Annetta behind the wheel.

Approximately an hour before dark the women would begin scouting for a good campsite, one that was not too near a ranch or pueblo, one without too much *cholla* cactus or too many rocks. While Annetta and Louise unloaded the truck and set up camp, Annie would collect firewood, most frequently dry mesquite. Dinner would be cooked over an open fire and was usually preceded by a drink. Dishes were generally washed by the light of a Coleman lantern. After dinner the women would sort and mark their specimens. On a good day, when they had collected extensively, the plants in their presses might create a stack 6 feet high. They would then write up their field notes or accounts for family and friends before retiring, up at dawn the next day to repeat the entire routine.

Although they carried tents and cots in the truck, most evenings the women preferred to sleep in the open. On one occasion, Annetta awoke in the middle of the night and saw a large rattlesnake passing through camp. Grabbing the long-handled shovel, she brought the blade down quickly, severing the snake's head. The experience caused her to pull out the sleeping cot for several nights thereafter; but not so Annie and Louise. They kept their mattresses and sleeping bags on the ground: "They weren't going to let any snake change their sleeping habits."[8]

South of Punta Prieta, at Rosarito, the dirt highway they were traveling skirted the edge of the mountain foothills, rejoining Highway 1 approximately forty-five miles south of Guerrero Negro. The women lunched by a mature grove of Joshua trees a short distance beyond Mesquital, a cattle ranch situated near the base of a high lava mesa. By evening they were camped on open granite ground about seven miles north of El Arco, a small village of adobe houses. The following night found them camped on a flat of limy soil mixed with fine sand. Their collecting was sporadic, good at some points, poorer at others, depending upon the geology and the geography of the varied landscape, and on the incidence of recent rain.

They stopped briefly in San Ignacio, a village nestled amidst a canyon of running water and date palms. In its center a distinguished old church graced a large plaza. Crowds of children shyly approached them, fascinated by their unusual orange vehicle and the three *gringas* driving it. Further south, in Santa Rosalía, they obtained the name of a trustworthy guide living in Loreto who would pack them into the Sierra de la Giganta, a rugged volcanic mountain range stretching almost 200 miles south to La Paz. The women were anxious to climb the peak they called Cerro de la Giganta. At almost 5,800 feet, it was the highest mountain in the range but had been relatively little explored and its flora remained virtually undescribed.

The road south out of Santa Rosalía ran through flat, dry country as far as Mulegé, a well-watered, picturesque village of palm groves mixed with orange, banana, and mango trees. The town seemed marred only by its reputation as a malarial enclave. From that point on, the road south began to climb. From the top of a range of hills, the women had their first view of Bahía de la Concepción, before the road dipped, following the bank of a deep canyon to their right. Upon reaching level ground they made camp on a nearby flat. A drive of five more miles brought them close to the beach and "nothing could restrain the girls from stripping and plunging into the limpid waters of the bay."[9]

On reaching Loreto, their first order of business was to seek out the guide who had been recommended to them. Arrangements were made for a pack trip and while their guide went about gathering the necessary animals, the women headed south on the dirt track that served as the main highway out of Loreto, the same trail that Steinbeck described when writing about his stay in Loreto several years earlier.[10] Colleagues at home had told them of the beauty of Escondido Bay and they decided to spend what little time they had exploring its environs.

The departure point for their pack trip was a relatively remote spot eighteen miles north of Loreto, then seven miles west along a branch road that could hardly be called a road at all. Annetta did the driving while their guide navigated. Annie and Louise sat perched above the cab, hanging on to a rope that had been stretched from door to door. At their inauspicious rendezvous point, this party of four was met by two additional guides, two horses, four mules, and five small pack burros.

After transferring their gear to the animals and mounting their burros, the group began a six-hour trail ride that followed around the northeastern base of the mountain, wending in and out of deep arroyos and, at times, ascending steep rocky ridges through plains and across mesas. In places, the steepness of the trail forced them to dismount and walk. At approximately 2,000 feet in elevation, they set up camp in a huge north-facing amphitheater formed by two jagged ridges running out from the main peak of the mountain that loomed above them.

The site proved excellent for collecting, but as their main objective was to gather plants from the upper, unexplored slopes of the peak, after several days they decided that an attempt should be made to reach the higher elevations. Few men and no women had ever scaled La Giganta. As the youngest botanist in the party, Annetta was designated to make the first bid. With a pack on her back and two of the guides to accompany her, she set off. The slopes were steep and rough, there were no real trails, and the three were continually plagued by an *Acacia* plant with horribly hooked spines that tore at their clothes and skin.[11] When they finally reached the peak, the trio was rewarded with a clear view of the Pacific Ocean to the west and the Mexican mainland to the east. Pausing only briefly to view the magnificent panorama, they hurriedly gathered ferns from the crest of the ridge that had not been found elsewhere, plus several species of tall grasses, an oak, and a tree lily previously known to occur only in the southern part of the state. The high winds near the summit made collecting dangerous and that night they slept poorly, returning to base camp the next morning having achieved their objective.

In 1947 there was no passable road heading south from Loreto. Instead, the women backtracked almost to Bahía de la Concepción before turning southwest on the road that ran through Comondú. The roughness of this stretch of track had not been exaggerated. The road crossed lava mesas and followed up and down canyons. But the area around Comondú proved to be a subtropical oasis, a valley full of palms and papayas fed by perennial springs welling up from the deep canyon. They did not follow Highway 1 south as they left Comondú. Instead, they were confined to the road that ran through the *llanos,* the Magdalena plains, great open stretches of land covered predominantly by a species of *Sphaeralcea* bearing orange-colored flowers. The nights that they spent camping in this region were cold and damp.

In time this unbroken and unwavering stretch of road, "the road that knows no turning"[12] *did* turn, bringing them out toward the sea where it paralleled the coastline for many miles. Finally a ranch house, identified simply as Arroyo Seco, marked the juncture in the road that would take them east across the peninsula toward La Paz.

A little over a month after they had left Tijuana, the women arrived in La Paz, at that time a town still paved entirely with cobblestones. The approach to the city was novel, a muddy road intersected by a series of concrete bridges that extended across numerous sloughs. The women spent several days in the city, cleaning up and organizing their specimens.

Leaving La Paz, they continued south to San José del Cabo, then west to Cabo San Lucas, collecting feverishly along the way in the wake of recent rains. The road out of La Paz proved surprisingly good, mostly granite sand. They climbed through wooded hills, dropping down a long grade that ended at the mission village of San Antonio, then across a wide plain to a long, sandy wash where they encountered a profusion of oak trees. They passed San Bartolo situated picturesquely on the edge of a deep canyon, and the sandy beaches at Los Barriles and Buena Vista. At Santiago they enjoyed oranges that were for sale in the market. San José del Cabo, a sizable village, was no more to their liking than the crowds that gathered about them, and rumors of malaria provided an additional excuse to move on. Five and a half miles from town, they camped near the top of a ridge on an old branch road.

The women reached the tip of the Baja peninsula, the village of Cabo San Lucas, forty-four days after setting out from Tijuana. They had not passed up any opportunity to collect interesting plants or swim in the ocean.

Alexander had been told in La Paz that the road up the western coast to Todos Santos had been washed out and was impassable, but the residents of Cabo San Lucas denied this account. Nonetheless, they chose to return the way they had come. The area between the two towns at the tip proved extraordinarily productive from a botanical perspective and, while Annie and Louise scrambled among the rocks collecting specimens, Annetta crouched on her knees in the wash preparing them.

Heading north again toward La Paz, they turned southwest at San Pedro and followed the road in the direction of Todos Santos. Here they arranged for a pack trip into the Sierra de la Laguna. From a ranch twenty-five miles east, they began their climb up a steep rocky trail. After several hours the landscape became more open and offered extended views of the lowlands and the ocean beyond. At the top of the spur, the trail wound around slopes that provided them glimpses of peaks to the south. It then gradually descended into the *laguna*, a long, open, sandy meadow enclosed by a forest of pine, oak, and madrona. Their camp was made in the woods above the meadow and, at almost 5,300 feet, Alexander likened the area to the foothills of the Sierra Nevada. In the mornings the temperature approached freezing. One day a thin film of ice formed on the water in their bucket. On the day after Christmas, they hiked to the top of Cerro de la Laguna, at 6,450 feet (according to Alexander) the highest peak in the range. Through woods and brush they followed a ridge until they reached a rocky promontory from which they could view the ocean on one side and the waters of Bahía de La Paz on the other.

On December 29 the party came down from the mountains. After paying their guides, the women drove thirteen miles down the road and made camp in a wash within earshot of the ocean. That evening they organized their plant collections, wrote up their field notes, and retired. In most respects, the day had been much like any other. In one respect it differed. Although Alexander and Kellogg did not say a word to Carter, Annie had celebrated her eightieth birthday.

January 1 marked the start of their return trip. It took four days to cover 150 miles. In one place, they had to widen the road using their long-handled shovel. Several nights the temperature fell below 30° F, but by noon the next day it might be 90° F in the sun. At Bahía de San Juanico a local fisherman sold them four lobsters, which they boiled for lunch and ate with great relish. After reaching Socorro, they backtracked along the route they had followed going south, stopping to collect at sites overlooked on the trip down or ones with plants now in flower. On January 21, 1948, three weeks after leaving La Paz for the second time, they arrived in Tijuana. Together, they had prepared 4,608 sheets for the herbarium, representing over 700 numbers, many of which were undescribed species or significant range extensions of previously known taxa.

Throughout, the Power Wagon performed just as Annie had hoped. Given her penchant for naming vehicles, it is surprising that she did not christen the truck when they set out. Perhaps it had not yet acquired a distinctive personality. In any event, by the end of their expedition they affectionately referred to it as "El Carro" and Annetta was addressing Louise as "Señorita Luisa," although she continued to refer to Annie as "Miss Alexander."[13]

To say that three women driving a large red truck (the color seemed to depend on who was describing it) through Baja California was a highly unusual sight fails to convey the intense level of interest that the trio generated in each of the small, dusty towns through which they passed. The people they met were kindly and responsive, but local women tended their homes and families; they were not accustomed to driving trucks unescorted across the Mexican desert or seeing other women do so. On numerous occasions, villagers asked the botanists if they were afraid of making such a journey without a man to guide them. In time they coined the response, *Somos tres mujeres sin miedo*—we are three women without fear—paraphrasing the title of a children's book that Annetta recalled having read.[14] Appropriately enough, when the women were asked to write a popular article about their adventure for *California Monthly* magazine a year later, Carter chose that English phrase for its title.

28

Investing in the Future

At almost the same time that Alexander began funding MVZ fieldwork in Baja California, she expanded her patronage to encompass graduate student research. She had long ago dropped her objections to Grinnell's involvement in undergraduate teaching and had come to view the museum's graduate students as its future. Evidence of this was her establishment of a two-year research assistantship in 1926 to provide support for an exceptionally promising individual. Her contributions to the university from this time forth were directed almost exclusively toward research and she began to insist that the institution assume a greater measure of responsibility for all other aspects of the MVZ's program. Thus in 1924, when Grinnell approached her for permission to purchase additional casework with residual funds from the museum's budget, his request was denied.[1]

Alexander now saw her mission as one of supporting and protecting the museum's research program by exacting financial commitments from the university for all other functions. Her next target was to suggest that the university compensate the museum for its teaching services to the Department of Zoology. While Alexander had come to accept the training of graduate students by museum staff as integral to their degree programs, she continued to view teaching as "a matter quite distinct from Museum activities," and one for which the university was profiting at the MVZ's expense.[2] The museum's prime function, as she viewed it, was and always would be research. If her annual donation entirely paid the salaries of museum staff, then the museum was entitled to compensation equaling the amount that the university saved when museum staff taught courses under the auspices of the department. This battle was not immediately resolved.

Ever since the end of World War I, placement of the museum's graduate students in university positions, federal and state agencies, and curatorships in other museums nationwide had borne out the MVZ's position as a center of authority for the study of vertebrate evolution. In 1931 Grinnell wrote proudly to Alexander, "Bryant, of course, has carried what I think may justly be called MVZ's point of view into the administrative headquarters of the National Park Service. This means a greatly improved attitude toward wild life, even such carnivores as mountain lions and wolves—which Bryant declares will no longer be campaigned against in any of the larger National Parks, such as Yellowstone. That is a great gain."[3]

Two years later he wrote, "I think further of the group of younger people centering here now—eager, enthusiastic—and each one with opportunity provided to aid him to develop along the line of his prime interest. There was practically no such opportunity 25 years ago; and I don't believe there would be such opportunity today except for the existence of M.V.Z."[4]

As the museum's program expanded and flourished, Alexander adopted an increasingly maternal attitude toward its graduate students. She enjoyed promoting them as individuals and took pride in their achievements. For Alexander, students were missionaries of "pure science." These young men (and later women) carried forth the museum's philosophy and its approach to the study of evolutionary biology in the positions that they accepted after graduation. Their influence became evident in the greater numbers of students applying for advanced degrees in vertebrate zoology, in the diversity of the curricula being offered by the university and, eventually, in the number of colleges and universities nationwide offering similar graduate programs.[5]

Alexander's contributions and commitment to the graduate program in paleontology were similarly strong and remained so after Merriam's departure in 1920. In 1915 he wrote, "I attribute to your personal interest and your generous support such success as we may have attained in paleontological work. In stating this view now, with even stronger emphasis, I wish to add my conviction that you have not only made possible the realization of considerable scientific results, but through students engaged in the work an influence has been exerted reaching beyond any limits that seemed attainable."[6]

Alexander came to know every graduate student by name. Through her voluminous correspondence with Grinnell and periodic visits to the museum, she familiarized herself with their research programs and followed with interest their progress and maturation. Grinnell often wrote to her before an upcoming oral comprehensive examination and, again, upon its completion. If the student performed exceptionally well, Alexander shared Grinnell's pride in his academic progeny.

Based on her own field experience, Alexander frequently recommended collecting sites to students. She also provided funds for their research projects apart from her regular appropriation, both for specimen acquisition and for fieldwork, and she funded trips that would train graduate students in collecting techniques and specimen preparation, for example, Roy Reinhart received $800 to acquire sirenian specimens (dugongs and manatees) for his dissertation, whereas Warren Wagner received a stipend and logistic support for several years to study the ferns of Hawaii. For one student interested in Africa, Alexander simply brought in an album of her 1904 travels for him to review. Hilda Grinnell was fond of recounting that when Alexander was approached by a wealthy friend from Hawaii seeking to interest her in a lucrative investment, she brought him to the museum and, pointing to a group of busy students, stated simply, "Here are my investments."[7]

In scientific disciplines where fieldwork played a prominent role in the graduate program, the presence of women was generally rare until after World War II. Alexander seemed particularly pleased when women were finally accepted into the MVZ program and when their numbers and participation in museum activities increased markedly in the early 1940s. The first female graduate student was Mary Erickson in 1931. The second was Barbara Blanchard, who was accepted in 1933. Within the MVZ, women were not permitted to go on museum-sponsored field trips during Grinnell's tenure as director and for a number of years thereafter. A man of excruciating shyness and strict Victorian correctness, he doubtless balked at the very idea of a field expedition that included women. He and Hilda might go camping and collecting by themselves for a weekend, or even a week, but on official museum trips even wives were excluded. Propriety aside, Grinnell felt that women were incapable of enduring the hardships of camp life.[8] Alexander and Kellogg, of course, were exceptions. As Swarth had remarked in 1910, these women were clearly in a class by themselves.

Oddly enough—for a woman who once wrote, "I've been wondering if a knowledge of the Carboniferous, Triassic and Cretaceous periods wouldn't be better for the girls than a study of Beowolf!"—Alexander never questioned Grinnell's policy.[9] She certainly had met other women who were capable field biologists, although most had chosen to marry and had abandoned their careers. Did she view training such women as wasteful? Or did she merely feel that it was not her place to interfere in the academic aspects of the museum program or Grinnell's operation of museum affairs?

Both Erickson and Blanchard arrived at the museum wanting to work on birds. During Blanchard's first meeting with Grinnell he suggested that she might wish to change her dissertation topic and work on worms, a less-taxing subject in his view. Much to her amazement, when she replied, "But I'm not interested in worms. I want to work on birds," the matter was dropped and she received no further opposition.[10] Blanchard, who by happenstance became the last student to finish a Ph.D. under Grinnell, credited Erickson with easing the way for her. She did not recall any overt discrimination once she was accepted into the museum program, but an anecdote provides perspective on life for these women.

Erickson and Blanchard were planning their first collecting trip to Humboldt County along the northern coast of California in the spring of 1935. Erickson was to trap five female pocket gophers at each site that Grinnell had designated, in addition to bringing back any male gophers that were caught. Blanchard thought it would be wise to carry a letter of introduction from Grinnell as they were likely to be trapping on private land much of the time. Ordinarily, the contents of such a letter would have been fairly perfunctory but, according to Blanchard's recollection, Grinnell wrote something on the order of: "These two women are of cheerful disposition. I bespeak the cooperation of landowners for the collection of gophers."[11] Reacting with wry understanding, she and Erickson subsequently referred to themselves as "the Museum Sisters, of Cheerful Disposition."

As Grinnell's protégé, E. R. Hall shared his mentor's attitudes regarding women as field biologists, but he was mortified when Alexander was made privy to an expression of his feelings on this matter. Both men clearly placed her outside the circumscribed individuals. When Alexander called the curator on a slight that had been attributed to him, Hall replied defensively, albeit humorously,

> I am still cogitating over your remarks about what a "bunch of women students would have done had they been with you at the Hoover Dam Ferry when it was 112° F." While I might say several things in reply I shall limit myself to voicing the suspicion that some one who I have regarded as a friend has basely betrayed my trust by recounting to you difficulties met with by me when attempting to select personnel of the past season's field party. May this person, whoever he or she is, all together and simultaneously some day get a cactus thorn in his knee, sunburn the back of his neck, and have all of his gophers spoil before he can skin them![12]

Much protestation, but no denial of the denigrating remark.

The presence of women in paleontology at the university preceded that in the MVZ by several decades. In the same year that the MVZ was founded, Edna Wemple became the first woman on campus to complete a master of science in paleontology.[13] Merriam did not seem to harbor Grinnell's prejudice, to judge by his willingness to suggest a female companion for Alexander as early as 1902.

William Diller Matthew acceded to the directorship of the UCMP in the late 1920s, a time when the MVZ's policy of discrimination against women was still well entrenched, yet he immediately showed a greater willingness to have women participate in fieldwork than did either Grinnell or Hall. In 1930 he took the initiative and wrote to Alexander about an enthusiastic student of his, Natasha Smith, who he felt would benefit greatly by the opportunity to take part in a fossil-hunting expedition with the women. Alexander immediately wrote to Smith, who accepted the invitation "with alacrity."[14]

Smith seemed unperturbed by the hardships of camp life or by conditions in the field, in this instance the intense heat and the alkali dust that permeated their stay in the barren hills of Nevada. Alexander suggested that she try to find a specimen of *Ilingoceros*, a graceful antelope believed to have been a precursor to modern pronghorns. Instead, her first morning out, the young woman uncovered the lower jaw of an *Aplodontia*, the mountain beaver, with all its teeth intact, as well as a smattering of camel, horse, and elephant bones. This was a creditable beginning. Pleased with their new companion, Annie wrote to Edna, "Miss Smith . . . has fitted right into the camp life and the dirt. She reminds us a little bit of Lorna [Armine von Tempsky's sister] in blue denim jeans and is husky and willing—does not seem impatient that we do not find more fossils, takes things as they come and knows a bone when she sees it."[15] In fact, the only thing that put Alexander off about Smith was her diet. The young woman was a vegetarian, a culinary style that neither Alexander nor Kellogg was accustomed to accommodating.

Their enjoyment of Natasha's company, coupled with a realization of the opportunity they had afforded the young woman, may have prompted Alexander to invite Mary Erickson to accompany them on a two-month collecting trip during the winter of 1933–34. The women planned to travel from the Palm Springs Desert area, along the Colorado River and east into western Arizona. Erickson intended to collect bird specimens for her dissertation research while Alexander and Kellogg planned to collect additional topotypes for the museum.

The women arrived in Palm Springs on Christmas Eve. The nearby canyon proved to be "an amazing bit of greenery in the midst of desert surroundings."[16] Only a short distance west of town, they walked in sandy soil among creosote bushes and horseweed. Alexander noted old mammal burrows but no fresh signs of animals. Her observations were confirmed when long trap lines set out on sandy tracts over the course of two nights secured only a single specimen of *Perognathus*, the pocket mouse. But the one specimen showed that these rodents remained active even late in the year.

After working a week in the Palm Springs area, the women headed east toward the Colorado River. Beginning at Yuma, they collected throughout southwestern Arizona, camping and trapping in the vicinity of Castle Dome, Tinajas Atlas, Tule Tank, Gila Bend, Ajo, Quitobaquito (a spring on the Mexican boundary at Monument 172), Bates Well, Wickenburg, the Harquahala Mountains, Salome, Ehrenburg, Parker, and La Paz Slough before returning home.

While the women were still in Palm Springs, Erickson's Ithaca 20-gauge shotgun disappeared from their parked car. Alexander felt personally responsible for the loss, and "humiliated beyond measure" when she wrote to Grinnell of the incident.[17] Realizing that Erickson attached a great deal of sentimental value to the gun, she asked Grinnell to purchase a new one immediately, as nearly identical to the original as possible, or if he could not, to get "the best on the market." Her letter was sent airmail and contained a check for $100. Grinnell could not find an exact replacement. But the L. C. Smith, superior grade he bought was much to Erickson's liking.

Overall, the trip seemed to have been a resounding success. According to Grinnell's calculations, the women averaged 13.25 specimens per day "and you were doing much traveling, reaching remote points, and doing *selective* trapping and hunting!" In his letter to Sproul reporting this latest gift, he noted that in two months' time the women had collected 3 reptiles, 112 birds, and 749 mammals. But perhaps the truer measure of their success came almost a decade later when Erickson confided to Alden Miller, "I'd not take anything for the trip I had with Miss Alexander."[18]

Given Alexander's desire to support graduate student research and her wish to encourage women to participate in fieldwork, her letter to Hilda Grinnell in April 1941—less than two years after her husband's death—is not surprising. Alexander suggested that she take the female graduate students in the museum on a field trip.[19] Alexander enclosed a $200 check for their expenses. Mrs. Grinnell would serve as their chaperone and teacher, and the women would study the nesting habits of birds and learn how to prepare study skins. Hilda was delighted, undoubtedly for the same reasons that Carlotta Hall had been thrilled with the opportunity to again spend time in the field. Like Virginia Miller, they had gone on field expeditions with their husbands, collecting specimens and preparing them for the museum collections. Hall and Miller had bachelor's degrees in zoology; Grinnell had a master's.

The women, four graduate students plus Hilda Grinnell, spent two weeks at Russian Gulch State Park, a stretch of the humid coastal belt along the northern coast in Mendocino County.[20] The area was only 163 miles from Berkeley but its environment was so different in terms of its lush vegetation and distinctive fauna that to the participants it seemed a world apart. In a sure sign of the times, the group was driven to the park by one of the male graduate students and, on the Sunday afternoon before their return, a second male student drove up to help the women break camp and drive them back to Berkeley.

Upon their return, the women wrote to Alexander expressing their gratitude for the opportunity that she had afforded them. Grinnell, undoubtedly accustomed to her husband's meticulous manner of dealing with expenses and museum accounts, also wrote to Alexander. Her letter detailed both the group's expenditures and their success, for example, the number of mammals trapped, birds observed, plants photographed, and it included a check to Alexander for the balance of the $200 that had not been expended.

It is difficult to interpret Alexander's reaction to receiving that check. She replied, "Thank you so much for your letter of June 8 enclosing record of expenses and check balance. I wish you had felt like keeping the check balance but I suppose one trip was enough for you."[21] The rejoinder seems out of character in a businesswoman who had always demanded accountability for every penny.

The situation obviously resolved itself, as the following spring Alexander sent a second check to Grinnell for the same purpose. This time two trips were taken and a greater number of students were able to participate. During the university's Easter vacation, one group of women drove south to Santa Cruz and Monterey counties, a dry and brushy region strikingly different from the humid northern coastal belt of the first trip, to collect specimens of the kangaroo rat, *Dipodomys*, for Jean Boulware's thesis research. Then, toward the end of the summer, five students plus Grinnell spent two weeks in the Cedar Grove area of King's Canyon National Park.[22] Two young men from the museum were sent ahead to set up camp for the women in an open stand of yellow pine near the mouth of Sheep Creek. Dixon, who had left the MVZ in 1931 to go to work for the National Park Service, happened to be staying at the rangers' headquarters in the park and helped in selecting the campsite and areas for study.

Alexander seemed to be of an entirely different mindset concerning these trips. Again, there seemed to be no demand for financial accountability, but there was also no hint of an unkind word in her correspondence with Grinnell. Quite to the contrary, she wrote a chatty letter to Hilda following the Cedar Grove trip which she concluded by noting sympathetically, "It was a sad time for you to be away for it must have taken you back to those rare happy times when you and Dr. Grinnell were off together in the mountains he loved so much and yet you could not have been in a better place, so peaceful and reassuring are the mountains."[23]

Paralleling Erickson's exhilaration and gratitude after her own trip with Alexander and Kellogg, a delighted Jean Boulware closed her long and enthusiastic letter from Russian Gulch State Park to a female colleague who remained behind in the museum with the emphatic salutation "Grinnell Expeditionary Force."[24] Her comment gives a sense of the import of those first Alexander-sponsored field trips to their participants.

But perhaps the most profound impact of these trips was not on the "girls" at all, but on Grinnell herself. Reminiscing in a letter to Alexander written in 1944, she confessed, "Those trips are going to bear dividends in the lives of the girls through all the years to come, but I do not see how they could mean more to them than they have, and will have, to me. I had felt so hopelessly cut off from the out of doors and you opened the way again for me in such a wonderful manner. I shall always be grateful."[25]

The trip to King's Canyon was the last that Alexander sponsored for women, perhaps because gasoline rationing soon became a limiting fact of life. And after the war relaxed many of the inhibitions that had prevented women from conducting fieldwork without escorts, there was no longer any reason why the museum should not sponsor expeditions conducted by women, as well as those conducted by men, if the specimens accumulated were to be deposited in its collections.

Following the war, Alexander began to think about establishing a permanent fund for graduate student research and, in 1948, she wrote to Alden Miller signaling her intent to establish a graduate student scholarship in the museum. That same year, Juliette passed away and Alexander was designated the beneficiary of her older sister's home, a lovely house located in the community of Piedmont next to Oakland. Alexander deeded the property to the regents, instructing them to sell it within a reasonable period of time and use the proceeds of the sale to create two scholarship funds of equal value, one in paleontology, the other in vertebrate zoology. Predictably, Alexander stipulated that no publicity was to attend either the creation of the fellowships or the name of the donor.[26]

As they had done many times before, the regents gratefully accepted Alexander's gift. What neither party realized was that it would be her last.

29

An Enduring Legacy

Through the winter and early spring of 1949 Alexander and Kellogg divided their time between the farm and their apartment in Oakland. Though she was now eighty-two, Alexander had barely altered her annual cycle of activity. Her only concession to age was to give up foreign travel. A year earlier, she and Kellogg had contemplated a trip to Costa Rica after receiving letters from E. R. Hall about his fieldwork there but did not pursue the idea. Similarly, their plans to join Alexander's nephew Jack on a trip to South America in 1947 had fallen apart and he had gone on without them. Instead, Alexander remarked to Hilda Grinnell, "While I live through the winter with the thought of getting away to the desert somewhere and resting my eyes on distant blue ranges, it is surprising how much of one's time is taken up with odds and ends of things to do."[1]

Most immediately, there were the matters of settling Juliette's estate and establishing the graduate fellowships. And the women now often hosted small dinner parties upon the return of a researcher after an extended field trip, often one that Alexander had financed. Alexander was both a delightful and an unconventional hostess, setting an attractive table with lovely china and then serving coffee from her old, beat-up camp pot.[2] Both she and Kellogg enjoyed hearing stories from the field and, after the dessert plates had been cleared away and the last of the coffee drained from their cups, they looked forward to viewing the slide presentations that inevitably served as after-dinner entertainment on such occasions. Alden Miller and Ruben Stirton had recently been collecting in Colombia and they and their wives were dinner guests on more than one occasion.

Since their trip to Baja California in 1947, Alexander and Kellogg had confined their collecting to the United States, but the number of plant specimens that they gathered annually remained prodigious and the variety of taxa that they collected diverse. An accounting by the herbarium put the total number of sheets donated at 17,851. In explaining her affinity for amassing large quantities of specimens, Alexander expressed her desire to create a record of the beauty and natural diversity from which she derived so much pleasure, but which she recognized was rapidly being lost in the wake of development and human population growth. Toward the end of her life she wrote to Willis Jepson:

> You ask what my objective is in making such painstaking collections. Well, I'm just a born collector and it is an excuse to get off into the mountains. When I was a child we had a summer home at 4000 ft. on the slope of Haleakala on the island of Maui and it was our greatest delight to go rambling in the woods and gulches for land shells and ferns. No doubt those times had their influence on my life but I am old now, 76, and I often feel with Wordsworth "that there hath passed away a glory from the earth."[3]

By the 1940s, the collections that Alexander and Kellogg had gathered were almost legendary within the natural history museum community on campus. In addition to the sheer quantity of specimens that Alexander collected, there are seventeen taxa named in her honor (see appendix). Undoubtedly, there would have been more had she not strenuously objected. Naturally, such assignations are often politically motivated, but far more frequently such action stems from a sincere desire to honor a mentor or a colleague who has made substantive contributions to the field. Before the MVZ's founding, three fossil taxa carried the specific epithet *alexandrae*. Two had been assigned by John Merriam and one by Edna Wemple. From Merriam's standpoint, such appellations must have seemed appropriate given Alexander's discovery of the ichthyosaur specimens that he was describing and her financial backing of the field expeditions associated with gathering the material. It was the sort of recognition that most patrons would have demanded. But Alexander was not most patrons and she soon dissuaded others from taking similar action.

It is unlikely that either Merriam or Wemple ever considered asking Alexander for permission to publish the description of a new species that had been named for her. After completing a manuscript on the mammals collected by the 1907 Alexander Expedition to Alaska, however, Edmund Heller was forced to write to Grinnell, "Miss A. apparently doesn't like to have species named for her, at all events, she objects to linking hers with a beaver for all time."[4]

Despite Heller's failure to secure her approval, for several years hence

Alexander continued to have new taxa named in her honor, presumably in instances where she was not consulted beforehand. She soon learned to stave off such honors by preempting the author's attempts. For example, in 1926 she wrote to Grinnell, "By the way, don't name the gopher from there [Toyabe Range, Nevada] after me. I have a funny feeling about things being named after me, just would rather they weren't."[5]

Acutely aware of this sentiment but nonetheless wishing to please, in 1932 E. R. Hall sent Alexander a manuscript that he had just completed in which he described a new and rare species of shrew. He proposed calling it *Sorex alexanderi*. His motivation in naming the species for Alexander stemmed in part from the fact that she and Kellogg had collected four of the eleven specimens then in existence. The magnitude of this feat was further impressed upon him by his own inability to trap even one, despite repeated efforts. Hall did not ask Alexander's permission to use the new name. He rightly assumed that her critique of the manuscript would indicate the tenor of her feelings. Her reply was brief and succinct, "We [she and Kellogg] were much interested in your description of the new shrew from Arizona (shall we not call it *Sorex grahamensis* for you know my aversion to having things named after me)."[6] He did as she requested.

In 1939 Hall reported to the office of the university president that the women had donated a minimum of 22,738 specimens to the MVZ, or about 12 percent of the total number of specimens in the museum at that time. And, he emphasized, 80 percent or more of the total MVZ collection resulted from Alexander's "sustained interest in the collections and her provision of the necessary means of collecting and using these specimens."[7]

Once Alexander and Kellogg began to collect plants in earnest, Alexander faced a new coterie of researchers to indoctrinate on this issue. "We asked the people at the Herbarium not to have any new species we might find, named after us," she wrote matter-of-factly to Jepson in 1944.[8] Perhaps the fanatical botanist had felt assured that this sort of obsequiousness would garner him the specimens he coveted.

Researchers during the 1940s were still discovering new species of plants with some frequency, but one find that Alexander made in 1949 was truly remarkable. The coming of spring that year found the women on the eastern side of the Sierra Nevada scouring the landscape in Eureka Valley for new and unusual plants, both wildflowers and grasses. They walked long distances, enjoying the warmth of the sunshine and the beauty and solitude that surrounded them in this somewhat remote locale. Their accumulated bounty from this excursion included one variety of grass that Alexander had collected on the lower slope of a huge sand dune and could not identify.

Unable to make a determination himself, Mason sent the specimen on to Jason Swallen at the Smithsonian. Swallen declared that the specimen represented an exceedingly rare find, a new genus of grass that he went on to name *Ectosperma alexandrae*, presumably without Alexander's knowledge and without worrying about her consent.

While Alexander awaited Swallen's identification of the mysterious grass, she and Kellogg postponed their visit to Hawaii to attend the Hitchcock lecture series that was to be given on campus from late October through early November that year by the noted evolutionary biologist George Gaylord Simpson. A curator of vertebrate paleontology at the American Museum, Simpson had written extensively on the intercontinental migrations of animal species through geologic time, particularly mammals. In addition, his published hypotheses on adaptive peaks and on accidental dispersal by species enriched the evolutionary debate around the time that a flourish of discoveries in genetics led to the emergence of the "modern synthesis." The topics that Simpson planned to address on successive Tuesday and Thursday evenings over the course of three weeks included the "Pre-history of Mammals," "The Paradox of the Ancient-Recent," "The Dawn of the Recent," "Island Continents," and "Progress and Failure in Evolution."[9]

Greatly anticipating Simpson's lectures, Alexander invited the paleontologist and his wife, as well as Ruben Stirton and his wife, to dinner Friday evening before the start of the series. Predictably, but disappointingly, the only written account of the evening was a single notation in her diary, "Talk about Frick!"[10]

In the week before the Simpson dinner, Alexander suffered from what she described as "indigestion." She stayed in the apartment all day Tuesday but by Wednesday morning felt well enough to run errands. She was also up and about on Thursday, but she stopped at the doctor's office for a shot, nose drops, and lozenges while she was out. Then, on Sunday following the dinner, she wrote in her diary, "Had a bad night." It was the last entry that she would record in the slim volume.

The following Tuesday Louise took Annie to Peralta Medical Center where her doctor ordered a cardiogram and a blood count. When she did not feel well enough to go out that evening, Martha Beckwith accompanied Louise to Simpson's first lecture. The next day, Annie began taking digitalis but had no appetite and remained in bed. The following Monday, Louise arranged to have her taken to the hospital by ambulance.

Stubbornly, Alexander refused to remain in the hospital. Her faith in the medical profession had not grown and she demanded to return home. She blamed a lack of nourishing food for her failure to recover and professed that all that she needed was for Louise to cook for her. On October 31, Louise capitulated and Annie came home. A week later, she was readmitted to the hospital and two days after that suffered a major stroke. On November 11, she lapsed into a coma and remained in that state for ten months.

Louise continued to make entries in Annie's diary for a week after her beloved partner and friend became comatose, noting Annie's slightest movements, optimistically hoping to glimpse signs of her recovery. She also indicated the cards and gifts received, as well as the names of those who stopped to visit or called to inquire about Alexander's condition. Then the pages of the diary were left blank. Months passed. Finally, on September 10, 1950, forty-six years to the day after Samuel had died, Louise picked up the diary once again and wrote simply, *Finis*.[11]

Following a private funeral, Kellogg arranged for Alexander's remains to be returned to Maui where they were interred in Makawao Cemetery, surrounded by the graves of many of her family and childhood friends, but where Louise's body was unlikely to lie. Even in Annie's death, Louise preserved the facade of separation that belied the intermingling of their lives and their love for more than forty years. It is difficult to imagine that either woman expected anyone to be sufficiently interested in the routine of their daily affairs to scrutinize the entries in their line-a-day diaries, yet with only one or two exceptions the entries are so generic, so impersonal, that only supplementary sources, for example, letters to friends or family, reveal the presence of two women, not just one. If the object of such discretion was to protect their families, or protect themselves from their families, then perhaps they succeeded, as neither set of relatives ever suggested that their relationship was anything more than that of close friends.

Everything that Alexander contributed financially to her museums and to the University of California, she gave during her lifetime. In a will dated June 7, 1949, she bequeathed a small portion of her estate to her sister Martha and divided the remainder equally among Martha's four surviving children.[12] Yet the programs made possible through her contributions survive as an enduring legacy.

The steadfastness with which Alexander pursued her vision for almost fifty years earned her the admiration and gratitude of even those in the administration with whom she clashed. Those with whom she worked most closely frequently expressed their appreciation to her in letters, carefully weighing their words so that the force and sincerity of their meaning would not become trite. On one occasion Grinnell wrote, "Again have decisions and appropriations on your part energized projects which, with small margin of hazards, carried out, will bring 'repercussions' of a beneficial sort far, far into the future. I don't know of any sort of investment that is more promising than in *intellectual* directions."[13]

Following Grinnell's death in 1939, a former MVZ graduate student, Walter P. Taylor, wrote to Alexander of the profound loss to science that Grinnell's passing represented. But apart from this expression of sympathy, Taylor simultaneously acknowledged Alexander's own significant contribution to Grinnell's accomplishments by noting, "There is no figure in the history of American vertebrate zoology who stands as superior to this unassuming man. With all this, I want to express to you my personal deep appreciation of the keen insight and understanding mind and generous spirit through which you have made it possible for Dr. Grinnell and his boys to do effective work all these years."[14] Taylor's words evidence the pervasiveness of Alexander's influence—in spite of her desire to remain in the background.

Alexander expressed her satisfaction in the fulfillment of goals that she and Grinnell set out to achieve when she wrote to him in 1926, "I am gratified that the Museum has its contribution to make to the solution of the great problems of evolution. That is the ultimate, if not the only goal, is it not, of our special kind of scientific work?" Reflecting her "sustained interest" in addressing such extraordinary questions, she commented on a batch of correspondence that Grinnell asked her to read a year later: "Mr. Sherwood's letter is a reminder that the size of an institution like the American Museum doesn't count for so much in itself. What counts is the ideals of its representatives and the actual output of scientific publications."[15]

The friendships that developed between Alexander and the early MVZ collectors were enduring and their mutual respect profound. In one of the chatty letters that Joseph Dixon and Allen Hasselborg continued to exchange for many years, the former MVZ staff member wrote to the veteran trapper and bear man, "I had a letter from Miss Alexander the other day and she asked to be remembered to you. She said a lot of nice things about you being the finest woodsman that she had ever met. But I think the thing she admired most is that you have lived your life in your own way, the way that you wanted it, without being dependent on others."[16] Such a statement might equally well apply to Alexander and I believe that she would most like to be remembered that way.

Epilogue

The friendship between Annetta Carter and Louise Kellogg that began during the 1947 trip to Baja California lasted many years. Early in the spring of 1950, while Alexander lay in a coma, Herbert Mason followed up on a comment made by Carter and wrote to Kellogg about her expressed interest in working at the herbarium.[1] Mason suggested that she might first like to prepare her own recently collected specimens, that is, laying the material out onto herbarium sheets and affixing the printed or typed data labels to them, an offer she happily accepted. The work was cathartic and it allowed Louise to focus her thoughts elsewhere while Annie languished.

Kellogg continued to pursue her own interests in fieldwork and collecting after Alexander's death. She and Carter made a second expedition to Baja California in the Power Wagon, driving south from Tijuana to the Sierra de la Giganta. They visited Mission San Borja and, after reaching Loreto, arranged for pack trips to Mission San Xavier, to Arroyo de Tabor, and to Cajon de Tecomaja. On the return trip they collected near Mission Santa Gertrudis and along the old coast road north of Miller's Landing to Punta Prieta, gathering and preparing approximately two thousand specimens for the herbarium.[2] They were natural as collecting partners, sharing botanical interests and a rhythm in the field that was conducive to gathering large quantities of material. With respect to herbarium business, the women always referred to each other in correspondence as "Miss Carter" and "Miss Kellogg," but their personal letters began with the salutations "Dear Annetta" and "Dear Señorita Luisa."

In October 1951 Louise joined Annetta on a third trip to Baja. This time the women flew from Tijuana to Loreto, taking a mail plane from Santa Rosalía as Loreto was not a regular commercial stop at that time. The following year they made a short trip to Punta Banda, south of Ensenada, to collect.

Their last trip together was in 1953, when they drove to Sonora and Sinaloa. After collecting in those states for several days they flew from Guaymas to Loreto to resume their work.[3]

When Louise was not on a collecting expedition or working in the herbarium, she continued to spend much of the year on the farm in Suisun. In 1927 the Baby Beef Company in Collinsville had begun buying up land on Grizzly Island to grow grain for cattle feed. The company's need to irrigate its crops, coupled with construction of Shasta Dam to the north, decreased the freshwater supply in Montezuma Slough and precipitated a decline in dairy farming on the island. Coincident with these events were several winters of below-average rainfall in which local crops did poorly. By 1934 an overexpansion of irrigation districts, increased farming in the region, and a lowering of the water table caused Solano County to feel that its very livelihood was threatened by an insufficient freshwater supply.[4]

In 1938 Annie deeded her interest in the ranch to Louise. That same year, the salinity of the water in Montezuma Slough was so high that a recommendation was put forth advocating that asparagus farming be discontinued on the island altogether. By that time, Innisfail Ranch remained the only sizable asparagus farm below the junction of the Sacramento and San Joaquin rivers. In 1945 as their plants reached maturity and production dropped off, the women dug up the roots and went back to raising oats and barley. When Kellogg died in 1967, under her will the property passed to her ranch manager, Dick Wilson, whose father, Don, had served the women so faithfully for many years.

The Suisun Marsh Preservation Act, passed in 1977, recognized the Grizzly Island area as a unique habitat, one of only two estuarine marshes in the world, and its designation coincided with the formation of the Suisun Resource Conservation District. For more than twenty-five years, increasing salinity of the irrigation water drawn from Montezuma Slough had contributed to a decline in productivity on the ranch. In 1977 Wilson sold the property to a fourth owner, who in turn sold it to the state of California. In conjunction with the California Wildlands Program, the California Department of Fish and Game acquired the farm as wetlands mitigation with the intent of restoring it to the marshy condition that prevailed when Alexander purchased the land in 1911—replacing the pastureland with seasonal ponds and tearing down the structures that Alexander and Kellogg had erected. In 1991 the ranch became part of the Grizzly Island Wildlife Area and flourishes as a wetland habitat, a designation that would undoubtedly have pleased both women.[5]

Appendix

TAXA NAMED IN HONOR OF ANNIE MONTAGUE ALEXANDER

Shastasaurus alexandrae Merriam 1902. University of California, Bulletin of the Department of Geology, 3:96. (fossil)—"In the fall of 1901, Miss A. M. Alexander generously contributed funds for another expedition into the Shasta region. Under the direction of Mr. H. W. Furlong, this party spent two months in the field and obtained some very valuable specimens."

Thalattosaurus alexandrae Merriam 1904. Extrait des Comptes rendus du 6e Congrès international de Zoologie. Session de Berne, 1904:247. (fossil)—No attribution given.

Acrodus alexandrae Wemple 1906. University of California Publications, Bulletin of the Department of Geology, 5:71. (fossil)—No attribution given.

Lagopus alexandrae Grinnell 1909. University of California Publications in Zoology, 5:204. (bird)—"This new species is named for the courageous organizer and leader of the expedition, Miss Annie M. Alexander, to whose energetic supervision much of its success was due."

Ilingoceros alexandrae Merriam 1909. University of California Publications, Bulletin of the Department of Geology, 5:320. (fossil)—"The species is named in honor of Miss Annie M. Alexander, through whose efforts the collections of Tertiary mammals from northwestern Nevada have been obtained and made available for scientific investigation."

Aplodontia alexandrae Furlong 1910. University of California Publications, Bulletin of the Department of Geology, 5:398. (fossil)—"During the past summer a continuation of the work begun in 1906 was carried on by a party from the University of California, organized and supported financially by Miss Annie M. Alexander."

Ursus alexandrae Merriam 1914. Proceedings of the Biological Society of Washington, 27:174. (mammal)—"Named in honor of Miss Annie M. Alexander of Oakland, California, whose collection of Alaska Bears is second only to that of the Biological Survey and National Museum."

Alticamelus alexandrae Davidson 1923. University of California Publications, Bulletin of the Department of Geology, 14:399. (fossil)—"Specimen found by Miss Annie M. Alexander and the species named in her honor."

Sitta carolinensis alexandrae Grinnell 1926. University of California Publications in Zoology, 21:405. (bird)—"It is fitting that the name of the one who has made possible this Museum's faunistic explorations in the San Pedro Martir region should be commemorated in connection with its vertebrate animal life; and I have chosen the designation for the new nuthatch accordingly."

Thomomys alexandrae Goldman 1933. Journal of the Washington Academy of Sciences, 23:464. (mammal)—"The species is named for Miss Annie M. Alexander whose own faunal investigations and generous support of the studies by others have contributed greatly to knowledge of the mammals of western states."

Hydrotherosaurus alexandrae Welles 1943. Memoirs of the University of California, 13:126. (fossil)—"The specific name is in honor of Miss Annie M. Alexander, who has contributed so much to the work on the vertebrates of the West."

Lupinus alexandrae Smith. Species Lupinorum, 1944:393. (plant)—No attribution given.

Ectosperma alexandrae Swallen. Journal of the Washington Academy of Sciences, 1950, 40:19. (plant)—"A very unusual and striking grass was collected by Miss Annie M. Alexander and Miss Louise Kellogg in the course of exploration in Inyo County, California, early last summer. It was submitted to the author for identification by Dr. H. L. Mason, of the University of California. It was evident at once that the grass did not belong to any species known from the United States, and on further examination it proved to be not only a new species but a new genus."

Bouvardia alexanderae Carter. Madroño, 1955, 13:142. (plant)—"Bouvardia Alexanderae is named in memory of Miss Annie M. Alexander, who in 1947 invited me to accompany her and Miss Louise Kellogg on an expedition covering the length of the peninsula, thus initiating my field work of successive years in Baja California. She assisted generously in the financing of the 1949 trip during the course of which I collected this plant. In 1947 Miss Alexander, a keen and painstaking collector and an inspiring field companion, contributed her full share of work to the expedition and, although in her eightieth year, endured with cheerful equanimity the rigors of three months of rough travel and camping in the peninsula."

Scaphiopus alexanderi Zweifel 1956. American Museum Novitates, No. 1762:16. (fossil)—"Collected by Annie M. Alexander and Louise Kellogg in 1925."

Mojavemys alexandrae Lindsay 1972. University of California Publications in Geological Sciences, 93:59. (fossil)—"Patronym for Annie M. Alexander in recognition of her interest and support for vertebrate paleontology in the Mojave Desert."

Eriogonum ochrocephalum S. Wats. var. *alexanderae* Reveal. Great Basin Naturalist, 1985, 45:276. (plant)—"The var. *alexanderae* is named for Annie M. Alexander (1867–1950) who discovered this plant during her last botanical expedition to Nevada with Louise Kellogg in 1947. Miss Alexander was then 80 years old. It is a pleasure to remember this fine Nevada collector by naming this variant in her honor."

TAXA NAMED IN HONOR OF LOUISE KELLOGG—
INCOMPLETE LIST

Lagopus rupestris kelloggae Grinnell 1910. University of California Publications in Zoology, 5:383. (bird)—"The new ptarmigan is given its name in honor of Miss Louise Kellogg, whose unfailing energy as a collector helped materially to make the 1908 Alaska Expedition a success."

Peridiomys kelloggi Wood 1936. American Journal of Science, 32:116. (fossil)—"I take great pleasure in naming this species for Miss Louise Kellogg, who has done much collecting of small rodents, and who, with Miss Annie Alexander, discovered the locality where this material was found and made the collections."

Acacia kelloggiana Carter & Rudd. Madroño, 1981, 28:221. (plant)—"*Acacia kelloggiana* is named in memory of Louise Kellogg, with whom, in company of Annie M. Alexander, the senior author made her first trip to Baja California in 1947, as well as a number of subsequent memorable trips following Miss Alexander's death in 1950. Alexander and Kellogg botanical specimens were collected in many remote parts of California and Nevada, and their collection numbers reached almost 6000. Many of their specimens serve as the bases for new taxa; duplicates have been distributed widely by UC, where the first set is deposited."

TAXA DESCRIBED FROM THE ALEXANDER
AND KELLOGG FARM AT SUISUN

Melospiza melodia maxillaris Grinnell 1909. University of California Publications in Zoology, 5:265. Suisun Song Sparrow—"collected by L. Kellogg" (MVZ #5476).

Sorex sinuosus Grinnell 1913. University of California Publications in Zoology, 10:187. Suisun shrew—"collected by Miss Annie M. Alexander" (MVZ #16470).

Lutra canadensis brevipilosus Grinnell 1914. University of California Publications in Zoology, 12:306. California river otter "[s]ecured from a local trapper [on Grizzly Island] by Miss AMA" (MVZ #20775).

Mustela vison aestuarina Grinnell 1916. Proceedings of the Biological Society of Washington, 29:213. California lowland mink—trapped by Luscomb for Alexander on Grizzly Island (MVZ #23660).

Telmatodytes paustris aestuarinus Swarth 1917. Auk, 34:310. Type specimen of the Suisun Long-billed Marsh Wren—collected by Grinnell (#3152). Alexander collected most of the specimens examined in conjunction with this determination (MVZ #25349).

Microtus californicus aestuarinus R. Kellogg 1918. University of California Publications in Zoology, 21:15. Tule meadow mouse—collected by Alexander on Grizzly Island (MVZ #18699).

Notes

Alaska State Library	Allen Hasselborg papers, ms 2, Alaska State Library
AMNH	American Museum of Natural History Special Collections
AQH	Alice Q. Howard
BANC mss C-B970	John Campbell Merriam papers, Bancroft Library, University of California, Berkeley
BANC mss C-B995	Joseph Grinnell papers, Bancroft Library, University of California, Berkeley
BANC mss C-B1003	Annie M. Alexander papers, Bancroft Library, University of California, Berkeley
BANC mss 67/121c	Annie M. Alexander additions, Bancroft Library, University of California, Berkeley
BANC mss 73/25c	Joseph Grinnell additions, Bancroft Library, University of California, Berkeley
CU-5	Records of the Office of the President, University Archives, University of California, Berkeley
CU-120	MVZ records, University Archives, University of California, Berkeley
Diary	Alexander and Kellogg's line-a-day diaries
KU	University Archives, University of Kansas, Lawrence
Letters to Martha	Letters to Martha Beckwith, Joseph Grinnell papers, Bancroft Library, University of California, Berkeley
Mission Children's Society	Hawaiian Mission Children's Society Library, Honolulu
MVZ	Museum of Vertebrate Zoology archives

UCH   University and Jepson Herbaria archives, University of California, Berkeley

UCMP   University of California Museum of Paleontology archives

INTRODUCTION

1. Several short articles have been written about Alexander. Those not cited elsewhere in this biography are M. M. Bonta, "Annie Montague Alexander: Intrepid Explorer," in *Women in the Field: America's Pioneering Women Naturalists* (College Station: Texas A&M University Press, 1992); T. S. Palmer, "Alexander, Annie Montague," in *Biographies of Members of the American Ornithologists Union* (Baltimore: Lord Baltimore Press, 1954), 7–8; R. G. Sproul, "A Tribute to Annie Montague Alexander" (University of California, 1958); and R. Williams, "Annie Montague Alexander: Explorer, Naturalist, Philanthropist," *Hawaiian Journal of History* 28 (1994):113–27.

2. The poem about Alexander's hair is as follows:

An apology for becoming a flapper of the advanced age variety

"Off with my little old lady's p—g!"
The grey haired spinster cried.
"Off with these scolding locks that have
So long my patience tried!"
So down she marched to the barber shop
And sat her in his chair
"Ten years I'll shave off of your age!"
The barber did declare.
So he clipped and snipped and snipped and clipped
And laid the tresses by.
A side-long glance discovered them,
She winced to see them lie.
Symbol of dignity—alas!
T'was such to call forth tears,
To think she should have cherished them
These many, many years!
"Do you want this hair?" the barber asked,
The lady shook her head
"Pathetic little faded wisps!"
Unto herself she said.
"Now lady come and see yourself!"
The barber beamed with joy.
She looked—"Great God!" the spinster gasped,
"I've turned into a boy!"

—May 1, 1929, to Martha Beckwith

3. Letters to Martha Beckwith are part of the Joseph Grinnell papers in the collection of the Bancroft Library, University of California, Berkeley (hereafter referred to as Letters to Martha).

4. Alexander and Kellogg's line-a-day diaries (hereafter referred to as Diary) are the personal property of Alice Q. Howard, Louise's great-niece by marriage (hereafter referred to as AQH); March 18, 1911, Diary.

5. See L. Faderman, *Odd Girls and Twilight Lovers: A History of Lesbian Life in Twentieth-Century America* (New York: Columbia University Press, 1991); and C. Smith-Rosenberg, "The female world of love and ritual: relations between women in nineteenth-century America," *Signs: Journal of Women in Culture and Society* 1 (1975):1–29.

6. Close friends of Alexander who might have spoken with me about her relationship with Kellogg were no longer living when I began this project. Because of the discretion that Alexander and Kellogg understandably exercised, there is almost nothing in writing to support the conclusion that the women were lovers. My sincere thanks to Bill Kaiwa, who was able to confirm it.

7. Warren Herb Wagner, Leslie Marcus, and David Mason spoke with me about the relationship; quote from Lois Chambers Taylor Stone oral history, Museum of Vertebrate Zoology archives (hereafter referred to as MVZ).

8. For a complete history of the company see A. L. Dean, *Alexander & Baldwin, Ltd. and the Predecessor Partnerships* (Honolulu: Alexander & Baldwin, 1950).

9. Presentation entitled "Annie Montague Alexander: An Invisible Woman?" in the symposium "The History of Women at Cal: The First 125 Years," University of California, Berkeley, April 1994.

10. Correspondence written by Alexander to her mother, to Edna Wemple, and to her cousin Mary Charlotte Alexander is in the private collection of AQH. The Joseph Grinnell papers and additions (hereafter referred to as BANC mss C-B995 and BANC mss 73/25c respectively) and the Annie M. Alexander papers and additions (hereafter referred to as BANC mss C-B1003 and BANC mss 67/121c respectively) in the Bancroft Library, University of California, Berkeley, contain the overwhelming majority of material referenced in this volume.

11. November 29, 1935, Alexander to Grinnell, MVZ records, University Archives, University of California, Berkeley (hereafter referred to as CU-120); December 1, 1935, Grinnell to Alexander, BANC mss C-B1003.

1. SAMUEL ALEXANDER AND HENRY BALDWIN

1. H. W. Grinnell, *Annie Montague Alexander* (Berkeley: Grinnell Naturalists Society, 1958) is the sole source for the avocado seedling anecdote, which simple math undermines. Uncle Henry was Henry Baldwin, her father's business partner.

2. "A&B, Land & Sea: One Hundred and Twenty-five Years Strong," *Ampersand* [a publication of Alexander & Baldwin] (1995). B. Millard, *History of the San Francisco Bay Region* (San Francisco: American History Society, 1924); vol. 3 gives a short biographical sketch of Wallace Alexander's business and professional affiliations.

3. M. C. Alexander, *William Patterson Alexander in Kentucky, the Marquesas, Hawaii* (Honolulu, 1934). William Patterson (1805–84) and Mary Ann McKinney Alexander (1810–88) are listed in the Family Record of April 17, 1982 (Alexander & Baldwin archives, Honolulu).

4. See two works by M. C. Alexander: *The Story of Hawaii* (New York: American Book, 1912) and *Dr. Baldwin of Lahaina* (Berkeley, 1953); as well as A. D. Baldwin, *A Memoir of Henry Perrine Baldwin 1842 to 1911* (Cleveland, 1915).

5. E. A. Kay, ed., *A Natural History of the Hawaiian Islands: Selected Readings*, 2 vols. (Honolulu: University of Hawaii Press, 1972–94).

6. Samuel's oldest brother, William, developed an interest in geology, becoming surveyor general of the Hawaiian kingdom and eventually mapping the entire island chain. John Thomas Gulick wrote extensively about the evolution of land snails in Hawaii, and Sanford B. Dole published a checklist of Hawaiian birds. The most prolific natural history writer among the missionary sons was David Dwight Baldwin, Samuel's brother-in-law. Baldwin's interests in natural history ranged widely and his publications included catalogues of land snails and lists of mosses, as well as descriptions of the indigenous woods, trees, and shrubs of his adopted home, and details of the eruption of the volcano Mauna Loa.

7. Alexander, *William Alexander in Kentucky*.

8. Ibid., 394.

9. In 1854 Samuel went to work for William H. Rice in Lihue, Kauai.

10. Alexander, *William Alexander in Kentucky*, 401.

11. Ibid., 440.

12. "Brief Sketch of Mr. Alexander's Life," published in conjunction with a memorial service for Samuel Alexander held at the First Congregational Church, Oakland, California, on September 18, 1904 (Hawaiian Mission Children's Society Library, Honolulu [hereafter referred to as Mission Children's Society]); and "A Biography of Samuel T. Alexander," in *Oahuan* (1908):9–10, the yearbook of Oahu College (Punahou School archives, Honolulu).

13. There is a discrepancy between Alexander's account of Samuel's education on the East Coast (in *William Alexander in Kentucky*) and that reported in the "Brief Sketch."

14. Family Record, April 17, 1982 (Alexander & Baldwin archives, Honolulu).

15. This purchase marked the beginning of their business partnership. For a detailed history see A. L. Dean, *Alexander & Baldwin, Ltd. and the Predecessor Partnerships* (Honolulu: Alexander & Baldwin, 1950).

16. I. L. Bird, *Six Months in the Sandwich Islands* (Honolulu: University of Hawaii Press, 1864), 198.

17. "Alexander and Baldwin: Early Hawaiian Pioneers," *Ampersand* (1990): 1–6; Baldwin, *Memoir*; and Dean, *Alexander & Baldwin* all describe construction of the aqueduct in great detail.

18. Their holdings had previously been referred to as the Alexander & Baldwin Plantation, the Sam T. Alexander & Co., and Haleakala Sugar Company.

19. Alexander, *William Alexander in Kentucky*, 402, 442.

20. Talk of annexation continued for almost a decade until 1898, when the U.S. Congress finally annexed Hawaii and the monarchy came to an end. For details see G. Daws, *Shoal of Time: A History of the Hawaiian Islands* (Honolulu: University of Hawaii Press, 1968); and J. Adler, *Claus Spreckels: The Sugar King in Hawaii* (Honolulu: Mutual Publishing Paperback Series, 1966).

21. Adler, *Claus Spreckels*.

2. LIFE IN OAKLAND

1. B. Bagwell, *Oakland: The Story of a City* (Oakland: Oakland Heritage Alliance, 1982).

2. For details of the house see "Residence of Mr. S. T. Alexander, Oakland," *S.F. News Letter* 1887, San Francisco, Mission Children's Society.

3. G. L. Bamford, *The Mystery of Jack London, Some of His Friends, and also a Few Letters: A Reminiscence* (Oakland, 1931). AQH provided information on the Kellogg family.

4. T. P. Martin, *The Sound of Our Own Voices: Women's Study Clubs 1860–1910* (Boston: Beacon Press, 1987); D. J. Winslow, *Lasell: A History of the First Junior College for Women* (Boston: Nimrod Press, 1987).

5. Lasell class records.

6. August 17, 1899, Alexander to Beckwith, Letters to Martha; and April 10, 1950, Louise Kellogg to George Pettitt, archives, University and Jepson Herbaria, University of California, Berkeley (hereafter referred to as UCH).

7. H. W. Grinnell, *Annie Montague Alexander* (Berkeley: Grinnell Naturalists Society, 1958) contains the sole reference to these events although Alexander's niece, Martha Hurd Kreuter, generally confirmed Grinnell's account.

8. M. C. Alexander, *William Patterson Alexander in Kentucky, the Marquesas, Hawaii* (Honolulu, 1934), 464; "Brief Sketch of Mr. Alexander's Life," published in conjunction with a memorial service for Samuel Alexander held at the First Congregational Church, Oakland, California, on September 18, 1904, Mission Children's Society.

9. Details of the trip are taken from October 11, 1944, Alexander to Herbert Mason, UCH; and M. W. Beckwith, "Lassen Buttes; from Prattville to Fall River Mills," *Sierra Club Bulletin* 3 (1901):288–97; August 1, 1899, Alexander to Mary Beckwith, Letters to Martha.

10. August 1, 1899, Alexander to Mary Beckwith, Letters to Martha.

11. August 7, 1899, Alexander to Beckwith, ibid.

12. August 17, 1899, ibid.

13. Undated, 1902, ibid.; November 2 and 24, December 3, 1902, ibid.; undated, 1904, ibid. Based on content, the letter was written between October and November 1904, after Alexander's return from Africa.

14. February 7 and May 2, 1901, ibid.

15. B. B. Peterson, ed., *Notable Women of Hawaii* (Honolulu: University of Hawaii Press, 1984).

16. August 17, 1899, Alexander to Beckwith, Letters to Martha.

17. August 12, 1899, ibid.; October 25, 1899, ibid. There is no indication that Alexander ever contracted malaria. I believe this is either an error in penmanship or a euphemism for "malaise."
18. October 25, 1899, ibid.
19. March 13, 1901, ibid.; November 10, 1901, ibid.; July 17, 1905, ibid.
20. December 14, 1899, ibid.
21. Personal communication from Martha Hurd Kreuter.
22. September 1, 1900, Alexander to Beckwith, Letters to Martha. "Miss Wilson" may be Emily Wilson, Alexander's former governess, or an acquaintance from Berkeley, Mary Wilson.
23. Ibid.

3. A PASSION FOR PALEONTOLOGY

1. R. A. Stirton, "The Role of Paleontology in the University of California: Honoring the Twenty-Fifth Presidential Year of President Robert Gordon Sproul" (University of California, Berkeley, 1955); and J. T. Gregory, "UCMP History: John Campbell Merriam," *UCMP News*, June 1995. In the early 1860s, legislation creating the University of California had stipulated that the collections of the California State Geological Survey should become an intrinsic part of the new institution at its first campus, Berkeley (which acquired the nickname of Cal). Accordingly, the survey's entire assemblage of both rocks and fossils was transferred to the university upon its founding in 1868 (V. A. Stadtman, *The University of California 1868–1968: A Centennial Publication of the University of California* [New York: McGraw-Hill, 1970]; and C. D. Wagner, "Paleontology at the University of California: the history of Bacon Hall," *Journal of the West* 8 [1969]:169–82). Merriam built up the collections from this core of material.
2. February 7, 1901, Alexander to Beckwith, Letters to Martha.
3. January 12, 1901, ibid.
4. Ibid. Those mentioned are Mary W. (presumably Mary Wilson who accompanied Alexander on the 1901 Fossil Lake expedition), Mr. Furlong, and a Mr. Cornish. Tule fog is a phenomenon peculiar to the Central Valley of California in winter; a dense overcast rises from wet ground and hangs in the air, providing a uniform gray light from all points of the sky and obliterating the surrounding landscape.
5. February 24, 1910, Alexander to Mary Beckwith, ibid.; December 21, 1905, ibid.
6. February 7, 1901, Alexander to Beckwith, ibid.
7. J. A. Shotwell, "Journal of first trip of University of California to John Day Beds of eastern Oregon by Loye Miller, 1899," *Bulletin of the Museum of Natural History, University of Oregon*, no. 19 (1972). In 1901 the university was just thirty-three years old, and its reputation for research excellence was not yet established.
8. May 2, 1901, Alexander to Beckwith, Letters to Martha.
9. Details of the expedition have been extracted from June 10 [?], July 7,

July 26, and November 10, 1901, ibid.; and from the Fossil Lake photo album compiled by Mary Wilson, University of California Museum of Paleontology archives (hereafter referred to as UCMP).

10. July 7, 1901, Alexander to Beckwith, Letters to Martha.
11. July 26, 1901, ibid.; June 10 [?], 1901, ibid. Mary Wilson and Herbert Furlong subsequently married.
12. April 23, 1902, ibid.
13. July 26, 1901, ibid.
14. Spring 1901, ibid.; undated, 1901 (before May 2, based on content), ibid.; July 7, 1901, ibid.
15. November 10, 1901, ibid.
16. February 7, 1901, ibid.
17. June 8 [?], 1902, ibid. A question mark and the word "Aug." have been added in pencil, but only pages 2 through 6 remain—the rest of the letter is missing. I, too, question the letter's date as it is a recap of the trip.
18. Details of the expedition have been extracted from "Notes of Miss Katherine Jones, Paleontological Expedition, University of California, Shasta County, June 16, to July 13, 1902," MVZ. See also J. L. Zullo, "Annie Montague Alexander: her work in paleontology," *Journal of the West* 8 (1969): 183–99.
19. June 8 [?], 1902, Alexander to Beckwith, Letters to Martha. J. C. Merriam, "Triassic Ichthyopterygia from California and Nevada," *University of California Publications, Bulletin of the Department of Geology* 3 (1902):63–108.
20. January 30, 1903, Alexander to Beckwith, Letters to Martha. The volume is probably W. H. Flower, *An Introduction to the Osteology of the Mammalia* (London: Macmillan, 1885).
21. No biographical background on Ray or Esterly was given in the account of the expedition.
22. June 7, 1903, Alexander to Beckwith, Letters to Martha.
23. March 24, 1903, Wheeler to Alexander, UCMP.
24. June 7, 1903, Alexander to Beckwith, Letters to Martha.
25. A type specimen is the single specimen, or series of specimens, which is examined and upon which the description of a new taxon is based. For the type description of *Thalattosaurus* see J. C. Merriam, "A new group of marine reptiles from the Triassic of California" (Extrait des Comptes Rendus du 6e Congrès international de Zoologie, Session de Berne 1904), 247–48.
26. June 8, 1902, Alexander to Beckwith, Letters to Martha.
27. August 11, 1899, ibid. This letter is a continuation of a letter written August 5, 1899. Because none of the letters from Beckwith to Alexander survive, the context for this remark is unknown.

4. AFRICA, 1904

1. December 7, 1903, Alexander to Beckwith, Letters to Martha.
2. October 17, 1903, Merriam to Alexander, UCMP.

3. December 7, 1903, Alexander to Beckwith, Letters to Martha. Thomas Gulick was the youngest brother of John Gulick, the well-known malacologist.

4. Details of the expedition are taken from a scrapbook of letters written by Alexander and her father to family and friends back home. They are included in BANC mss C-B1003.

5. May 28, 1904, Samuel to his son, Wallace, BANC mss C-B1003. Technically, the term "deer" refers to members of the family Cervidae, a group that does not occur in Africa.

6. Ibid. See P. G. Abir-Am and D. Outram, eds., *Uneasy Careers and Intimate Lives: Women in Science, 1789–1979* (New Brunswick: Rutgers University Press, 1989) on relationships that influence the development of women as scientists.

7. May 21, 1904, Alexander to her mother, BANC mss C-B1003; June 1, 1904, Alexander to Aunt Mary, ibid.

8. B. B. Peterson, ed., *Notable Women of Hawaii* (Honolulu: University of Hawaii Press, 1984).

9. June 24, 1904, Samuel to Pattie, BANC mss C-B1003; July 4, 1904, Samuel to Pattie, ibid.

10. *San Francisco Chronicle*, September 3, 1905, 1. Alexander's photographs are in the Bancroft Library, University of California, Berkeley (PIC 1966.013-ALB).

11. August 3, 1904, Samuel to Pattie, BANC mss C-B1003; undated, 1904, Alexander to Mary Beckwith, Letters to Martha.

12. Ibid.

13. Undated, 1904, Alexander to Mary Beckwith, Letters to Martha.

14. August 22, 1904, Alexander to her mother, BANC mss C-B1003.

15. Ibid.

16. Ibid.

17. Undated, 1904, Alexander to her family, reel 6 XXII, Alexander Family papers, Bancroft Library, University of California, Berkeley.

18. Today, all that remains of old Livingstone is this small cemetery (D. W. Phillipson, "The Early History of the Town of Livingstone," in *Mosi-oa-Tunya: A Handbook to the Victoria Falls Region*, ed. D. W. Phillipson [London: Longman Rhodesia, 1975], 88–104).

19. Undated, 1904, Alexander to her family, reel 6 XXII, Alexander Family papers.

20. October 10, 1904, Alexander to Martha and Mary Beckwith, Letters to Martha.

21. January 14, 1943, Alexander to Hilda Grinnell, BANC mss 73/25c.

5. MEETING C. HART MERRIAM

1. "Recent" animals are those that are living or have lived within the past 10,000 years.

2. March 1, 1903, Alexander to Beckwith, Letters to Martha.

3. Undated, 1904, ibid.

4. H. W. Grinnell, "Biographical sketch of Annie Montague Alexander," ms, MVZ. Native American artifacts were one of C. Hart Merriam's interests.

5. October 8, 1907, Alexander to Beckwith, Letters to Martha.

6. Undated, 1905, ibid.

7. H. Kuklick and R. E. Kohler, "Science in the field," *Osiris* 11 (1996): 1–265.

8. Alexander's field notes from the 1905 expedition have been reprinted in J. L. Zullo, "Annie Montague Alexander: her work in paleontology," *Journal of the West* 8 (1969):183–99. A delightful description of the intricacies of paleontological fieldwork can be found in G. G. Simpson, *Attending Marvels: A Patagonian Journal* (New York: Macmillan, 1934).

9. May 23, 1901, Alexander to Beckwith, Letters to Martha; July 17, 1905, ibid.

10. August 20, 1905, Alexander to John Merriam, John Campbell Merriam papers, Bancroft Library, University of California, Berkeley (hereafter referred to as BANC mss C-B970); Zullo, "Annie Montague Alexander."

11. Zullo, "Annie Montague Alexander."

12. W. H. Osgood, "Clinton Hart Merriam—1855–1942," *Journal of Mammalogy* 24 (1943):421–57; and K. B. Sterling, *Last of the Naturalists: The Career of C. Hart Merriam* (New York: Arno Press, 1977). In 1940 the Department of Interior took over the work of the Biological Survey and merged it with the former Fish Commission to form the U.S. Fish & Wildlife Service. Merriam's sister, Florence, was also an ornithologist who went on to her own notable career (H. Kofalk, *No Woman Tenderfoot: Florence Merriam Bailey, Pioneer Naturalist* [College Station: Texas A&M University Press, 1989]).

13. July 17, 1905 Alexander to Beckwith, Letters to Martha.

14. October 10, 1905, ibid.

15. On John Muir see M. Austin, *Earth Horizon* (New York: Houghton Mifflin, 1932); and on Hittell, G. W. Dickie, L. M. Loomis, and R. Pratt, "In memoriam: Theodore Henry Hittell," *Proceedings of the California Academy of Sciences* 8 (1918):1–25. Hittell's 1,563-page manuscript on Hawaii was never published.

16. February 22, 1906, Alexander to C. Hart Merriam, C. Hart Merriam correspondence, reel 28, Bancroft Library, University of California, Berkeley.

17. February 22, 1906, Alexander to Beckwith, Letters to Martha.

18. March 4, 1905, ibid. Bryan went on to serve as director of the Los Angeles County Museum of Natural History in the 1920s.

19. March 1, 1906, C. Hart Merriam to Alexander, MVZ.

20. A. E. Leviton and M. L. Aldrich, eds., *Theodore Henry Hittell's The California Academy of Sciences: A Narrative History, 1853–1906* (San Francisco: California Academy of Sciences, 1997).

21. February 24, 1911, C. Hart Merriam to Joseph Mailliard, MVZ.

22. March 11, 1932, Alexander to C. Hart Merriam, C. Hart Merriam correspondence, reel 28, Bancroft Library, University of California, Berkeley; and

April 30, 1939, Alexander to Grinnell, in Grinnell, "Biographical sketch." Elsewhere, the prize was referred to as the Roosevelt Memorial Medal and was awarded to Merriam "for excellence in his work as a naturalist" (K. B. Sterling, *Last of the Naturalists: The Career of C. Hart Merriam* [New York: Arno Press, 1977]).

6. ALASKA, 1906

1. July 17, 1905, Alexander to Beckwith, Letters to Martha.
2. October 10, 1905, ibid.
3. December 9, 1905, ibid.
4. Merriam eventually amassed 1,864 brown and grizzly bear specimens, mainly skulls. He assigned these to an enormous number of new species and subspecies, the resulting taxonomy being greeted with a fair bit of skepticism by researchers of the day, as well as those who later reexamined his work.
5. J. Burroughs, J. Muir, G. B. Grinnell, W. H. Dall, C. Keeler, and B. E. Fernow, *Alaska: The Harriman Expedition, 1899* (New York: Dover Publications, 1986); and W. H. Goetzmann and K. Sloan, *Looking Far North: The Harriman Expedition to Alaska 1899* (Princeton: Princeton University Press, 1983). Harriman subsidized the publication of thirteen volumes on the scientific results of the expedition.
6. J. A. Allen, "Mammals collected in Alaska by the Andrew F. Stone Expedition of 1903," *Bulletin of the American Museum of Natural History* 20 (1904):273–92.
7. February 22, 1906, Alexander to Beckwith, Letters to Martha.
8. March 8, 1906, Alexander to C. Hart Merriam, Allen Hasselborg papers, ms 2, Alaska State Library (hereafter referred to as Alaska State Library).
9. Seale was a Bureau of Fisheries employee in Washington, D.C. in January 1906 (and had been, coincidentally, a classmate of Joseph Grinnell's at Stanford). In 1907 the bureau donated to the Smithsonian marine specimens collected by its ship, the *Albatross*. Seale, Wemple, and Alexander may have been on board during the ship's 1906 voyage to Alaska, but there is no record of their presence (Smithsonian Institution archives, RU 192). Details of the 1906 expedition to Alaska are taken from H. W. Grinnell, "Biographical sketch of Annie Montague Alexander," ms; and uncataloged photographs, MVZ.
10. Grinnell, "Biographical sketch."
11. Ibid.

7. MEETING JOSEPH GRINNELL

1. P. F. Covel, *Beacons Along a Naturalist's Trail: California Naturalists and Innovators* (Oakland: Western Interpretive Press, 1988); F. Stephens, *California Mammals* (San Diego: West Publishing, 1906).
2. Throop Polytechnic Institute later changed its name to the California Institute of Technology, known today as Cal Tech.

3. J. K. Jones, Jr., "Genealogy of Twentieth-Century Systematic Mammalogists in North America: The Descendants of Joseph Grinnell," in *Latin American Mammalogy: History, Biodiversity, and Conservation*, ed. M. Mares and D. Schmidly (Norman: University of Oklahoma Press, 1991), 48–56. No biography of Grinnell has ever been written but summaries of his accomplishments appeared in scientific journals after his death, e.g., W. K. Fisher, "When Joseph Grinnell and I were young," *Condor* 42 (1940):35–38; H. W. Grinnell, "Joseph Grinnell: 1877–1939," *Condor* 42 (1940):3–34; E. R. Hall, "Joseph Grinnell (1877 to 1939)," *Journal of Mammalogy* 20 (1939):409–17; E. R. Hall, "Obituary: Joseph Grinnell," *The Murrelet* 20 (1939):46–47; J. M. Linsdale, "In memoriam: Joseph Grinnell," *Auk* 59 (1942):269–85; D. D. McLean, "Dr. Joseph Grinnell: February 27, 1877–May 29, 1939," *California Fish and Game* 26 (1940):174–77; and W. E. Ritter, "Joseph Grinnell," *Science* 90 (1939): 75–76. See also A. H. Miller, "Joseph Grinnell," *Systematic Zoology* 13 (1964): 235–42.

4. Grinnell, "Joseph Grinnell."

5. J. Grinnell, "Birds of the Kotzebue Sound region, Alaska," *Pacific Coast Avifauna* 1 (1900):1–80.

6. Grinnell, "Joseph Grinnell."

7. September 18 and October 15, 1907, Alexander to Grinnell, BANC mss 67/121c.

8. S. Eppenbach, *Alaska's Southeast: Touring the Inside Passage*, 4th ed. (Old Saybrook: The Globe Pequot Press, 1991). The Alexander Archipelago was named in 1867, the year in which the U.S. signed the Treaty of Purchase for Alaska. Although the region remained largely unexplored in 1907, that same year Congress created the 16-million-acre Tongass National Forest in southeastern Alaska in an attempt to control massive logging and regulate timber sales in the region.

9. Details about the expedition are taken from Alexander, Dixon, and Stephens's field notes, MVZ; and F. Stephens, "A summer's work: a natural history expedition to southeastern Alaska," *Forest and Stream* 71 (1908):568–70, and 608–9. Grinnell placed an announcement of the expedition in the May 1907 issue of *Condor* (9:94). The expedition became the first scientific party to explore the interior of Admiralty Island.

10. Littlejohn had previously hunted bears in the Aleutian Islands but it is unknown how Alexander came to contact him.

11. May 12, 1907, Alexander to Beckwith, Letters to Martha.

12. Ibid.

13. See J. R. Howe, *Bear Man of Admiralty Island: A Biography of Allen E. Hasselborg* (Fairbanks: University of Alaska Press, 1996); and J. M. Holzworth, *The Wild Grizzlies of Alaska* (New York: G. P. Putnam's Sons, 1930) for more on Hasselborg.

14. May 27, 1907 (a continuation of May 26, 1907), Alexander to Beckwith, Letters to Martha.

15. At the conclusion of the 1907 expedition, successful application was made

to the Geographical Society to have those names formalized. See D. J. Orth, "Dictionary of Alaska Place Names," Geological Survey Profession Paper no. 567, 1967.

16. Heller described a new subspecies of beaver from the lakes (J. Grinnell and E. Heller, "Birds and mammals of the 1907 Alexander Expedition to southeastern Alaska," *University of California Publications in Zoology* 5 [1909]: 171–264). See also "Beavers in Alaska," *Forest and Stream* 72 (1909):875–76, for a popular account of the discovery.

17. June 2, 1907 (a continuation of May 26, 1907), Alexander to Beckwith, Letters to Martha. Grinnell named a new species of ptarmigan in Alexander's honor (see appendix). Dixon collected two species of ptarmigan on this trip, rock ptarmigan (*Lagopus mutus dixoni*) and a new subspecies of willow ptarmigan, which Grinnell subsequently named after Alexander, *Lagopus lagopus alexandrae*.

18. Stephens's 1907 field notes, MVZ.

19. July 9, 1907, Alexander to Beckwith, Letters to Martha.

20. Ibid.

21. Ibid.

22. September 18, 1907, Alexander to Grinnell, BANC mss 67/121c; October 15, 1907, ibid. The specimens collected on the 1907 expedition, together with all field notes and accompanying photographs, comprised the first Museum of Vertebrate Zoology accession and formed the basis for the museum's first published report.

8. FOUNDING A MUSEUM OF VERTEBRATE ZOOLOGY

1. J. Grinnell, "The methods and uses of a research museum," *Popular Science Monthly* 77 (1910):163–69. For additional references on this topic see H. W. Greene and J. B. Losos, "Systematics, natural history, and conservation: field biologists must fight a public-image problem," *BioScience* 38 (1988):458–62; "Systematics Agenda 2000: Charting the Biosphere" (Washington, D.C.: Systematics Agenda 2000 in cooperation with the Association of Systematics Collections, 1994); E. H. Miller, "Biodiversity Research in Museums: A Return to Basics," in *Our Living Legacy: Proceedings of a Symposium on Biological Diversity*, ed. M. A. Fenger, E. H. Miller, J. A. Johnson, and E. J. R. Williams (Victoria, B.C.: Royal British Columbia Museum, 1993), 141–73; S. M. Goodman and S. M. Lanyon, "Scientific Collecting," *Conservation Biology* 8 (1994):314–15; and C. E. Bock, "The role of ornithology in conservation of the American West," *Condor* 99 (1997):1–6.

2. November 14, 1907, Grinnell to Alexander, BANC mss C-B1003; April 8, 1908, ibid.; February 18, 1908, ibid.

3. October 29, 1907, ibid.

4. November 2, 1907, Alexander to Grinnell, BANC mss 67/121c.

5. October 29, 1907, Grinnell to Alexander, BANC mss C-B1003; November 2, 1907, Alexander to Grinnell, BANC mss 67/121c.

6. October 28, 1907, Alexander to Wheeler, Records of the Office of the President, University Archives, University of California, Berkeley (hereafter referred to as CU-5). Details concerning the founding of the MVZ and the UCMP, and of Alexander's relationships with Grinnell and Merriam, were first published in B. R. Stein, "Annie M. Alexander: extraordinary patron," *Journal of the History of Biology* 30 (1997):243–66.

7. Grinnell "Report of activities of the California Museum of Vertebrate Zoology for the period from its inauguration, March 23, 1908, to July 1, 1910" (MVZ).

8. January 30, 1908, Alexander to Wheeler, BANC mss C-B1003.

9. In 1912 Grinnell wrote to David Starr Jordan, his former major professor, stating his desire to complete his Ph.D. He submitted "An Account of the Mammals and Birds of the Lower Colorado Valley with Especial Reference to the Distributional Problems Presented" toward fulfillment of the thesis requirement, took a written examination and, in May 1913, received his doctorate. Simultaneously he acquired the title of assistant professor of zoology at the university.

10. Undated, John C. Merriam correspondence, MVZ and CU-5.

11. A. B. Comstock, *Handbook of Nature Study* (Ithaca: Comstock Publishing, 1935). There is no explanation for the omission of amphibians as a vertebrate class worthy of study here or in Alexander's original proposal.

12. December 24, 1907, Alexander to Wheeler, CU-5.

13. January 26, 1908, Alexander to Grinnell, BANC mss 67/121c.

14. January 30, 1908, Alexander to Wheeler, BANC mss C-B1003.

15. February 11, 1908, Wheeler to Alexander, CU-5 (the letter says that the regents' opinion is attached but the attachment was missing); February 5, 1908, Alexander to Grinnell, BANC mss 67/121c.

16. February 14, 1908, Alexander to Wheeler, BANC mss C-B1003.

17. February 15, 1908, Wheeler to Alexander, ibid.

18. March 23, 1908, Alexander to Grinnell, BANC mss 67/121c.

19. November 6, 1908, ibid. The letter reiterates the sentiments she expressed in the contract she and Grinnell signed February 3 and February 6, 1908, respectively (ibid.). That agreement outlined the conduct of business upon Grinnell's appointment as museum director.

20. November 14 and 20, 1908, Grinnell to Alexander, BANC mss C-B1003; S. G. Herman, *The Naturalist's Field Journal: A Manual of Instruction Based on a System Established by Joseph Grinnell* (Vermillion, S. Dak.: Buteo Books, 1980); N. K. Johnson, "Ornithology at the Museum of Vertebrate Zoology, University of California, Berkeley," in *Contributions to the History of North American Ornithology*, ed. W. E. Davis, Jr., and J. A. Jackson (Cambridge, Mass.: Nuttall Ornithological Club, 1995), 183–221; J. R. Griesemer, "Modeling in the Museum: on the role of remnant models in the work of Joseph Grinnell," *Biology and Philosophy* 5 (1990):3–36; J. R. Griesemer and E. M. Gerson, "Collaboration in the Museum of Vertebrate Zoology," *Journal of the History of Biology* 26 (1993):185–203; and S. L. Star and J. R. Griesemer, "Institutional ecology, 'translations,' and boundary objects: amateurs and professionals in

Berkeley's Museum of Vertebrate Zoology, 1907–1939," *Social Studies of Science* 19 (1989):387–420.

21. January 20, 1907, Alexander to Beckwith, Letters to Martha. Bailey was married to C. Hart Merriam's younger sister, Florence.

22. November 14, 1907, Grinnell to Alexander, BANC mss C-B1003; J. Grinnell and E. Heller, "Birds and mammals of the 1907 Alexander Expedition to southeastern Alaska," *University of California Publications in Zoology* 5 (1909):171–264.

23. September 19, 1908, Alexander to Wheeler, CU-5.

24. October 15, 1907, Alexander to Grinnell, BANC mss 67/121c.

25. November 20, 1907, Grinnell to Alexander, BANC mss C-B1003.

26. Grinnell credited Alexander with stimulating the academy to focus on public exhibitions in rebuilding, noting, "Perhaps the threat or fear of competition has stirred him up" (November 8, 1907, ibid.). After seeing the sheep and sea lion groupings that Rowley had constructed for Alexander, Loomis, director of the Academy of Sciences, hired Rowley to create similar pieces for that museum.

27. January 6, 1911, Alexander to Grinnell, BANC mss 67/121c.

28. By 1930 the habitat groups had begun to deteriorate and the university agreed to offer them, free of charge, to any person or institution willing to pay for the cost of their removal. The sheep group was too badly damaged to salvage, but the San Diego Museum of Natural History acquired the sea lion groupings.

9. AN UNUSUAL COLLABORATION

1. J. R. Griesemer and E. M. Gerson, "Collaboration in the Museum of Vertebrate Zoology," *Journal of the History of Biology* 26 (1993):185–203; S. L. Star and J. R. Griesemer, "Institutional ecology, 'translations,' and boundary objects: amateurs and professionals in Berkeley's Museum of Vertebrate Zoology, 1907–1939," *Social Studies in Science* 19 (1989):387–420; and M. P. Winsor, *Reading the Shape of Nature: Comparative Zoology at the Agassiz Museum* (Chicago: University of Chicago Press, 1991).

2. August 30, 1910, Grinnell to Alexander, BANC mss C-B1003. The Tring Zoological Museum in Hertfordshire, England was established by the Hon. N. Charles Rothchild of Tring and eventually became a branch of the British Museum of Natural History. *Pure science* was the term used at that time to refer to basic research.

3. August 17, 1899, Alexander to Beckwith, Letters to Martha. Direct knowledge about Alexander's investment habits come from Diary entries. Alexander also subscribed to the *Wall Street Journal*.

4. February 12, 1929, ibid.

5. April 4, 1921, memorandum written by Grinnell, BANC mss C-B1003; January 16, 1930, Alexander to Grinnell, CU-120; and November 20, 1931, Alexander to UC Comptroller L. A. Nichols, BANC mss 67/121c.

6. January 31, 1947, Alexander to Miller, CU-120.

7. March 18, 1920, Alexander to Grinnell, BANC mss 67/121c. A "Statement of Proposed Appropriations and Expenditures for 1909," dated early March, summarizes these details, while other letters between Alexander and Grinnell written throughout the year substantiate this agreement (BANC mss 67/121c and C-B1003).

8. February 16, 1908, Grinnell to Alexander, BANC mss C-B1003; February 6, 1908, ibid.

9. October 29, 1907, ibid.; December 6, 1907, ibid.

10. February 28, 1908, ibid.; March 27, 1911, ibid.

11. C. H. Merriam, "The Museum of Vertebrate Zoology of the University of California," *Science* 60 (1914):703.

12. January 15, 1922, Alexander to Grinnell, CU-120.

13. March 29, 1908, Alexander to Grinnell, BANC mss 67/121c, commenting on J. Grinnell, "A new museum," *The Condor* 10 (1908):95. A year later, Grinnell announced construction of the new building with a notice in the journal *Science* (29 [1909]:254).

14. P. G. Abir-Am and D. Outram, eds., *Uneasy Careers and Intimate Lives: Women in Science, 1789–1979* (New Brunswick: Rutgers University Press, 1989); and M. W. Rossiter, *Women Scientists in America: Struggles and Strategies to 1940* (Baltimore: Johns Hopkins University Press, 1982).

15. References to her participation in such organizations are contained in letters to Beckwith and to Grinnell, as well as in the membership lists of the various scientific societies listed. See also May 14 and 21, 1920, assistant secretary to Alexander, American Museum of Natural History Special Collections (hereafter referred to as AMNH). On at least one occasion she attended the annual meeting of the Society for Vertebrate Paleontology.

16. August 7, 1925, memorandum written by Grinnell, BANC mss 67/121c.

17. A. M. Alexander, "A Further Chronicle of the Passenger Pigeon and of methods employed in hunting it," *The Condor* 29 (1927):273; A. M. Alexander, "'Control, not extermination,' of Cynomys ludovicianus arizonensis," *Journal of Mammalogy* 13 (1932):302.

18. March 23, 1911, Alexander to Grinnell, BANC mss 67/121c.

19. Not only did Alexander make this stipulation upon Grinnell's appointment as director, but she reiterated it in her letter to the regents when she endowed the museum more than a decade later (November 15, 1919, BANC mss C-B1003; and December 5, 1919, Alexander to the regents, MVZ).

20. October 28 and 30, 1923, Grinnell to Alexander, BANC mss C-B1003.

21. October 30, 1923, ibid.

22. April 1, 1908, Alexander to Grinnell, BANC mss 67/121c.

10. LOUISE AND PRINCE WILLIAM SOUND

1. November 2, 1907 and November 13, 1908, Grinnell to Alexander, BANC mss C-B1003; undated, Alexander to Grinnell (H. W. Grinnell, "Biographical sketch of Annie Montague Alexander," ms, MVZ).

2. April 16, 1908, Grinnell to Alexander, BANC mss C-B1003.
3. April 16, 1908, Alexander to Grinnell, BANC mss 67/121c. "Mrs. Bailey" was Florence Merriam Bailey, a respected ornithologist, sister of C. Hart Merriam and wife of the Smithsonian mammalogist Vernon Bailey.
4. April 22, 1908, Grinnell to Alexander, BANC mss C-B1003.
5. April 25, 1908, Alexander to Grinnell, BANC mss 67/121c.
6. May 2, 1908, ibid.
7. May 5, 1908, Grinnell to Alexander, BANC mss C-B1003.
8. See D. W. Ryder, *Men of Rope* (San Francisco: Historical Publications, 1954) for a history of the Tubbs Cordage Co.; A. Arnold, *Suisun Marsh History: Hunting and Saving a Wetland* (Marina, Calif.: Monterey Pacific Publishing, 1996) offers a perspective on the Cordelia Duck Club.
9. Background on Louise Kellogg and Kellogg family history from Alice Q. Howard, personal communication.
10. Details of the 1908 Alaska expedition are from the field notes of Alexander, Kellogg, Dixon, Hasselborg, and Heller, MVZ; and from the Alexander and Kellogg Diary, AQH. See also J. Grinnell, "Birds of the 1908 Alexander Alaska Expedition with a note on the avifaunal relationships of the Prince William Sound District," *University of California Publications in Zoology* 5 (1910): 361–428.
11. April 1, 1908, Alexander to Grinnell, BANC mss 67/121c.
12. May 29, 1908, Heller to Grinnell, MVZ.
13. Hasselborg had left the two cubs in Yakutat. Alexander shipped the animals to San Francisco and cabled John Merriam, asking that he make arrangements for their transfer to the zoo in Golden Gate Park (see also T. I. Storer and L. P. Tevis, Jr., *California Grizzly* [Berkeley: University of California Press, 1955]).
14. June 14, 1908, Dixon to Grinnell, MVZ.
15. July 18, 1908, Heller to Grinnell, ibid.
16. August 7, 1908, ibid.
17. August 6, 1908, Alexander to Beckwith, Letters to Martha.

11. SUPPORT FOR PALEONTOLOGY

1. October 31, 1905, Alexander to Wheeler, UCMP. Alexander also outlined the nature of her proposal in a letter to John Merriam (October 23, 1905, BANC mss C-B970).
2. October 31, 1905, Alexander to Wheeler, UCMP.
3. Based on the Consumer Price Index, which did not begin until 1913, at which date $1.00 was the equivalent of $17.27 in January 2000.
4. With Alexander's support, the Department of Paleontology at Berkeley began to amass what would become the second largest collection of fossil specimens from La Brea in the world.
5. Details of Alexander's 1909 paleontological expedition can be found in her field notes, MVZ; and in Kellogg's diary entries, AQH.

6. August 4, 1909, Alexander to Beckwith, Letters to Martha.
7. Ibid.
8. December 29, 1909, ibid.
9. R. A. Stirton, "The Role of Paleontology in the University of California: Honoring the Twenty-Fifth Presidential Year of President Robert Gordon Sproul" (Berkeley: University of California, 1955).
10. December 13, 1910, Alexander to Wheeler, BANC mss C-B970.
11. December 23, 1910, Wheeler to Alexander, UCMP.
12. July 4, 1914, Merriam to Alexander, ibid.
13. December 8, 1915, Alexander to Wheeler, ibid.

12. HEARST, SATHER, FLOOD

1. February 7, 1901, Alexander to Beckwith, Letters to Martha.
2. V. A. Stadtman, *The University of California 1868–1968: A Centennial Publication of the University of California* (New York: McGraw-Hill, 1970).
3. Wheeler was a Greek scholar and former professor of philology at Cornell University. He had been offered at least six presidencies prior to the one that he received from the University of California in 1899. He accepted the UC post, believing that the fledgling institution was the best and most promising university west of the Mississippi (Benjamin Ide Wheeler papers, Bancroft Library, University of California, Berkeley, BANC mss C-B1044).
4. Stadtman, *University of California;* S. Birmingham, *California Rich: The Lives, the Times, the Scandals and the Fortunes of the Men & Women who Made & Kept California's Wealth* (New York: Simon and Schuster, 1980); and J. Ruyle, ed., "Ladies Blue and Gold," *Chronicle of the University of California* 1 (1998):1–10.
5. January 6, 1911, Alexander to Grinnell, BANC mss 67/121c.
6. March 26, 1909, Alexander to Beckwith, Letters to Martha. As pressure grew within the university for Grinnell to become involved in teaching, Alexander became increasingly insistent that the university assume responsibility for salary increases of the museum staff.
7. Records of Donation, vol. 1, University Archives, University of California, Berkeley; and "President Announces Beneficent Gifts," *California Alumni Weekly* 4 (1912):1–2.
8. P. G. Clark, "Big Game Bagged by a California Girl in Africa," *San Francisco Chronicle*, September 3, 1905, 5–6; August 21, 1905, Alexander to Merriam, BANC mss C-B970.
9. M. C. Alexander, *William Patterson Alexander in Kentucky, the Marquesas, Hawaii* (Honolulu, 1934); and Samuel's obituaries.
10. March 16, 1919, Alexander to Beckwith, Letters to Martha; and December 11, 1919 through December 23, 1924, Alexander correspondence, Vassar College archives, Poughkeepsie, N.Y.; September 14, 1919, Alexander to Beckwith, Letters to Martha; M. Beckwith, *The Kumulipo, a Hawaiian Creation Chant*, trans. and ed. Martha Beckwith (Chicago: University of Chicago Press, 1951).

11. October 23, 1901, Alexander to Beckwith, Letters to Martha; November 10, 1901, ibid.

12. January 13, 1913; July 29, 1915; August 6, 1915; November 13, 1915, Alexander to Grinnell, BANC mss 67/121c. See also MVZ and UCMP correspondence, e.g., Benson and Welles.

13. INNISFAIL RANCH

1. L. Kellogg, "Innisfail (Beautiful Isle) Ranch," a three-page history of the farm written by Kellogg after Alexander's death (AQH). The cousin referred to was never specified.

2. Personal communication from Alice Howard.

3. Photocopy of manuscript page, California Historical Society, San Francisco, AQH; and A. Arnold, *Suisun Marsh History: Hunting and Saving a Wetland* (Marina, Calif.: Monterey Pacific Publishing, 1996); December 19, 1910, Alexander to Beckwith, Letters to Martha.

4. J. Skinner, "An Historical Review of the Fish and Wildlife Resources of the San Francisco Bay Area," Water Projects Branch Report no. 1, California Department of Fish and Game, Sacramento, 1962; December 25, 1911, Alexander to Beckwith, Letters to Martha.

5. April 8, 1908, Grinnell to Alexander, BANC mss 67/121c.

6. J. Grinnell, "Description of river otters in California, with description of a new subspecies," *University of California Publications in Zoology* 12 (1914):305–10; the specimen was cataloged as MVZ no. 20775. In 1925 Grinnell presented the Smithsonian Institution with five individuals representing three species of small mammals that Alexander and Kellogg had trapped on the island after recalling that the National Museum lacked representative specimens of these taxa (September 1, 1925, Grinnell to Alexander Wetmore; September 18, 1925, Wetmore to Alexander, Smithsonian Institution archives, accn. no. 88707); December 3, 1915, Grinnell to Alexander, BANC mss 67/121c.

7. J. Frost, *A Brief Pictorial History of Grizzly Island* (San Francisco: Arthur V. Fay, The Trade Pressroom, 1978).

8. The origin of the ranch name Innisfail is lost. It is assumed that Alexander and Kellogg chose it themselves as a translation of "beautiful isle." Innisfail, Canada states that its name is a derivative of the Gaelic *InnisVail*, meaning "isle of destiny," but there is no presumed relationship between the two (Jack Schafer, pers. comm.).

9. J. Grinnell, H. C. Bryant, and T. I. Storer, *The Game Birds of California* (Berkeley: University of California Press, 1918).

10. The tractor that Alexander purchased was the forerunner of the now-ubiquitous caterpillar tractor developed by Best and Holt. See also R. L. Adams and C. M. Haring, "Grisly [sic] Island Project," College of Agriculture, University of California, Berkeley, 1921.

11. February 18, 1912, Dixon to Hasselborg, Alaska State Library.

12. December 18, 1918, ibid.

13. January 30, 1918, ibid.
14. October 2, 1918, Alexander to Grinnell, BANC mss C-B1003.
15. August 30, 1918, ibid.; October 18, 1919, Alexander to M. C. Alexander, Mission Children's Society; and November 6, 1919, March 20, 1922, Dixon to Hasselborg, Alaska State Library.
16. November 11, 1921, Alexander to Beckwith, Letters to Martha.
17. May 23, 1922, ibid.
18. Adams and Haring, "Grisly Island Project."
19. Schmidt Lithography, San Francisco.
20. February 15, 1927, Dixon to Hasselborg, Alaska State Library.
21. March 30, 1919, Alexander to Beckwith, Letters to Martha.
22. Alexander's cost for the approaches totaled $2,267.40 (Diary). See also *Solano Republican*, September 17 and 21, 1920.

April 5, 1920, Solano County Road Minutes, Solano County Courthouse, Fairfield; the minutes' first mention of the road occurs on May 3, 1915; see also deeds relating to Road 578, Solano County Administrative Office.

14. VANCOUVER ISLAND AND THE TRINITY ALPS

1. June 11, 1910, Grinnell to Alexander, BANC mss 67/121c. Details of the 1910 Vancouver expedition are taken from the field notes of Alexander, Kellogg, and Swarth, MVZ. See also H. S. Swarth, "Report on a collection of birds and mammals from Vancouver Island," *University of California Publications in Zoology* 10 (1912):1–124.
2. May 12, 1910, Alexander to Grinnell, BANC mss C-B1003; May 29, 1910, ibid.
3. June 11, 1910, Grinnell to Alexander, BANC mss 67/121c.
4. May 29, 1910, Alexander to Grinnell, BANC mss C-B1003. An aux is a metal sleeve that fits in a shotgun barrel, allowing the hunter to fire a smaller shot (e.g., in a .38-caliber shell) than the gun ordinarily permits. The smaller gauge was necessary to collect songbirds without obliterating them.
5. June 27, 1910, Swarth to Grinnell, MVZ.
6. June 29, 1910, ibid.
7. The vole, collected on Coronation Island, was renamed *Microtus coronarius* (H. S. Swarth, "Birds and mammals of the 1909 Alexander Alaska Expedition," *University of California Publications in Zoology* 7 [1911]:9–172). It is assumed that specimens subsequently named for Alexander (see appendix) were done without her consent or any concern for her disapproval.
8. H.S. Swarth, "Report on a collection of birds and mammals from Vancouver Island," *University of California Publications in Zoology* 10 (1912): 1–124.
9. L. Kellogg, "Rodent fauna of the late Tertiary beds at Virgin Valley and Thousand Creek, Nevada," *University of California Publications, Bulletin of the Department of Geology* 5 (1910):411–37; and L. Kellogg, "A fossil beaver from Kettleman Hills, California," *University of California Publications, Bulletin of the Department of Geology* 6 (1911):401–2.

10. December 11, 1910, Alexander to Grinnell, BANC mss C-B1003.
11. December 14, 1910, Grinnell to Alexander, BANC mss 67/121c.
12. February 10, 1911, Alexander field notes, MVZ. Details of the 1911 Trinity Alps expedition are taken from the field notes of Alexander and Kellogg. See also J. Grinnell, "An analysis of the vertebrate fauna of the Trinity region of northern California," *University of California Publications in Zoology* 12 (1916):399–410; and L. Kellogg, "Report upon mammals and birds found in portions of Trinity, Siskiyou and Shasta counties, with description of a new *Dipodomys*," *University of California Publications in Zoology* 12 (1916): 335–98.
13. February 18, 1911, Alexander field notes, MVZ; February 22, 1911, ibid.
14. February 21, 1911, Alexander to Grinnell, BANC mss C-B1003.
15. July 19, 1911, Alexander field notes, MVZ.
16. June 23, 1911, Alexander to Grinnell, BANC mss C-B1003; July 17, 1911, Alexander to Beckwith, Letters to Martha.
17. June 23, 1911, Alexander to Grinnell, BANC mss C-B1003.
18. Grinnell, "An analysis of the vertebrate fauna of the Trinity region of northern California."
19. Kellogg, "Report upon mammals and birds." See B. R. Stein, "Women in mammalogy: the early years," *Journal of Mammalogy* 77 (1996):629–41, for relevance of this publication in another context. The three additional publications authored by Kellogg between 1910 and 1916 are "A collection of winter birds from Trinity and Shasta counties, California," *The Condor* 13 (1911): 118–21; "Pleistocene rodents of California," *University of California Publications, Bulletin of the Department of Geology* 7 (1912):151–68; and "*Aplodontia chryseola*, a new mountain beaver from the Trinity region of Northern California," *University of California Publications in Zoology* 12 (1914):295–96.
20. August 29, 1948, Alexander to Hall, University Archives, the University of Kansas, Lawrence (hereafter, KU).

15. THE TEAM OF ALEXANDER AND KELLOGG

1. August 26, 1910, Alexander to Beckwith, Letters to Martha.
2. August 30, 1918, ibid.; October 2, 1918, Alexander to Grinnell, BANC mss C-B1003.
3. Personal communication from Henry Fitch.
4. June 17, 1932, Grinnell to Alexander, BANC mss 67/121c. Seth Benson was a graduate student in the museum who became associate curator of mammals.
5. J. Grinnell, "The methods and uses of a research museum," *Popular Science Monthly* 77 (1910):163–69.
6. Records of Donation, vol. 1, University Archives, University of California, Berkeley, and MVZ correspondence; January 25, 1925, Alexander to Grinnell, BANC mss C-B1003.
7. Ward Russell oral history, MVZ.

8. May 22, 1912, Grinnell to Alexander, CU-120; and November 14, 1931, Grinnell to Alexander, BANC mss 67/121c.

J. Grinnell, "Geography and evolution in the pocket gophers of California," *University of California Chronicle*, 1926, 247–62; J. Grinnell, "A new lake-side pocket gopher from south-central California," *University of California Publications in Zoology* 38 (1932):405–10; J. Grinnell, "Differentiation in pocket gophers of the *Thomomys bottae* group in northern California and southern Oregon," *University of California Publications in Zoology* 40 (1935):403–6; and J. Grinnell and J. E. Hill, "Pocket gophers (*Thomomys*) of the lower Colorado Valley," *Journal of Mammalogy* 17 (1936):1–10.

April 25, 1911, Alexander to Grinnell, BANC mss C-B1003. Alexander was referring to *Thomomys leucodon navus* Merriam 1901. *Proceedings of the Biological Society of Washington* 14:112 (= *T. umbrinus navus* = *T. u. nanus*).

9. J. Grinnell and T. I. Storer, "Animal life as an asset of national parks," *Science* 44 (1916):375–80; J. Grinnell, "Recommendations concerning the treatment of large mammals in Yosemite National Park," *Journal of Mammalogy* 9 (1928):76; A. Runte, *Yosemite: The Embattled Wilderness* (Lincoln: University of Nebraska Press, 1990); A. Runte, "Joseph Grinnell and Yosemite: Rediscovering the Legacy of a California Conservationist," in *Yosemite and Sequoia: A Century of California National Parks*, ed. R. J. Orsi, A. Runte, and M. Smith-Baranzini (Berkeley: University of California Press, 1993), 85–95; C. E. Bock, "The role of ornithology in conservation of the American West," *The Condor* 99 (1997):1–6; and M. V. Barrow, Jr., *A Passion for Birds: American Ornithology after Audubon* (Princeton: Princeton University Press, 1998).

10. August 1, 1928, Alexander to Beckwith, Letters to Martha.

11. M. Grant, "Saving the redwoods: an account of the movement during 1919 to preserve the redwoods of California," *New York Zoological Society Bulletin* (1919):90–118; and J. C. Merriam, "The task ahead of the Save-the-Redwoods League: a message to members from the League's president, Dr. John C. Merriam of Washington, D.C.," Save-the-Redwoods League, Berkeley, 1934 [?].

16. FROM "A FRIEND OF THE UNIVERSITY"

1. February 21, 1931, Alexander to Grinnell, BANC mss 67/121c.
2. June 25, 1914, Grinnell to Alexander, BANC mss C-B1003.
3. March 27, 1911, ibid. See November 3, 1910, Alexander to Grinnell, BANC mss 67/121c with respect to Alexander's chagrin with the university administration.
4. J. Grinnell, J. S. Dixon, and J. M. Linsdale, *Fur-Bearing Mammals of California: Their Natural History, Systematic Status, and Relations to Man* (Berkeley: University of California Press, 1937).
5. December 11, 1912, Alexander to Grinnell, BANC mss 67/121c.
6. October 22, 1915, ibid.

7. October 6, 1915, ibid.

8. November 15, 1919, Alexander to UC regents, CU-120; December 5, 1919, Alexander to UC regents, MVZ; and April 4, 1921, Grinnell memorandum, BANC mss C-B1003.

9. "Gift to the Museum of Vertebrate Zoology of the University of California," *Science* 50 (1919):585; December 15, 1919, Grinnell to President Barrows, and December 16, 1919, Barrows to Grinnell, MVZ. Included in the collection were twenty-seven type specimens, numerous individuals representing geographic records, and specimens of at least three species of birds that had already gone extinct. In true Grinnellian fashion, there were also several large series of birds that dramatically demonstrated geographic variation.

10. September 15, 1922, Alexander to Grinnell, BANC mss 67/121c; October 17, 1922, Grinnell to Alexander, BANC mss C-B1003; October 31, 1922, Alexander to Sproul, and November 16, 1922, Sproul to Grinnell, MVZ; September 15, 1922, Alexander to Grinnell, BANC mss 67/121c.

11. J. Grinnell and T. I. Storer, *Animal Life in the Yosemite: An Account of the Mammals, Birds, Reptiles, and Amphibians in a Cross-Section of the Sierra Nevada* (Berkeley: University of California Press, 1924).

12. January 25, 1925, Alexander to Grinnell, BANC mss 67/121c; January 3, 1934, Grinnell to Alexander, BANC mss C-B1003.

13. December 10, 1910, Grinnell to Alexander, BANC mss C-B1003. The money Alexander provided in this instance was used to purchase bird specimens obtained by a Mr. Kleinschmidt in Siberia and Alaska.

14. April 4, 1921, memorandum written by Grinnell, BANC mss C-B1003; January 16, 1930, Alexander to Grinnell, BANC CU-120; and November 20, 1931, Alexander to UC Comptroller L. A. Nichols, BANC mss 67/121c.

15. June 1, 1929, Grinnell to UC President Sproul, MVZ; January 16, 1930, Alexander to Grinnell, CU-120; November 19, 1930, Grinnell to Alexander, BANC mss C-B1003; and November 20, 1931, Alexander to UC Comptroller Nichols, CU-120. See December 17, 1920, Sproul to Grinnell, MVZ, for documentation that original endowment was to be pooled with the university's general fund.

16. July 13, 1932, Grinnell to Alexander, BANC mss C-B1003.

17. July 22, 1932, Alexander to Grinnell, CU-120.

18. April 19, 1933, Grinnell memorandum, BANC mss C-B1003.

19. September 9, 1935, Alexander to UC regents, BANC mss 67/121c.

20. By the time that Alexander proposed a faunal survey of the Colorado River area in 1934, the museum's bird collection was the seventh largest in the country (T. I. Storer, "The California Museum of Vertebrate Zoology," *California Alumni Monthly* 15 [1922]:257–59). The MVZ's amphibian and reptile collection did not begin to grow appreciably until 1945 when Robert Stebbins was hired as curator.

21. October 15, 1909, C. H. Merriam to Alexander; July 1, 1910, C. H. Merriam to Grinnell, MVZ; C. H. Merriam, "Descriptions of thirty apparently new grizzly and brown bears from North America," *Proceedings of the Biological*

Society of Washington 27 (1914):173–96; and C. H. Merriam, "The Museum of Vertebrate Zoology of the University of California," *Science* 60 (1914):703.

22. Undated, 1908, Alexander to Beckwith, Letters to Martha; and W. H. Osgood, "Clinton Hart Merriam—1855–1942," *Journal of Mammalogy* 24 (1943): 421–57. H. W. Grinnell, "Edmund Heller: 1875–1939," *Journal of Mammalogy* 28 (1947):209–18, states that Akeley recommended Heller to Roosevelt.

23. December 7, 1909, Grinnell to Alexander, BANC mss C-B1003.

24. Although Roosevelt's Charter Day speech (CU-5) clearly states that he is offering the entire elephant to the university (skin, skull, and skeleton), only the tanned hide was ever accessioned into the MVZ collections (accn. no. 242). It was later deaccessioned because of its poor condition. The hunt is described in T. Roosevelt, *African Game Trails: An Account of the African Wanderings of an American Hunter-Naturalist* (New York: Charles Scribner's Sons, 1923), 306–9.

25. March 27, 1911, Grinnell to Alexander, BANC mss C-B1003. See also February 15, 1911, Grinnell to Wheeler, MVZ.

26. T. Roosevelt and E. Heller, *Life-Histories of African Game Mammals* (New York: Charles Scribner's Sons, 1915).

27. April 22, 1939, C. H. Merriam to Grinnell, MVZ.

17. FOUNDING A MUSEUM OF PALEONTOLOGY

1. January 1, 1918, Merriam to Alexander, UCMP; November 10, 1917, Alexander to Beckwith, Letters to Martha. Alexander borrowed this quote from an article that she had read in *Science* (46 [1917]:331) and recommended to Grinnell. It stresses that for universities to fulfill their highest mission, i.e., research, investigators should not be burdened with administrative duties or the demands of public service.

2. July 27, 1920, Alexander to Grinnell, CU-120.

3. October 12, 1920, Alexander to Beckwith, Letters to Martha. Browning wrote the poem "The Lost Leader" upon learning of Wordsworth's appointment as poet laureate. Browning felt similarly betrayed, thinking that the award reflected the older poet's abandonment of political liberalism.

4. December 6, 1920, Alexander to UC President Barrows; and December 18 and 23, 1920, Barrows to Alexander, UCMP.

5. V. A. Stadtman, *The University of California 1868–1968: A Centennial Publication of the University of California* (New York: McGraw-Hill, 1970).

6. December 23, 1920, UC President Barrows to Alexander, UCMP.

7. J. Grinnell, *Joseph Grinnell's Philosophy of Nature: Selected Writings of a Western Naturalist* (Berkeley: University of California Press, 1943); and March 22, 1921, Alexander to Grinnell, BANC mss 67/121c.

Merriam's paper ("The research spirit in the life of an average man," reprinted in *Science* 52 [1920]:473–78), was originally delivered as an address marking his retirement as president of the Pacific Division of the American Association for the Advancement of Science, Seattle, Washington, June 17, 1920.

November 29, 1920, Alexander to Beckwith, Letters to Martha.

8. March 30, 1921, Grinnell to Alexander, BANC mss C-B1003.

9. February 7, 1921, Alexander to Barrows ("A Proposal for the Establishment of a Museum of Paleontology"); March 11, 1921, Bruce Clark to Alexander; March 14, 1921, Barrows to Alexander; and March 14, 1921, Alexander to Clark, UCMP.

10. J. T. Gregory, "UCMP History: Bruce L. Clark, first director of the UCMP," *UCMP News*, March 1996, 2; and J. T. Gregory, "UCMP History: Charles Lewis Camp, third director of UCMP," *UCMP News*, September 1996, 2. Alexander erroneously assumed that Chester Stock, a former student of Merriam's and later his colleague, was—along with all other paleontology department faculty—also a member of the UCMP. This assumption was corrected for her by Bruce Clark in a letter dated February 29, 1924 (AQH). For a brief synopsis of the history of paleontology at Cal see C. L. Camp, "Paleontology," in *The Centennial Record of the University of California*, ed. V. A. Stadtman and the centennial publications staff (Berkeley: University of California, 1968), 95–96.

11. November 16, 1921, Grinnell to Alexander, BANC mss C-B1003.

12. July 12, 1924, Alexander to Beckwith, Letters to Martha. Alexander raised this same issue with Alden Miller following his appointment as the MVZ's director. In this instance, however, her accusations would prove groundless as the circumstances stemmed from a misunderstanding on her part. Nonetheless, Miller went to great lengths to point out that the MVZ byline had become a cherished symbol of one's affiliation with a well-known and well-respected institution.

13. J. T. Gregory, "UCMP History: William Diller Matthew, a new start," *UCMP News*, May 1996, 2; W. K. Gregory, "William Diller Matthew," *Natural History* 30 (1930):664–66; and W. K. Gregory, "William Diller Matthew, Paleontologist (1871–1930)," *Science* 72 (1930):642–45.

14. June 23, 1927, Alexander to Beckwith, Letters to Martha. Directorship of the UCMP had been offered previously to William K. Gregory, a noted paleontologist also at the American Museum, but he had declined (see May 16, 1926, Alexander to Grinnell, BANC mss 67/121c); April 9, 1927, Alexander to Campbell, UCMP.

15. G. A. Pettitt, *Twenty-Eight Years in the Life of a University President* (Berkeley: University of California, 1966).

16. January 28, 1931, Grinnell to Sproul, MVZ.

17. May 6, 1931, Alexander to Sproul, UCMP.

18. See C. D. Wagner, "Paleontology at the University of California: the history of Bacon Hall," *Journal of the West* 8 (1969):169–82.

19. November 27, 1930, Alexander to Beckwith, Letters to Martha.

20. December 16, 1933, Alexander to Sproul, UCMP. Alexander suggested that the money be held in the university's general endowment pool, with interest earned to be applied to the salary of the UCMP's director. She would continue her ongoing support of the museum, merely deducting the amount of interest earned by the endowment from her monthly contribution.

21. March 21, 1934, Alexander to Sproul; undated memo written by Alexander, ibid. (Its context places it in 1939.)

22. September 21, 1934, Grinnell to Alexander, BANC mss C-B1003; and April 16, 1934, Paul Dorman to Sproul, UCMP.

23. September 27, 1934, Alexander to Sproul; and November 23, 1934, UC Finance Committee memorandum, ibid. Alexander requested that this gift (1,000 shares of common stock in California Packing Corporation, a Hawaiian producer of canned pineapple) be consolidated with the original $30,000 and administered under the same terms and conditions. With the endowment at $90,000, she hoped that the interest earned would equal her $1,000 monthly contribution to the museum (and would donate whatever money was necessary to bring the amount up to $12,000 annually). Any surplus from the endowment was to be applied to the museum's general support.

24. April 25, 1938, Alexander to Sproul; and April 19, 1938, Sproul to Alexander, BANC mss 67/121c.

25. May 17, 1938, Camp to Sproul, UCMP. Joseph Gregory, professor emeritus and a paleontology graduate student in the 1930s, was not aware that students were used as pawns until many years later (pers. comm.).

26. August 15, 1938, Alexander to Camp, ibid.

27. Memorandum, ibid. Comptroller's records were not consulted for verification.

28. May 1, 1939, Clark to Camp, ibid.

29. May 1, 1939, Camp to Sproul, ibid.

30. In her letter, Alexander charged that Merriam had used his position as head of the Carnegie Institute to secure Sproul the presidency of the university and that Sproul, in turn, had compensated Merriam by appointing Chaney as department chair.

31. May 26, 1947, Alexander to Alden Miller, CU-120.

32. April 29, 1945, Alexander to Camp; July 1, 1948, Alexander to UC regents, UCMP. Under the terms of this agreement, Alexander directed the university to combine the gift (2,000 shares of Pacific Lighting, 1,000 shares each of Bralone Mines and Honolulu Oil stock) with the UCMP endowment moneys that she had previously donated, remove the endowment from the university's general pool to a separate account, and, if any shares were ever sold, reinvest the amount gained in common stock earning a high rate of return, ideally at least 5 percent. In the trust agreement drawn up, she promised to supplement the endowment income by $1,200 annually for the duration of her life.

33. July 1, 1948, Alexander to UC regents, ibid. Citing the terms of her 1945 endowment agreement (which she wished to keep unchanged), Alexander transferred almost 1,400 shares of common and preferred stocks, worth approximately $100,000, to the university in July 1948.

34. December 2, 1947, Miller to Alexander, CU-120.

35. May 26, 1947, Alexander to Miller, ibid.

18. A RESTLESS DECADE

1. February 7, 1917, Alexander to Beckwith, Letters to Martha.
2. September 17, 1918, ibid.
3. October 19, 1919, ibid.
4. July 7, 1918, ibid.
5. August 30, 1918, ibid.
6. August 28, 1919, ibid.
7. October 8, 1919, ibid.
8. Details of the trip are taken from Alexander's field notes, UCMP.
9. June 28, 1920, ibid.
10. August 17, 1899, Alexander to Beckwith, Letters to Martha.
11. November 24, 1902, ibid.
12. November 5, 1940, Diary.
13. May 24, 1921, Alexander to Beckwith, Letters to Martha.
14. Details of the trip are taken from Alexander's field notes, UCMP. After receiving her degree at Cal, Davidson worked at UCLA until she married, at which point she apparently abandoned paleontological research.
15. March 5 and 6, 1922, ibid.
16. March 14, 1922, ibid.
17. April 22, 1922, Alexander to Beckwith, Letters to Martha.
18. July 24, 1924, Osborn to Alexander, AMNH. Alexander's July 14, 1924 letter to Osborn referenced within is not available.
19. April 22, 1923, Alexander to Beckwith, Letters to Martha.

19. EUROPE, 1923

1. October 18, 1919, Alexander to Mary Charlotte Alexander, Mission Children's Society.
2. H. F. Osborn, *Men of the Old Stone Age: Their Environment, Life and Art* (New York: Charles Scribner's Sons, 1915); and H. F. Osborn, *The Origin and Evolution of Life: On the Theory of Action, Reaction and Interaction of Energy* (New York: Charles Scribner's Sons, 1917).
3. Details of the journey were gained from Kellogg's trip diary, AQH.
4. April 26, 1923, Clark to Osborn; May 2, 1923, Osborn to Clark; May 2, 1923 Osborn to Nelson; May 2, 1923, Osborn to Alexander; May 7, 1923, Osborn to all relevant European colleagues; May 12, 1923, Clark to Osborn; May 13, 1926, Alexander to Osborn, June 18, 1923, Nelson to Osborn; June 25, 1923, Alexander to Osborn, AMNH; and May 12, 1923, Clark to Alexander, UCMP.
5. Emergency Preparation Fund 1924–25 donor list, AMNH; and November 29, 1926, Andrews to Alexander, UCMP. UC paleontologist Ralph Chaney participated in one of Andrews's Asiatic expeditions, providing another connection between the two institutions.
6. November 29, 1926, Andrews to Alexander, UCMP.
7. July 3, 1923, Kellogg's trip diary, AQH.

8. July 5, 1923, ibid.

9. In 1953 development of fluorine tests for bone aging showed that the dates of the animal teeth and bones supposedly found with the Piltdown skull fragments were not at all similar to them, and the specimen was pronounced a hoax.

10. July 6, 1923, Kellogg's trip diary, AQH.

11. For a picture of this unusual vehicle see: http://www.ccc-uk.demon.co.uk/launches.htm and click on "Pre-1930."

12. July 21, 1923, Kellogg's trip diary, AQH.

13. July 25, 1923, ibid.

14. July 31, 1923, ibid.

15. E. Peyrony, *Les Eyzies and the Vézère Valley: An Illustrated Guide for Scholars and Tourists*, trans. P. Smith (Montignac: Imprimerie de la Vézère, 1959).

20. THE TEMPLE TOUR

1. February 27, 1924, Alexander to Stock, AQH.

2. The discovery of Tut's tomb was publicly announced on February 16, 1923. See A. C. Brackman, *The Search for the Gold of Tutankhamen* (New York: Mason/Charter, 1976); P. Cone, ed., *Wonderful Things: The Discovery of Tutankhamun's Tomb* (New York: Metropolitan Museum of Art, 1976); and N. Reeves, *The Complete Tutankhamum: The King, the Tomb, the Royal Treasure* (London: Thames and Hudson, 1990).

3. August 21, 1924, Alexander to Temple Tours, AQH. Details of the tour are taken from Kellogg's trip diary, ibid.; supplemented by letters from Alexander to Grinnell, BANC, and to Edna Wemple, AQH.

4. January 19, 1924, Alexander to Wemple, AQH.

5. January 24, 1924, Alexander to Herrick Bros., ibid.

6. Jerboas are Old World jumping rodents with elongated hind limbs and long tails, generally similar in form to kangaroo rats that inhabit the deserts of the American Southwest.

7. February 27, 1924, Alexander to Stock, AQH.

8. February 7, 1924, Kellogg's trip diary, ibid.

9. February 24, 1924, Alexander to Wemple, ibid. Kellogg's diary has no mention of the visit.

10. Ibid.; February 16, 1924, Kellogg's trip diary, ibid.; February 26, 1924, ibid.

11. February 24, 1924, Alexander to Wemple, ibid.

12. February 22, 1924, Kellogg's trip diary, ibid.

13. May 20, 1924, Alexander to Grinnell, BANC mss 67/121c.

14. Ibid.

21. THE "AMOEBA TREATMENT"

1. October 12, 1920, Alexander to Beckwith, Letters to Martha.

2. September 20, 1925, Alexander to Mary Beckwith, ibid. Piedmont is the

small community adjacent to Oakland where Pattie lived after Samuel's death. She lived in a large house across the street from Alexander's brother, Wallace, who had settled in Piedmont after his marriage. Kellogg's mother died in the spring of 1922.

3. October 25, 1924, Mary Charlotte Alexander to her family, Mission Children's Society. The trip took place October 9–November 20, 1924.
4. October 25, 1924, ibid.
5. November 7, 1924, ibid.
6. January 25, 1925, Alexander to Grinnell, BANC mss 67/121c.
7. September 8, 1925, Grinnell to Campbell, MVZ.
8. April 25, 1925, Grinnell to Alexander, BANC mss C-B1003.
9. June 6, 1925, Alexander to Grinnell, BANC mss 67/121c.
10. September 28, 1925, Diary.
11. May 16, 1926, Alexander to Grinnell, BANC mss 67/121c.
12. May 25, 1926, Alexander to Beckwith, Letters to Martha.
13. August 10, 1926, ibid.
14. November 6, 1926, Alexander to Grinnell, BANC mss 67/121c; November 15, 1926, Grinnell to Alexander, BANC mss C-B1003.
15. December 4, 1926, Alexander to Beckwith, Letters to Martha; December 28, 1926, ibid.

22. FIELDWORK—THE LATER YEARS

1. Use of the polymerase chain reaction (PCR) now allows researchers to extract and analyze DNA from many museum specimens in order to address questions concerning the population genetics, speciation, and historical evolution of organismal clades.
2. October 21, 1931, Alexander and Kellogg to Edna Wemple McDonald, AQH.
3. September 15, 1931, ibid. This statement inextricably contradicts the fact that Alexander doubled the size of the MVZ's endowment during the depression; June 20, 1931, Alexander to Grinnell, CU-120.
4. July 18, 1930, Alexander and Kellogg to Mary Charlotte Alexander, AQH.
5. September 8, 1930, Alexander to Miss Wilson, ibid.
6. July 22, 1930, Alexander to Grinnell, ibid.
7. August 20, 1930, Alexander to Grinnell, ibid. On Grinnell's work ethic, see E. R. Hall, "Joseph Grinnell (1877 to 1939)," *Journal of Mammalogy* 20 (1939):409–17; and E. R. Hall, "Obituary: Joseph Grinnell," *The Murrelet* 20 (1939):46–47.
8. July 11, 1928, Alexander to Nora Perley, MVZ.
9. June 11, 1928, Alexander to Beckwith, Letters to Martha.
10. July 11, 1928, Alexander to Nora Perley, MVZ.
11. Ibid. Their letters do not explain why the women were driving the Franklin on this particular trip rather than Blundie or ever again refer to the Ford that had been parked in a garage in Wabuska, Nevada.

12. July 7, 1929, Alexander to Grinnell, BANC mss 67/121c.
13. October 16, 1929, Alexander to Beckwith, Letters to Martha; October 15, 1929, Alexander and Kellogg to Edna Wemple McDonald, AQH. On interest in the outdoors, see F. E. Brimmer, "Autocamping—the Fastest-Growing Sport," *Outlook*, July 16, 1924, 437–40.
14. October 16, 1929, Alexander to Beckwith, Letters to Martha.
15. October 19, 1929, Alexander to Paul Dorman, Oakland Bank Trust Dept., AQH.
16. October 22, 1929, Alexander and Kellogg to Edna Wemple McDonald, ibid.
17. November 20, 1929, Alexander to Grinnell, BANC mss 67/121c; October 31, 1929, Alexander and Kellogg to Edna Wemple McDonald, AQH.
18. June 10, 1930, Alexander and Kellog to Edna Wemple MacDonald, AQH.
19. July 1, 1930, ibid. Alexander never names the shrew species of interest.
20. August 23, 1929, Grinnell to Alexander, BANC mss C-B1003; September 30, 1930, Alexander to Grinnell, BANC mss 67/121c.
21. July 9, 1931, Alexander to Grinnell, CU-120.
22. July 17, 1931, Grinnell to Alexander, ibid.
23. June 28, 1931, Alexander and Kellogg to Edna Wemple McDonald, AQH.
24. Ibid.
25. Ibid.; July 18, 1931, ibid.
26. September 13, 1938, Alexander and Beckwith, Letters to Martha.
27. August 19, 1930, Alexander to Abercrombie and Fitch, AQH.
28. July 18, 1931, Alexander and Kellogg to Edna Wemple McDonald, ibid. Alexander's description of both the beast and its burrow is characteristic of *Dipodomys spectabilis*, the bannertail kangaroo rat; October 2, 1929, Diary ("dipo" = *Dipodomys*; *Microtus* is a meadow vole).
29. August 19, 1931, Grinnell to Alexander, BANC mss 67/121c. See T. R. Dunlap, *Saving America's Wildlife* (Princeton: Princeton University Press, 1988); and A. Runte, "Joseph Grinnell and Yosemite: Rediscovering the Legacy of a California Conservationist," in *Yosemite and Sequoia: A Century of California National Parks*, ed. R. J. Orsi, A. Runte, and M. Smith-Baranzini (Berkeley: University of California Press, 1993), 85–95, for more on Grinnell's views about destruction of predators and related issues.
30. In 1931 Harold Bryant, a former student of Grinnell's, was appointed assistant director of the National Park Service and Joseph Dixon permanently left the MVZ to accept a position under Bryant. Both men carried Grinnell's philosophy on predator control into their new jobs (see December 16, 1930; July 17 and August 4, 1931; and January 11, 1932, Grinnell to Alexander, BANC mss C-B1003).
31. S. B. Benson, "Concealing coloration among some desert rodents of the southwestern United States," *University of California Publications in Zoology* 40 (1933):1–70. Benson concluded that the species represented the best example of adaptive coloration in mammals to be found in the United States.

32. March 14, 1932, Grinnell to Alexander, CU-120.
33. December 18, 1929, Alexander to Beckwith, Letters to Martha; September 30, 1930, ibid.
34. April 13, 1932, Alexander to Grinnell, CU-120.
35. June 19, 1933, Alexander to Grinnell, BANC mss 67/121c.
36. July 9, 1933, Grinnell to Alexander, BANC mss C-B1003.

23. SALINE VALLEY

1. December 11, 1936, Alexander to Grinnell, CU-120. In addition to the field notes kept by both Alexander and Kellogg, details of the Saline Valley trip come from a journal letter that Alexander wrote to her sister Martha, beginning December 30, 1936, Mission Children's Society, and from Ward Russell oral history, MVZ.
2. December 17, 1936, Grinnell to Alexander, BANC mss C-B1003.
3. January 10, 1937, ibid.
4. January 14, 1937, ibid.
5. January 22, 1937, Alexander to Grinnell, CU-120.
6. July 6, 1984, John Davis to Alice Howard, MVZ; January 20, 1937, Alexander to Russell, CU-120.
7. *Daily Californian,* January 19, 1937; *Los Angeles Times,* January 27, 1937, 1; and *Sacramento Bee,* January 27, 1937, 2; February 19 and 20, 1937, Grinnell to Alexander, BANC mss C-B1003.
8. February 7, 1938, Grinnell memorandum, BANC mss C-B1003.
9. July 6, 1984, John Davis to Alice Howard, MVZ.

24. THE END OF AN ERA

1. June 2, 1939, Alexander to Beckwith, Letters to Martha.
2. Ibid.
3. June 4, 1939, Alexander to Hilda Grinnell, BANC mss 73/25c.
4. March 23, 1911, Alexander to Grinnell, BANC mss 67/121c.
5. President/chancellor's reports, MVZ; September 30, 1935, Grinnell to Alexander, BANC mss C-B1003.
6. October 13, 1935, Alexander to Grinnell, CU-120.
7. November 8, 1935, Grinnell to Sproul, MVZ.
8. March 8, 1946, Miller to Sproul, ibid.
9. Personal communication from Frank Pitelka. Miller's father was Loye Holmes Miller (1874–1970), a much beloved naturalist and biologist on the faculty of UCLA.
10. F. J. Graham, "Hall's Mark of Excellence," *Audubon,* July 19, 1984, 88–102; and obituary, "Noted KU Scientist E. Raymond Hall Dies," *Lawrence Journal World,* April 3, 1986 (Lawrence, Kans.).

May 17, 1946 and May 13, 1947, KU Chancellor Deane W. Mallot to Alexander; April 15, 1948, Alexander to Hall; and April 8, 1949, Hall to Irvin Youngberg, executive secretary, Endowment Association, KU. E. R. Hall, *Mammals of Nevada* (Berkeley: University of California Press, 1946). July 11, 1946, Hall to Alexander, UCMP.

11. June 1, 1939, Alexander to Sproul, CU-120.
12. June 16, 1939, Sproul to Alexander, BANC mss 67/121c.
13. October 27, 1939, ibid.
14. February 16, 1940, Sproul to Alexander, UCMP.
15. March 13, 1939, Grinnell memorandum, CU-120. Alexander contributed $2,000 toward publication of the volume. February 18 and 26, 1944, UC Press manager Samuel T. Farquhar to Alexander; March 1, 1944, Alexander to Hall; March 29, 1944, Hall to Alexander, KU; March 1, 1944, Alexander to Miller; and March 9, 1944, Miller to Alexander, CU-120. Throughout the years Alexander subsidized the publication of other MVZ volumes over and above her regular appropriation, but none under such unusual circumstances as these.
16. November 4, 1940, Miller to Alexander, CU-120; February 3, 1943, Hall to Alexander, UCMP.
17. November 20, 1943, Alexander to Miller, CU-120.
18. Undated, 1944, Hall to Alexander, MVZ.
19. May 23, 1944, Alexander to Hall, KU.
20. April 12, 1946, Hall to R. C. Ploss, assistant manager, UC Dept. of Insurance & Retirement Systems, UCMP; September 7, 1944, Miller to Alexander, CU-120; June 30 and September 16, 1945, Hall to Alexander, UCMP.
21. Undated, 1944, Hall to Alexander, MVZ. Hall's research did, in fact, culminate in the 1951 publication of a monumental two-volume set entitled *Mammals of North America*, coauthored by Keith Kelson (New York: Ronald Press). A second edition was published thirty years later—E. R. Hall, *The Mammals of North America* (New York: John Wiley & Sons, 1981).
22. December 26, 1945, Hall to Alexander, UCMP; September 28, 1946, Hall to Alexander, KU. Stirton remained at Cal for the duration of his career.
23. E. R. Hall, "Ralph Ellis," *Proceedings of the Linnaean Society of London* 159 (1947):158–59; and R. Vosper, "A Pair of Bibliomanes for Kansas: Ralph Ellis and Thomas Jefferson Fitzpatrick" (Lawrence: University of Kansas Libraries, 1982), 1–19; February 23, 1946, Alexander to Miller, CU-120.

25. HAWAII—"MY ONLY REAL HOME"

1. March 1, 1944, Dixon to Hasselborg, Alaska State Library.
2. Personal communication from Martha Hurd Kreuter.
3. February 24, 1910, Alexander to Mary Beckwith, Letters to Martha; May 17, 1934, Alexander to Grinnell, BANC mss 67/121c.
4. March 21, 1920, Alexander to Beckwith, Letters to Martha.
5. December 28, 1920, Alexander to Kellogg, AQH.
6. October 8, 1907, Alexander to Beckwith, Letters to Martha.
7. "In memoriam: George C. Munro," *Elepaio* 24 (1964):37–38; G. C. Munro,

Birds of Hawaii (Honolulu: Tongg Publishing, 1944); July 17, 1935 and July 21, 1938, Alexander to Grinnell, CU-120.

8. February 18, 1926, Alexander to Stanley C. Ball, curator of collections, Bishop Museum Vertebrate Zoology archives.

9. February 23, 1926, Munro to Ball; March 6, 1926, Ball to Munro; March 18, 1926, Ball to Alexander; and March 18, 1926, memorandum from the acting director to the trustees with Memorandum "A," Bishop Museum Vertebrate Zoology archives. Other donations to the Bishop Museum were specimens that Alexander herself collected—accession nos. 1922.153, a group of entomological specimens; 1926.025 and 1936.004, small collections of land shells from Maui; 1925.023, nearly complete human skeletal remains of six persons from the Paia battlefield on Maui; 1947.031 and 1948.053, botanical specimens from Maui and Kauai. Based on the Consumer Price Index, $5,000 in 1926 is the equivalent of $48,850 in January 2000.

10. April 6, 1944, Munro to Alexander (H. W. Grinnell, "Biographical sketch of Annie Montague Alexander," ms, MVZ).

11. Grinnell felt that California's faunal diversity presented more evolutionary problems than either he or the museum staff could hope to unravel in a lifetime. Because the accumulation of specimens was intimately tied to publications, there was little incentive to acquire material from outside the state if no one would be available to work on it. In later years, exceptions were made for areas coming under heavy development, such as Baja California, where opportunities for future collecting might not exist.

12. January 28, 1910, Grinnell to Alexander, BANC mss C-B1003; February 1, 1910, Alexander to Grinnell, BANC mss 67/121c. Alexander's letter explicitly mentions spoonbills, but since the MVZ has no spoonbills from the islands in its collections and these distinctive-looking birds are not known to occur in Hawaii, it is difficult to know the species to which Alexander was in fact referring.

13. The 1911 purchase was the Chester E. Blacow collection. A student of Loye Miller's, Blacow had learned from him how to prepare scientific study skins of birds.

The lei was MVZ accn. no. 4419; the Ne-ne specimens were MVZ accn. nos. 3815 and 3916. See A. H. Miller, "Structural modifications in the Hawaiian goose (*Nesochen sandvicensis*): a study of adaptation," *University of California Publications in Zoology* 42 (1937):1-80.

14. January 13, 1919, Alexander to Grinnell, BANC mss 67/121c. See also Alexander field notes, MVZ.

15. January 13, 1919, Alexander to Grinnell, BANC mss 67/121c.

16. Alexander field notes, MVZ; February 14, 1919, Alexander to Beckwith, Letters to Martha. The following year, Alexander and Kellogg collected approximately 150 species of Hawaiian shells for Bruce Clark in the UCMP.

17. February 5, 1933, Alexander to Grinnell, CU-120; August 28, 1939, Alexander to Hilda Grinnell, BANC mss 73/25c.

18. February 19-24, 1927, Diary.

19. Personal communication from Bill Kaiwa.

20. Ibid.

21. Diary; personal communication from Martha Hurd Kreuter. According to Kreuter, Alexander could not resist taking a short flight in a hydroplane while in Florida but, much to Kellogg's dismay, invited one of the cousins rather than Kellogg to accompany her on the flight.

22. March 25, 1934, Grinnell to Alexander, BANC mss C-B1003. The trip took place from June 9–July 4. Details are extracted from journals kept by all three individuals (AQH).

23. June 20, 1934, Alexander trip journal, AQH.

24. June 16, 1934, Kellogg trip journal, ibid.

25. Unfortunately, no diary or journal from the trip has been found.

26. W. B. Alexander, *Birds of the Ocean, a Handbook for Voyagers Containing Descriptions of All the Sea-Birds of the World, with Notes on Their Habits and Guides to Their Identification* (New York: G. P. Putnam's Sons, 1928). The trip lasted through August 8, after which the trio sailed to New Zealand before returning to Hawaii. Letters dated June 16, July 7, and August 10, 1936, Alexander to Grinnell, BANC mss 67/121c; and July 9, 1936, Alexander to Grinnell, CU-120, provide limited details of the trip. Rollo Beck was an ornithologist who spent much of his professional life traveling the world collecting and preparing bird specimens for research institutions. The Academy of Sciences in San Francisco employed Beck for many years and Alexander contracted with him ca. 1910 and 1911 to collect and prepare specimens of water birds for the MVZ.

27. July 2, 1942, Alexander to Mason, UCH.

28. Personal communication from Warren Herb Wagner. Wagner confirmed the exact location from which the fern had been collected. Alexander financed his subsequent doctoral thesis on the Hawaiian flora, research that launched his career as a well-known systematic botanist.

26. THE SWITCH TO BOTANY

1. May 13, 1944, Alexander to Willis Jepson; December 15, 1944, Alexander to Lincoln Constance; and September 17, 1948, Alexander to Mason, UCH. The UC Herbarium contains sheets of specimens that Alexander collected as early as 1901 on her expedition with Mary Wilson.

2. Personal communication from Robert Hobdy.

3. July 17, 1935, Alexander to Grinnell, BANC mss 67/121c.

4. May 18, 1939, ibid.; and June 3, 1939, Alexander to Mrs. Pettigrew and friends, AQH. See also M. Reisner, *Cadillac Desert: The American West and Its Disappearing Water* (New York: Viking, 1986); July 23, 1945, Alexander to Annetta Carter, UCH; and November 2, 1947, Alexander to Alden Miller, CU-120.

5. H. M. Hall and C. C. Hall, *A Yosemite Flora: A Count of the Ferns and Flowering Plants, including Trees, of the Yosemite National Park, with Simple Keys for Their Identification, Designed to be of Use throughout the Sierra Nevada Mountains* (San Francisco: P. Elder, 1912).

6. June 3, 1939, Alexander to Mrs. Pettigrew, UCH. Details taken from "Car-

lotta Case Hall's account of field trip to Charleston Mts. Nevada with Annie M. Alexander and Louise Kellogg," AQH.

7. "Carlotta Case Hall's account."

8. December 11, 1944, Howell to Alexander, UCH.

9. L. Constance, *Botany at Berkeley: The First Hundred Years* (Berkeley: University of California, 1978).

10. January 8, 1944, Mason to Sproul; and January 3, 1945, Alexander to Mason, UCH; C. Holleuffer, "Annetta Carter, UC Herbarium botanist, collector and interpreter of Baja California plants, an oral history conducted in 1985," IV. Baja California trips with Alexander and Kellogg. Regional Oral History Office, Bancroft Library, University of California, Berkeley, 1987, BANC mss 98/57.

11. August 15, 1942, Alexander to Miss Crum, UCH.

12. Unlike animals for which a single individual is usually captured at each specific locality, more than one plant of a given taxon is usually collected at a site.

13. August 15, 1942, Alexander to Miss Crum, UCH; W. L. Jepson, *A Manual of the Flowering Plants of California* (Berkeley: University of California Press, 1925), considered the definitive botanical handbook of the day, was reissued in 1993 as *The Jepson Manual: Higher Plants of California*, ed. J. Hickman.

14. December 24, 1944, Alexander to Annetta Carter; January 10, 1945, Carter to Alexander, UCH; and personal communication from Richard Beidleman. *A Flora of California*, 3 vols. (Berkeley: Cunningham, Curtis and Welch, 1909–43). Through a provision in Jepson's will, the Jepson Herbarium was founded in 1950, four years after he died. It carries out the botanist's life work of studying California plants and publishing works that will lead to a greater understanding and appreciation of the California flora. Remaining true to his proprietary obsession even after death, the collections of the Jepson and University herbaria remain adjacent to, but physically separate from, one another within the same building on the UC Berkeley campus.

15. August 17, 1944, Alexander to Mason, UCH.

16. August 23, 1944, Mason to Alexander; September 10, 1944, Alexander to Mason; February 7, 1947, Alexander to Carter, ibid.

17. September 21, 1942, Eastwood to Alexander, ibid.

18. January 28, 1941, Mason to Alexander, ibid.

19. August 22, 1941, ibid.

20. January 14, 1945, Alexander to Constance, ibid.

21. H. W. Grinnell, "Biographical sketch of Annie Montague Alexander," ms, MVZ. Alexander was seventy-two at the time.

22. The concept of life zones was first put forth by C. Hart Merriam as a way to explain observed floral and faunal distribution patterns ("Laws of temperature control of the geographic distribution of terrestrial animals and plants," *National Geographic Magazine* 6 [1894]:229–41). Accumulated data and hypothesis testing subsequently discredited the theory.

23. July 18, 1945, Alexander to Mason, UCH; September 16, 1945, Hall to Alexander, UCMP; and Alexander field notes, UCH.

24. May 4, 1947, Alexander to Mason, UCH.

25. October 3, 1947, Mary Charlotte Alexander to her family, Mission Children's Society.
26. For a detailed and entertaining description of what maté is and how it is brewed, see G. G. Simpson, *Attending Marvels: A Patagonian Journal* (New York: Macmillan, 1934).
27. October 3, 1947, Mary Charlotte Alexander to her family, Mission Children's Society.
28. Record of Donations, vol. 1, University Archives, University of California, Berkeley. See also correspondence files in the MVZ and BANC for numerous letters between Alexander and Grinnell, E. R. Hall, and Seth Benson regarding the nature and funding of these expeditions. Charles Lamb was the MVZ's principal collector in Baja California during the initial years of fieldwork there. Later, Benson and others worked in Sonora under Alexander's sponsorship.

27. BAJA CALIFORNIA—*TRES MUJERES SIN MIEDO*

1. E. W. Nelson, "Lower California and Its Natural Resources," *Memoirs of the National Academy of Sciences* 16 (1921):1–194; D. Lamb, *Enchanted Vagabonds* (New York: Harper and Brothers, 1938).
2. December 18, 1934, Alexander to Nora Perley, MVZ.
3. March 4, 1948, Mason to Sproul, UCH.
4. B. Ertter, "Obituary: Annetta Mary Carter (1907–1991)," *Madroño* 39 (1992):245–50. Following her trip with Alexander and Kellogg, Carter vowed to return to Baja California every year to collect plants. With rare exceptions she did, becoming an expert on the flora of that region.
5. "Suggested Equipment for Baja California Expedition," AQH.
6. Details of the expedition are taken from Alexander's trip journal, ibid.; M. Waters, "Trio finds rare flora in Baja California," *The Christian Science Monitor*, April 5, 1950; A. Carter, "Three women without fear: how three botanists drove 4,200 miles in Baja California," *California Monthly*, June 1949, 30–31, 78–81; and C. Holleuffer, "Annetta Carter, UC Herbarium botanist, collector and interpreter of Baja California plants, an oral history conducted in 1985," IV. Baja California trips with Alexander and Kellogg. Regional Oral History Office, Bancroft Library, University of California, Berkeley, 1987, BANC mss 98/57. There is also a short account of the trip written by Alexander after *California Monthly* asked the women for an article (AQH).
7. Carter, "Three women without fear"; and Holleuffer, "Annetta Carter."
8. Holleuffer, "Annetta Carter"; and "Annie Alexander—the intrepid collector," *PG&E Progress*, July 1980, 8.
9. Alexander's trip journal, AQH.
10. J. Steinbeck and E. F. Ricketts, *The Log from the* Sea of Cortez: *The Narrative Portion of the Book,* Sea of Cortez (New York: Viking, 1951).
11. When Carter was eventually able to collect the *Acacia* in flower, she named it as a new species, *Acacia kelloggiana*, in honor of Kellogg (see appendix).
12. Alexander's trip journal, AQH.

13. September 30, 1948, Carter to Alexander, UCH.

14. The book to which Carter referred was probably R. C. DuSoe, *Three Without Fear* (New York: D. McKay, 1947). Its publication date suggests that she read it as an adult.

28. INVESTING IN THE FUTURE

1. April 14, 1926, Alexander to Grinnell, BANC mss 67/121c; and August 6, 1932, Grinnell to Alexander, BANC mss C-B1003; note appended to July 7, 1924, ibid.

2. April 8, 1929, Alexander to Grinnell, ibid. An example of the changing nature of relations with the university can be gleaned in December 19, 1931, Grinnell to Alexander, ibid.

3. July 17, 1931, Grinnell to Alexander, ibid. Bryant was Harold C. Bryant.

4. January 1, 1933, Grinnell to Alexander, ibid. The graduate program and museum collection that E. R. Hall developed at the University of Kansas is perhaps the most outstanding example of the promulgation of the MVZ philosophy and practices that can be cited.

5. J. K. Jones, Jr., "Genealogy of Twentieth-Century Systematic Mammalogists in North America: The Descendants of Joseph Grinnell," in *Latin American Mammalogy: History, Biodiversity, and Conservation*, ed. M. Mares and D. Schmidly (Norman: University of Oklahoma Press, 1991), 48–56; and F. A. Pitelka, "Academic family tree for Loye and Alden Miller," *The Condor* 95 (1993):1065–67. See also October 17, 1944, Miller to Alexander, CU-120.

6. July 1, 1915, Merriam to Alexander, UCMP.

7. May 22, 1946, Alexander to E. R. Hall, KU; and April 7, 1949, Alexander to Miller, CU-120; June 2, 1947, Miller to Don Tappe, MVZ; H. W. Grinnell, "Biographical sketch of Annie Montague Alexander," ms, MVZ.

8. Personal communication from Virginia Miller Russell; E. R. Hall, "Joseph Grinnell (1877 to 1939)," *Journal of Mammalogy* 20 (1939):409–17; and E. R. Hall, "Obituary: Joseph Grinnell," *The Murrelet* 20 (1939):46–47; Virginia Miller Russell and Lois Stone oral histories, MVZ.

9. Undated, but after 1908, Alexander to Beckwith, Letters to Martha.

10. Personal communication from Barbara Blanchard DeWolfe.

11. Barbara Blanchard DeWolfe written recollections, MVZ. In actuality, Grinnell's letter was slightly more professional than Blanchard remembered (see May 27, 1935, MVZ).

12. July 28, 1932, Hall to Alexander, CU-120.

13. R. A. Stirton, "The Role of Paleontology in the University of California: Honoring the Twenty-Fifth Presidential Year of President Robert Gordon Sproul" (Berkeley: University of California, 1955).

14. July 8, 1930, Alexander to Edna Wemple McDonald, AQH.

15. July 29, 1930, ibid.

16. December 27, 1933, Alexander to Grinnell, CU-120. Details excerpted from Alexander, Kellogg, and Erickson's field notes, MVZ.

17. December 29, 1933, Alexander to Grinnell, CU-120.
18. March 15, 1934, Grinnell to Alexander, BANC mss C-B1003; September 29, 1942, Erickson to Miller, MVZ.
19. Correspondence between Alexander and Hilda Grinnell is in Joseph Grinnell papers, BANC mss 73/25c; April 14, 1941, Alexander to H. Grinnell, ibid.
20. The graduate students were Jean Boulware, Josephine Crowley, Viola Memler, and Frances Prack (H. Grinnell field notes, ibid.).
21. June 17, 1941, Alexander to H. Grinnell, ibid.
22. April 9, 1942, ibid.; those on the first trip were Jean Boulware, Viola Memler, and Hilda Grinnell; on the second, Josephine Crowley, Jean Rusick, Ida DeMay, Florence Ogilvie, Marie Redox, and Hilda Grinnell.
23. June 16, 1942, ibid.
24. Undated, Boulware to Janet Failla, MVZ.
25. June 27, 1944, H. Grinnell to Alexander, BANC mss 73/25c.
26. June 5 and August 29, 1948, Alexander to Miller, CU-120. Subsequent correspondence between Miller and the university administration (MVZ) clarified that what Alexander wished to establish were fellowships exclusively for graduate student research. Alexander hoped that the property would bring $60,000, but the university received only $32,000 from the sale.

29. AN ENDURING LEGACY

1. April 30, May 21, and June 23, 1947, Alexander to E. R. Hall, KU; H. W. Grinnell, "Biographical sketch of Annie Montague Alexander," ms, MVZ.
2. Personal communication from Barbara Toschi.
3. March 15, 1966, Annetta Carter to James Kantor, university archivist, UCH; August 19, 1944, Alexander to Jepson, ibid. Wordsworth's quote is taken from his poem "Ode: Intimations of Immortality" (1807).
4. May 29, 1908, Heller to Grinnell, MVZ. The beaver was ultimately named *Castor canadensis phaeus* (see E. Heller, "Birds and mammals of the 1907 Alexander expedition to southeastern Alaska," *University of California Publications in Zoology* 5 [1909]:250).
5. September 25, 1926, Alexander to Grinnell, BANC mss 67/121c.
6. February 27, 1932, Alexander to Hall, CU-120.
7. June 13, 1939, E. R. Hall to George Pettitt, assistant to the president, MVZ. The figure 22,738 is the cumulative number of specimens represented by the 170 separate accessions for which she was wholly or partly responsible as of that time. Whereas Alexander's financial support of the UCMP was approximately equal to that of the MVZ, her specimen contributions were considerably less. She and Kellogg donated ca. 1,540 specimens to the UCMP based on current database records. Different bookkeeping practices, and the fact that a great deal of fossil material is not catalogued until a researcher has need for it, partially account for the low figure.
8. May 13, 1944, Alexander to Jepson, UCH.

9. October 9, 1949, Alexander to E. R. Hall, KU; and October 13, 1949, Susan Chattin to Alexander, CU-120.

10. October 14, 1949, Diary.

11. By one reckoning, Alexander's last spoken words were, "I can't go yet, I'm not finished" (Barbara Toschi, pers. comm.).

12. On January 2, 1958, the president's office reported to Alden Miller that Alexander had "made approximately 945 contributions to the University of California, scarcely missing a month . . . from December 31, 1903, to September 30, 1949." The aggregate value of these monetary contributions was $1,387,494.42. In addition, Alexander wrote many checks that were never recorded in the university's ledger (BANC and MVZ). February 23, 1951, Kellogg to Alden Miller, MVZ. At the time of her death, Alexander's estate was valued at almost $900,000.

13. See especially November 3, 1947, Sproul to Alexander, UCMP; December 9, 1936, Grinnell to Alexander, BANC mss C-B1003.

14. June 7, 1939, Walter P. Taylor to Alexander, BANC mss 73/25c.

15. December 17, 1926, Alexander to Grinnell, BANC mss 67/121c; November 14, 1927, ibid. See also March 26, 1909, Alexander to Beckwith, Letters to Martha.

16. December 2, 1944, Dixon to Hasselborg, Alaska State Library.

EPILOGUE

1. March 24, 1950, Mason to Kellogg; and March 28, 1950, Kellogg to Mason, UCH.

2. January 15, 1950, Mason to Sproul, ibid. (probably misdated, should be 1951); October 9, 1950, Mason to Lee Bonar, chair, botany dept., ibid.; G. Lindsay, "Notes concerning the botanical explorers and exploration of Lower California, Mexico," a paper prepared for Biology 199, Stanford University (Belvedere Scientific Fund, 1955); and C. Holleuffer, "Annetta Carter, UC Herbarium botanist, collector and interpreter of Baja California plants, an oral history conducted in 1985," IV. Baja California trips with Alexander and Kellogg. Regional Oral History Office, Bancroft Library, University of California, Berkeley, 1987, BANC mss 98/57.

3. September 10, 1951, Carter to Kellogg, UCH; and Lindsay, "Botanical explorers."

4. "Solano County to Lose Water Benefit," *The Solano County Republican* (Fairfield), November 8, 1934; also October 9 and 22, 1929; November 3, 1929; and July 28, 1931, Alexander to Wilson, AQH.

5. Personal communication from Jack Schafer.

Index

ABCFM (American Board of Commissioners of Foreign Missionaries), 4
Abu Simbel temple (Egypt), 210
Acacia kelloggiana, 355n11
Academy of Sciences. *See* California Academy of Sciences
Adamana (Arizona), 185
Admiralty Island (Alaska), 66, 67 fig., 70, 71, 331nn9,15, 332nn16,17
African expedition. *See* British East African expedition
African Game Trails (Roosevelt), 163, 343n24
AK Brand, 134
Akeley, Carl, 86, 163, 334n26
Alaskan expeditions: Alexander's interest in, 52, 54, 58–59; and L. L. Bales incident, 60–61; bear collecting objective of, 59, 66, 106, 330n4; complaints on, 104; female companion for, 98–100; Glacier Bay campsite of, 73–74; Grinnell's endorsement of, 65, 66, 96, 331n9; Grinnell's manuscript on, 82–83; of Edward Harriman, 59–60, 330n5; on Hasselborg River, 71–72; with Louise Kellogg, 100, 101; to Kenai Peninsula, 61–62; members of, 61, 63–64, 66, 69, 72–73, 330n9, 331n10; Mole Harbor campsite of, 69–70; MVZ's publication on, 84; poor wildlife diversity on, 103–4, 105; to Prince William Sound, 96, 98, 102–6; routes of, 67 fig., 102–3; of Smithsonian, 330n9; to southeastern Alaska, 65–74, 67 fig.; specimens from, 62, 70, 71, 75, 103, 332nn16,17,22; Windfall Harbor campsite of, 68
Albatross (Smithsonian's ship), 330n9
Alberni (Vancouver Island), 140–41
Alexander, Annie Montague: anonymity preference of, xvii, 117–18, 119, 249; automobile needs of, 196, 197, 204–5, 215, 216, 217, 292, 347n11; Bay Area apartment of, 214–15, 347–48n2; and Martha Beckwith, 16–17, 18, 118–19; birth/family of, 6, 8 fig.; cattle breeding by, 128, 131, 190, 193–94; conservation concerns of, 152–53, 158, 219, 239, 275–76, 309; death of, 311–12, 358n11; diaries of, xv, 323n4; dinner parties of, 308; education of, 14, 19, 22, 29; Egyptian tent of, 212, 215–16; family's relations with, 268–69, 272, 353n21; father's influence on, 3, 34, 39, 40–41, 117–18, 148–49,

359

Alexander, Annie Montague *(continued)*
323n1; female companions of, 25–26, 30, 32, 35–36, 63–64, 98–100; financial expertise of, 88–90, 334n3; and Grinnell, xvii, 65, 88, 90–91, 93–96, 155–56, 161, 168–69, 236, 253–54; Grizzly Island purchase by, 120–22; Hawaii ties of, 23, 49–50, 185, 262–63, 267–68; health problems of, xiv, 15, 18, 19, 214, 215, 217, 220–21, 325n7; honorary degree for, 176; hunting by, 35, 36–37, 42–43, 45; intellectual insecurities of, 19, 33, 73, 91–92, 110–11, 222; introversion of, 27–28; and Louise Kellogg, xv–xvi, 101–2, 106, 312, 323nn6,7; lake named for, 70, 331n15; on marriage, 19–21; on the medical profession, 18, 20, 217, 220, 311; nomadic disposition of, 18–19, 50, 51, 58–59, 149, 185, 189, 222, 225–26; Oakland's confinement of, 12, 18–19, 21, 148; organizational memberships of, 93, 153, 335n15; on patriarchal restrictions, 33–34, 327n27; photography interest of, 14, 38, 40, 41, 43; physical appearance of, xiv, 217, 322n2; politics of, 184–85; published articles of, 93; research priority of, 32, 88, 111–13, 155–56, 157–58, 166–67, 179–80, 299, 334n2; on self-perception 19, 29, 33, 69, 80, 90, 91–92, 110, 148, 185, 215, 267–68, 285; taxa named for, 31, 33, 71, 141, 242, 309, 317–19, 327n25, 332n17, 339n7. *See also* Fieldwork; Innisfail Ranch

Alexander, Annie Montague, benefactions of: in Alexander's will, 157, 312; to American Museum of Natural History, 191; anonymity of, xvii, 117–18, 119; for Baja California fieldwork, 150, 160, 355n28; to Martha Beckwith, 118–19, 181–82; to Bishop Museum, 264, 265, 352n9; to UC Department of Paleontology, 107, 111–13, 167, 172–73, 336nn1,3; during the depression, 161–62, 348n3; to Joseph Dixon, 119; to graduate students, 299, 300–301, 304–7, 357nn20,22,26; to E. R. Hall, 256, 257, 351n15; to Allen Hasselborg, 90; to John Merriam, 23–24, 29–30, 65, 107, 327n17, 336n3; to MVZ, 78, 80–82, 85–86, 94–95, 117, 158, 159, 160–62, 179–80, 335n19, 337n6; to purchase bird specimens, 160, 342n13; to Frank Stephens, 90, 98; sugar's impact on, 89–90, 157; to Harry Swarth, 90, 109; total amount of, to UC, 358n12; to UCMP, 169, 172–73, 174, 175, 176, 177–78, 179, 344n20, 345nn23,32; to University Herbarium, 281; to Warren Wagner, 273, 353n28; Wheeler's response to, 32–33

Alexander, Annie Montague, collecting trips: for fossils, 23, 24–29, 30–34, 50–51, 108–110, 183–88, 215–18, 287–89; for mammals, 218–20, 221–22, 225, 226–243, 254, 278; for plants, 16–17, 227, 242, 273, 277–79, 285, 286–87. *See also* Alaskan Expeditions; Baja, California, expeditions; Botanical specimens; Egypt; Fossil Specimens; Samoan trip; Topotype specimens; Trinity Alps expedition; Vancouver Island expedition; Vertebrate specimens

Alexander, Annie Montague, correspondence of: with Mary Charlotte Alexander, 190, 225; archived collections of, 322n3, 323n10; with David Barrows, 168; with Martha Beckwith (*See* Letters to Martha); with Mary Beckwith, 16, 215, 262; with William Campbell, 120–21; with Lincoln Constance, 285; with Hilda Grinnell, 253–54, 268, 305, 306; with Joseph Grinnell (*See* Grinnell, Joseph, Alexander's letters to); with E. R. Hall, 147, 258–59; with W. L.

Jepson, 309; with Kellogg, xv; with Herbert Mason, 273, 283; with C. Hart Merriam, 54–55, 61; with Alden Miller, 257–58; with mother, 45, 182; with Robert Sproul, 175, 176, 256; with Chester Stock, 203; with Walter Taylor, 313; volume of, xv, 226; with Edna Wemple, 209, 210, 224–25, 231, 232, 234, 237–38, 303; with Benjamin Wheeler, 78, 80

Alexander, Annie Montague, foreign travels of: to Altamira caves, 198–99; to British East Africa, 35–39, 40, 41–45; to Castillo Cave, 199–200; to England, 191–96; with family members, 15–16, 21; to France, 196–98, 200–202; to Mexico City, 290–91; to Middle Eastern temples/tombs, 203–4, 206–10, 212; newspaper exposé on, 117; to Palestine, 210–11; with relatives as a young woman, 15–16; to Samoa, 269–71; to South Pacific, 271–72, 353n26; with tour company, 204; to Victoria Falls, 45–46

Alexander, Clarence (Alexander's brother), 3, 6, 8 fig.

Alexander, Ellen Charlotte (Alexander's Aunt Lottie), 8 fig., 15

Alexander, Emily Whitney (Alexander's aunt), 7, 8 fig.

Alexander, James (Alexander's uncle), 5–6, 8 fig., 10, 15–16, 263

Alexander, Juliette (Alexander's sister), 3, 6, 8 fig., 20, 306

Alexander, Martha Eliza (née Cooke, Alexander's mother), 5–6, 8 fig., 40, 45, 74, 182, 347–48n2

Alexander, Martha Mabel. *See* Waterhouse, Martha Mabel

Alexander, Mary (Alexander's aunt), 8 fig., 58

Alexander, Mary Charlotte (Alexander's cousin): Alexander's correspondence with, 190, 225, 323n10; Alexander's expeditions with, 215, 216–17, 288–89; and Carlotta Hall, 277

Alexander, Samuel Thomas (Alexander's father): African safari of, 35–39, 36 fig., 40, 41–43, 117; attitudes of, toward women, 34, 35–36, 39, 40–41; and Henry Baldwin, 6–7; death/burial of, 46–47, 328n18; education of, 6, 324n13; family of, 3–4, 6, 8 fig.; Hamakua ditch project of, 9; hunting interest of, 35, 39, 42–43, 44–45; marital engagement of, 5–6; Oakland relocation by, 9–10, 13–14; personal traits of, 3, 5, 117–18, 324n9; politics of, 184; sugar business of, 10–12; travels of, 15, 21

Alexander, Wallace McKinney (Alexander's brother), 25, 39, 47, 89, 347–48n2; and Alexander-Kellogg relationship, 268; birth/family of, 6, 8 fig.; occupation of, 3

Alexander, W. B., 271

Alexander, William De Witt (Alexander's uncle), 7, 8 fig., 10, 15, 324n6

Alexander, William Patterson, Rev. (Alexander's grandfather), 3–4, 5, 8 fig., 15, 275

Alexander Archipelago (Alaska), 65–66, 331n8

Alexander & Baldwin (firm), 10, 11

Alexander Lake (Admiralty Island, Alaska), 70, 71, 331–32n15

Algiers, Arab quarter of, 204

Altamira caves (Spain), 198–99

Amenhotep III, 208

American Board of Commissioners of Foreign Missionaries (ABCFM), 4

American Canyon expedition, 50–51

American Museum of Natural History (New York), 60, 93, 162, 172, 191, 313

American Ornithologists' Union (AOU), 52, 93

American Refinery, 11

American Society of Mammalogists, 93

American University (Beirut), 212

ʿAmr mosque (Cairo), 207

"Analysis of Functions" (Grinnell), 254–55, 256
Andrews, Roy Chapman, 191, 346n5
Animal Life in the Yosemite (Grinnell and Storer), 160
Anklin River (Alaska), 103
Antelope ground squirrels, 246
Anza Borrego Desert (California), 252
AOU (American Ornithologists' Union), 52, 93
Aplodontia (mountain beaver), 146, 303
Apolima (Western Samoa), 270
Archaeopteryx (ancestor of living birds), 194
Arizona striped skunk, 233
Arrowsmith, Mount (Vancouver Island), 140
Arsinoitherium (fossil mammal), 194
Asparagus farm, 133–34, 277, 316
Aswân (Egypt), 209–10
Austin, Mary, 244
Auto camps, 231
Automobiles, purchase of, 272. *See also* Citroën automobile; Dodge Power Wagon; Ford field vehicle; Franklin automobile; Herrick Bros.; Kégresse automobile
Azilian culture, 201

Baby Beef Company (Collinsville, California), 316
Badlands region (Anza Borrego Desert, California), 252
Bahía de la Concepción (Baja California), 294, 295
Bahía de la Paz (Baja California), 297
Bailey, Florence Merriam, 99, 329n12, 336n3
Bailey, Vernon, 243, 336n3
Baja California expeditions: Alexander's funding of, 150, 160, 289, 355n28; botanical specimens from, 295, 297, 355n11; Carter's introduction to, 291, 355n4; equipment/provisions for, 291–92; of Kellogg and Carter, 315–16; road travel on, 292–97

Baldwin, Abigail Charlotte (Alexander's aunt), 7, 8 fig.
Baldwin, David Dwight (Alexander's uncle), 272, 324n6
Baldwin, Dwight, 6–7
Baldwin, Henry (Alexander's uncle): and Alexander family, 6–7, 8 fig., 323n1; Hamakua ditch project of, 9; marriage of, 7; philanthropy of, 117–18; sugar business of, 10–12
Baldy, Mount (Arizona), 236–37
Bales, L. L., 60–61
Bannertail kangaroo rats *(Dipodomys spectabilis)*, 349n28
Baranof Island (Alaska), 66, 67 fig., 74
Baringo, Lake (British East Africa), 41
Barrel cacti, 278
Barrows, David P., 94–95, 167–68, 170, 342n9
Barstow Formation (San Bernardino County, California), 165, 187–88
Bartlett Cove (Alaska), 73
Bear Bay (Baranof Island, Alaska), 67 fig., 74
Bear Cove (Shasta County, California), 31
Beardslee islands (Alaska), 67 fig., 73
Bears: from Alexander Archipelago, 66, 70; Alexander's collection of, 62, 69, 91; Alexander's interest in, 58–59; Arizona grizzly, 240; and bear cubs episode, 103, 336n13; C. Hart Merriam's interest in, 54, 59, 162, 330n4; from Montague Island, 106; from Vancouver Island, 139
Beaver: on Admiralty Island, 70, 332n16; Heller's new species of, 309, 357n4; scarcity of, 150
Beaver Lake (Admiralty Island, Alaska), 70, 71, 331–32n15
Becco (nonalcoholic beverage), 234
Beck, Rollo, 271, 353n26
Beckwith, Martha, xvii, 190, 277, 311; Alexander's affection for, 16–17; Alexander's benefaction to, 118–19, 181–82; education/vocation of, 18,

118; magazine proposal to, 181.
  See also Letters to Martha
Beckwith, Mary, 14, 16, 47, 63, 215, 262
Bégouen (comte of Toulouse), 200
Beira (Mozambique), 45
Beirut (Lebanon), 212
Benson, Seth, 149–50, 234, 239–40, 280, 340n4, 349n31, 355n28
Bermuda, 21
Bernice Pauahi Bishop Museum. *See* Bishop Museum
Biarritz (France), 197–98
Big Pine (Inyo County, California), 245, 247, 248, 249, 250
Billings Gap (Arizona), 185
Biological Survey. *See* U.S. Biological Survey
Bird, Isabella, 7
"Birdie." *See* Franklin automobile
Birds. *See* Alexander, Annie Montague, collecting trips
Birds of Hawaii (Munro), 264, 265
Birds of the Oceans (Alexander), 271
Bird specimens: from Admiralty Island, 70–71, 332n17; at Bishop Museum, 264; of Ralph Ellis, 259; exotic, 276; funds for purchase of, 160, 342n13; of Joseph Grinnell, 64, 158–59, 342n9; on Grizzly Island, 122–23; from Hawaii, 264–66, 276, 352nn12,13; of George Munro, 264–65; national ranking of MVZ's, 342n20; at Papeete museum, 271–72; preparation of, 68; from Prince William Sound, 103, 104; at Puente Viesgo, 198; from Samoa, 271; from Vancouver Island, 139–40, 141
Birkat Qârûn (Lake Morris, Egypt), 206
Bishop Museum (Honolulu), 55, 97, 264, 265, 352n9
Black Oak Canyon (Shasta County, California), 33
Black-tailed prairie dogs, 239
Blacow, Chester E., 352n13
Blanchard, Barbara, 301–2

Blue Forest (Arizona), 185
Blue salvia, 278
Bluff (Utah), 233
"Blundie." *See* Ford field vehicle
Bly (Oregon), 27
Boalt, Elizabeth, 115, 116
Boas, Franz, 18
Bole, P. J., Jr., 285
"Boston marriages," xv–xvi, 323n6
Botanical specimens: from Baja California, 295, 297, 355n11; disappearance of native, 276; evolutionary importance of, 227, 285; of fossil plants, 287–88; of Hawaiian ferns, 272, 273, 353n28; to Herbarium, from Alexander, 242, 275, 309, 353n1; Herbarium's identification of, 281–82, 283–84, 354n12; of W. L. Jepson, 282–83, 354n14; from Mount Grant, 286, 354n22; of new grass variety, 310–11; praise of Alexander's, 281, 284–85; reasons for switch to, 274–75; from southern California/Nevada, 278; from Sweetwater Mountains, 279–80. *See also* UC Herbarium
Boulder Dam, 278
Boulware, Jean, 306, 357nn20,22
Boyton, Bill, 50
Brandy City (California), 16
British East African expedition: camping arrangements on, 38–39; hunting aspect of, 36–37, 39, 40, 42–43, 44–45, 328n5; members of, 35–36; photographic documentation of, 38, 40, 41; route of, 35, 36 fig.
British Museum of Natural History (London), 88, 194, 195–96, 334n2
Britts Meadow (Utah), 229
Brooks, Allan, 125
Browning, Robert, "The Lost Leader," 343n3
Bryan, William A., 48, 55, 97, 264
Bryant, Harold, 300, 349n30
Bulawayo (Zimbabwe), 45
Bunker Hill mine (eastern Sierra Nevada), 246–47, 249, 250

364 / Index

Cabo San Lucas (Baja California), 296
Cairo (Egypt), 205, 207, 212
California Academy of Sciences (San Francisco): acquisition/loan policy at, 57, 78; destruction of, 57; Fossil Lake specimens at, 26; public exhibitions at, 87, 334n26; specimen determinations at, 284
California and Hawaiian Sugar Company (C&H Sugar), 11
California Associated Societies for the Conservation of Wildlife, 153
California Fish and Game Commission, 150, 153
California Magazine, 298
California Mammals (Stephens), 63
California Museum of Vertebrate Zoology. *See* Museum of Vertebrate Zoology
California Save-the-Redwoods League, 153
California State Fair, xiv, 131
California State Geological Survey, 280, 326n1
California Sugar Refining Company, 11
California Wildlands Program, 316
Camel riding, 205–6
Camp, Charles, xvi, 170, 221; on Ralph Chaney's interference, 178; on paleontology schism, 176–77, 345n25; UCMP appointment to, 174
Campbell, William W., 170, 171–72, 173, 174, 175
Cape Town (South Africa), 45
Carnegie Institute of Washington, 166, 171
Carter, Annetta, 284; *Acacia* species of, 355n11; Baja expeditions of, 291, 292–98, 355n4, 356n14; education/career of, 291; and Kellogg, 315–16; mechanical skills of, 292
Carter, George, 265
Cascade Range (California), 141
Castillo Cave (Puente Viesgo, Spain), 198, 199–200

Castor canadensis phaeus (a beaver species), 357n4
Cat Creek (Mineral County, Nevada), 286
Cattle breeding, 128, 131–32, 190, 193–94
Cedar Grove (Kings Canyon National Park, California), 305–6
Central Asiatic expeditions (sponsored by American Museum of Natural History, New York), 191, 346n5
Cerro de la Giganta (Baja California), 294, 295
Cerro de la Laguna (Baja California), 297
Chamberlain Tract (Sacramento River Delta), 121
Chaney, Ralph, 167, 172, 174, 178, 184, 345n30, 346n5
Charleston Peak (Nevada), 278
Chatham Strait (Alaska), 67 fig., 74
Chattin, John, 251–52
Chichagof Island (Alaska), 66, 67 fig., 73, 74
Chipmunks, 234, 237
Cholla cactus, 278
C&H Sugar, 11
Church of the Holy Sepulchre (Jerusalem), 211
Citroën automobile, 196, 197, 347n11
Civet cat (spotted skunk), 231–32
Clark, Bruce, 170, 171, 177–78, 190–91
Clark Mountain (California), 151, 278
Clethrionomys (red-backed vole), 233
Cliff chipmunk *(Eutamias dorsalis)*, 237
Clyde, Norman, 249, 250, 251
Coal Valley (California), 288
Coast Range (California), 141, 146
Collecting trips. *See* Alexander, Annie Montague, collecting trips
Colobus monkeys, 44–45
Colorado River projects, 160, 276
Colorado wood rat, 233
Colossi of Mennon (Egypt), 209
Comondú (Baja California), 295

The Condor (journal), 92, 331n9
Conservation: Alexander's interest in, 219, 239, 275–76, 309; Grinnell's focus on, 122, 123, 150, 239, 300, 349n30; MVZ's role in, 158, 159–60, 299–300, 356n4; by National Park system, 152–53
Constance, Lincoln, 285, 290
Contra Costa County (California), 23
Cooke, Amos Star (Alexander's grandfather), 4, 5
Cooke, Joseph P. (Alexander's cousin), 10
Cooke, Martha Eliza. *See* Alexander, Martha Eliza
Cooke, Will (Alexander's cousin), 15, 271
Cooper Ornithological Society, 93, 99
Coppermine Cove (Alaska), 67 fig., 73
Cordelia Shooting Club (San Francisco), 101, 121
Cordova Bay (Prince William Sound, Alaska), 103
Crater Lake (Oregon), 16, 28, 275
Creosote bush, 278
Crockett (California), 11
Cro-Magnon art, 197, 199–200, 201
Crowley, Josephine, 357nn20,22
Cugullière (cave explorer), 201

Daily Californian (UC student newspaper), 250–51
Damascus (Syria), 212
Davidson, Pirie, 183, 187, 188, 346n14
Davis, John, 251–52
Dawson, Bill, 62
Dawson, Charles, 194–95
Death Valley (California), 243
Deep Spring Lake (eastern Sierra Nevada), 244
Deer mice, 246
DeMay, Ida, 357n22
Dendur (Egypt), 210
Depression, the, 161–62, 348n3
Desman *(Myogalea [Galemys pyrenaicus])*, 200
Despard, Edward, 138, 139

Diablo, Mount (Berkeley), 23, 122
Dipodomys (kangaroo rat), 227, 238–39, 246, 305; *spectabilis* (bannertail), 349n28
Disenchantment Bay (Alaska), 61–62
Dixon, Joseph S., 65, 73, 74; Alaskan expeditions of, 66, 84, 98, 102, 104; Alexander's benefaction to, 119; bird collecting by, 64, 69, 70, 71, 332n17; at Cedar Grove, 305–6; letter to Hasselborg, on Alexander, 313; on Innisfail Ranch activities, 127, 129, 130; on preservation, 349n30
Dodge Power Wagon, 288, 289, 292–93, 297, 315
Dole, Sanford B., 324n6
Dordogne (France), 202

Earthquake (San Francisco, 1906), 116
East Bay hills (Berkeley), 23
Eastwood, Alice, 53, 284
Ectosperma alexandrae, 311
Edna (Alexander's boat), 105
Egypt: collecting specimens in, 205, 207, 210, 332n22, 347n6; illness in, 212–13; riding camels in, 205–6; temples/tombs of, 203, 208–9, 210, 347n2; tents from, 206, 207, 212, 215–16; voyage to, 204, 205
"El Carrol." *See* Dodge Power Wagon
Eldama Ravine (British East Africa), 42
Elephants, 45
Elgon, Mount (British East Africa), 36 fig., 42, 43
Ellis, Ralph, 259
El Paso (Texas), 224–25
Enchanted Vagabonds (Lamb), 290
English sparrows, 265–66
Entebbe (Uganda), 44
Eocene fossils, 25 fig., 205
Erickson, Mary, 301–2, 303–4
Escondido Bay (Baja California), 294
Estanove, M. Jacques, 200
Esterly, W. B., 32
Eucalyptus trees, 275

Eutamias dorsalis (cliff chipmunk), 237
Evolutionary studies: of botanical specimens, 227, 285; Grinnellian approach to, 76, 83, 152; of Mongolian "missing link," 191, 346n5; in perturbed habitats, 224–25, 348n1; of Piltdown man, 194–95, 347n9; of pocket gophers, 218; by G. G. Simpson, 311; of Trinity Mountains fauna, 141–42, 146–47; of Vancouver Island fauna, 138
Excelsior Mountains (Nevada), 226–27
Exhibits. *See* Public exhibitions at museums
Les Eyzies national museum of (France), 202

Fairfield (California), 123, 125, 185
Faiyûm (Egypt), 206–7
Fall River Mills (California), 16
Farm. *See* Innisfail Ranch (Grizzly Island)
Ferry, to Grizzly Island/Innisfail Ranch, 123, 136–37, 339n22
Field Museum of Natural History (Chicago), 86, 97
Field notes: Alexander's before meeting Grinnell, 51–52; Grinnell's method of, 77, 83; score keeping, 219; transcription of, 144, 151–52
Fieldwork: Alexander's passion for, 148–49, 235–36; attire for, 238; from auto camps, 231; and automobile misadventures, 221, 228, 230–31, 232–33, 288, 348–49n11; automobile's impact on, 153–54; with Pirie Davidson, 183, 187–88; and driving/cooking, 288–89; with Mary Erickson, 303–4; governmental restrictions on, 285–86; Grinnell's methods of, 76–77, 83, 97–98, 152, 243; guns used in, 25, 229, 238, 304, 339n4; with Carlotta Hall, 277; messiness of, 139; partnership requirements of, 17–18, 98–99; in perturbed areas, 76–77, 149–50, 160, 224–25, 239, 348n1; and skeleton preparation, 48–49, 68, 143–44, 151; with Natasha Smith, 303; traps used in, 143; during wartime, 261–62; and weather complaints, 234, 237–38; by women biologists, 303–6, 357nn20,22; women's exclusion from, 99, 301. *See also* Botanical specimens; Fossil specimens; Topotype specimens; Vertebrate specimens
Fifteen Club (book club), 277
Fisher, Walter, 64–65
Flat Island (Kailua Bay, Oahu), 266–67
Flint, Edward, 101
Flint, James P., 100
Flood, Cora, 114, 115
Flood, James, 115–16
Flying squirrel, 146
Ford field vehicle, 204–5, 215, 216, 217, 219, 348–49n11
Fossil Lake (Oregon) expedition, 24–29, 326n7
Fossil specimens: at American Canyon, 50–51; at Barstow Formation, 187–88; in Bay Area, 22–23, 326n4; at Billings Gap, 185; of California State Geological Survey, 326n1; at Castillo Cave, 199–200; comparative vertebrate specimens for, 56, 76, 79; in Egypt, 205–6; excavation of, 50–51; at Fossil Lake, 24–29, 25 fig.; at High Rock Canyon, 183; in Humboldt County, 50–51, 108–9; in Kern County, 215–17; at Mint Canyon, 186–87; named for Alexander, 31, 33, 309, 317, 327n25; new UCMP building for, 174–75; on Oahu, 23; from Petrified Forest, 186; at Piltdown, 194–95, 347n9; at Quaternary sites, 24, 25 fig.; at Rancho La Brea excavations, 107–8, 336n4; in Shasta County, 30–31, 33; at Thousand Creek Formation, 109, 183–84; to UCMP, by Alexander,

357n7; in Virgin Valley, 108–9. *See also* UC Department of Paleontology
Francis I (king of France), 197
Franklin automobile: dilapidated condition of, 287; fieldwork misadventures in, 221, 228, 230–31, 232–33, 288, 348–49n11; name for, 183; on Saline Valley expedition, 245, 246–47
French Creek (Vancouver Island), 140
Frick, Childs, 187, 188–89, 191, 216, 311
Fur-Bearing Mammals of California (Grinnell, Dixon, and Linsdale), 156
Furlong, Eustace: Alexander's expeditions with, 30–31, 32, 50, 108, 110; and Ford truck, 215, 216; UCMP appointment to, 170
Furlong, Herbert, 24, 25, 26

Geological Society of London, 195
George W. Elder (steamship), 59
Gîza (Egypt), 205
Glacier Bay (Alaska), 73–74, 104
Glass Peninsula (Alaska), 67 fig., 68
Goddard, Malcolm, 50
Goethe, *Wilhelm Meister's Apprenticeship*, 225
Goldman, E. A., 242
Graduate students: Alexander's benefaction to, 299, 300–301, 304–7, 357n26; Alexander's field trips with, 303–4
Graham Mountains (Arizona), 235
Grand Canyon, 185
Grant, Mount (Mineral County, Nevada), 286
Great Basin, 222, 279–80
Great Pyramid, 208
Greely, Will, 26
Green, Morris, 241
Greene, Edward L., 280
Greenville (California), 16
Gregory, Joseph, 345n25
Gregory, William K., 344n14
Grinnell, Hilda, 301; Alexander's correspondence with, 253–54, 268, 305, 306, 308; book club of, 277; women's field expeditions with, 304–6, 357n22
Grinnell, Joseph: Alaskan expeditions of, 64, 66; on Alexander's contributions, 93, 155–56, 160, 168–69, 175–76, 312–13; Alexander's relationship with, xvii, 65, 88, 90–91, 93–96, 161, 236, 253–54; on Arizona topotypes, 236; bird collections of, 64, 158–59, 342n9; cataloguing method of, 83–84; conservation activism of, 122, 123, 150, 152–53, 239, 300, 349n30; death of, 253, 278; debate over successors to, 255–57, 350n9; on directorship of UCMP, 173; doctorate of, 333n9; early education of, 64–65; fieldwork methodology of, 76–77, 83, 97–98, 152, 243; on Hawaiian bird species, 265–66, 352n11; on Kellogg's manuscript proposal, 142; MVZ functions document of, 254–55; MVZ proposal to, 75, 332n22; MVZ role/contributions of, 79, 82, 93–96, 157–58, 313, 333n19; on MVZ site, 77; nonresearch activities of, 117–18; offer from Biological Survey, 156; on perturbation's impact, 149–50, 340n4; pocket gophers study by, 152, 218–19, 221–22, 251, 302; on preservatives, 235; on public exhibitions, 334n26; resignation contemplated by, 94–95; on Theodore Roosevelt's visit, 163; on Saline Valley expedition, 244–45, 248–49; Trinity fauna manuscript of, 146–47; on Vancouver fauna, 138, 139–40; on Vancouver song sparrows, 139–40; on women naturalists, 99, 100, 301, 302
Grinnell, Joseph, Alexander's letters to: on ammunition needs, 140, 339n4; on Bluff, Utah, 233; on

Grinnell, Joseph *(continued)*
    eastern Sierras route, 244; on Egypt, 212–13; on female partner, 99, 100; on E. A. Goldman, 242; on Morris Green, 241; on Grinnell's accomplishments, 157–58; on Hawaii, 263, 267–68, 276; on Hawaiian fauna collections, 265–67, 352n12; on intestinal ailments, 217, 220; on John Merriam, 166; on MVZ functions document, 254; with natural history details, 235–36; on Saline Valley rescue, 249; on self-perception, 92
Grizzly Island (Suisun Bay, California): asparagus from, 133–34; landscape/wildlife of, 120–22; means of access to, 123, 136–37, 339n22; as MVZ field site, 122–23, 338n6; as wetlands habitat, 316. *See also* Innisfail Ranch
Grizzly Island Wildlife Area, 316
Ground sloth, 24, 108, 194
Gulick, John Thomas, 324n6, 328n3
Gulick, Thomas, 36, 39, 41–42

Habitat perturbation. *See* Perturbed habitats
Haiku (Maui), 262–63, 275; irrigation of, 9; landscape of, 7
Haleakala volcano (Maui), 7, 262, 309
Halfa (Sudan), 210
Hall, Carlotta, 277–78, 305
Hall, E. R., xvi, 147, 253, 279, 285, 356n4; Alexander's benefaction to, 256, 257, 351n15; on Alexander's gate episode, 286; at Kansas museum, 259–60, 356n4; personal traits of, 255–56; rejected as MVZ director, 256–58; taxa appellations by, 310; on women naturalists, 302
Hall, Harvey Monroe, 277
Hamakua ditch project (Maui), 9
Hanalei (Kauai), 272
Hanning Bay (Montague Island), 105
Harding, Warren, 184

Harriman, Edward H., Alaskan expedition of, 59–60, 330n5
Hasselborg, Allen, 74, 138; Alexander's admiration of, 69–70, 73, 313; Alexander's benefaction to, 90; bear killed by, 103, 106, 336n13; Joseph Dixon's correspondence with, 127, 129, 130, 313
Hasselborg Lake (Admiralty Island, Alaska), 70, 71, 331–32n15
Hasselborg River (Admiralty Island, Alaska), 71–72
Hawaii: Alexander-Kellogg's stays in, 265, 268; Samuel Alexander's childhood in, 4–5; Alexander's familial ties to, 23, 49–50, 185, 262–63, 267–68; Alexander's grave in, 312; bird specimens from, 264–66, 352nn12,13; habitat perturbation in, 275; Hamakua ditch project in, 9; Theodore Hittell's book on, 53–54; missionary activities in, 4; native ferns of, 272, 273; nonnative species in, 275–76; sugar industry in, 10–12; trips to, with nephew Jack, 272; Waterhouse home in, 262. *See also specific locations*
Hawaiian Commercial Co., 10
Hawaiian Commercial & Sugar Co., 112, 118
Hawaiian ferns, 272, 273
Hawaiian rats, 266–67
Hawkins Island (Prince William Sound, Alaska), 103–4
Hawley, Colonel, 192
Hearst, George, 115
Hearst, Phoebe Apperson, 114, 115
Hearst, William Randolph, 115
Heather vole *(Phenacomys)*, 233
Heindl, A. J., 108, 110
Heller, Edmund: Alaskan expedition of, 102, 104–5; on Louise Kellogg, 97, 98, 102, 106; and Theodore Roosevelt, 162–63; on taxa appellations, 309, 357n4
Herbarium. *See* UC Herbarium
Herrick Bros. (Maine), 204–5, 215

High Rock Canyon (Washoe County, Nevada), 183
Hinchinbrook Island (Prince William Sound, Alaska), 102, 103, 104
Hipparion (fossil horse), 29, 184
"History of the Hawaiian Islands" (Theodore Hittell), 53–54
Hittell, Katherine, 53, 284
Hittell, Theodore, 53–54
Hog-nosed skunk, 237
Holy Land sites, 211
Honey creepers, 264
Hopi pueblos (Arizona), 241–42
Hospital for Tropical Diseases (London), 220
Howard, Alice Q., 323n4
Howard, Jack, 286
Howell, John Thomas, 280
Hubert (French museum curator), 197
Humboldt County (Nevada) expeditions, 50–51, 108–9
Hunting: for bears, 58–59, 66, 330n4; for ducks, 121; for food rations, 39, 40, 42–43; for sport, 35, 36–37, 44–45, 328n5
Huntoon Valley (Nevada), 228
Hurlburt family, 44

Ichthyosaurs: American Canyon expedition for, 50–51; named for Alexander, 31, 33, 309, 327n25; Shasta County expeditions for, 30–31, 32, 33; South Kensington Museum, 194
Ictonyx libyca (zorilla), 207
Icy Strait (Alaska), 73
Idaho Inlet (Chichagof Island, Alaska), 74
Ilingoceros (fossil ungulate), 303, 317
Indian Snake Dance, 241–42
Innisfail Ranch (Grizzly Island, California): anniversary party at, 132; asparagus reputation of, 133–34, 277, 316; broken levee on, 136; cattle breeding on, 128, 131–32; coat-of-arms at, 125; farm machinery, 125–27, 338n10; as a haven, 120–21; infrastructure at, 128–29); irrigation of, 126–27, 132–33; Kellogg's contributions to, 129, 134; Kellogg's ownership of, 316; labor management at, 129–30, 134–35; map of, 124 fig.; means of access to, 123, 136–37, 339n22; as MVZ field site, 122–23, 338n6; origin of name, 338n8; plants grown at, 130–31, 275; plowing problems on, 125–26, 338n10; taxa described from, 319–20; as wetlands habitat, 316
Inside Passage (southeastern Alaska), 65, 66, 102
Irrigation projects: at Innisfail Ranch, 126–27, 132–33; on Maui, 9
Island Slough (Grizzly Island, California), 125

Jepson, Willis Linn, 282–83, 309, 354nn13,14
Jepson Herbarium, 354n14
Jerboas, 205, 207, 347n6
Jericho (Jordan), 211
Jerusalem, 211
Jinja (British East Africa), 36 fig., 44
John Day Formation (Oregon), 24
Johnston, Ivan M., 284
Joice Island (Sacramento River Delta), 123
Jones, Katherine, 30, 327n18
Jordan, David Starr, 87, 333n9
Joshua trees, 278
Juneau (Alaska), 61, 66, 67 fig.

Kahala (Honolulu), 268
Kahului Railroad (Maui), 11
Kailua Bay (Oahu), 266–67, 268
Kamehameha, King, 4
Kangaroo rats *(Dipodomys)*, 227, 238–39, 246, 305, 349n28
Kanzler (steamer), 36
Karnak, Temple of (Egypt), 208
Kauai (Hawaii), 4, 5, 272
Keck, David D., 285
Kégresse automobile, 196, 197, 347n11

Kellogg, Albert, 101
Kellogg, Anita, 13, 100, 101
Kellogg, Charles W., 13, 100–101, 121
Kellogg, Louise: ailments of, 100, 101, 102, 212, 237–38, 272; and Alexander family, 268–69, 353n21; and Alexander's death, 311–12; Alexander's relationship with, xv–xvi, 101–2, 106, 110, 312, 323n7, 323nn6,7; on Altamira caves, 198–99; asparagus farming by, 133–34; automobile misadventures of, 228, 230–31, 232–33, 288, 348–49n11; Baja California expeditions of, 292–98, 315–16; Bay Area apartment of, 214–15; botanical collecting by, 277–80, 279, 282, 284–85, 286; and Annetta Carter, 315–16; at Castillo Cave, 199–200; cattle breeding by, 190, 193–94; cooking responsibility of, 289; diaries of, xv, 323n4; in England, 192–96; family background of, 100; fieldwork attire of, 238; financial status of, 120; fishing/hunting interests of, 101, 225; fossil expeditions of, 183–84, 185–88, 215–16; in France, 196–98, 200–202; in Hawaii, 263, 265, 268; Edmund Heller's praise of, 102, 106; manual skills of, 101, 129, 286; mastodon find by, 109; on Menna's wall paintings, 209; Middle Eastern tour of, 204, 205, 206–11; publications of, 142, 147, 340n19; Saline Valley trip of, 245–51; Samoan trip of, 269–71; shrew find by, 242; and skunk incident, 232; South Pacific trip of, 271–72, 353n26; specimen preparation by, 149, 151–52; taxa named for, 319, 355n11; topotype collections by, 224–25, 226, 227, 229, 234; wartime activities of, 261–62
Kellogg, Martin, 13, 101, 114–15
Kenai Peninsula (Alaska), 61, 62, 102
Kenya expedition. *See* British East African expedition

Kern County (California) expedition, 215–17
Keys sisters, 190
Kijabe village (British East Africa), 41, 44
Killisnoo (Admiralty Island, Alaska), 67 fig., 71, 72
Kingman (Arizona), 186
Kings Canyon National Park (California), 305–6
Klamath Falls (Oregon), 16
Kodiak Island (Alaska), 59, 62
Kokee (Kauai), 272
Konig (steamer), 45
Kootznahoo Inlet (Admiralty Island, Alaska), 67 fig., 71, 72
Kreuter, Martha Hurd (Alexander's niece), 325n7, 353n21
The Kumulipo (Beckwith), 118

Lagopus (ptarmigan), 71, 103; *lagopus alexandrae* (willow), 332n17; *mutus dixoni* (rock), 332n17
Lahaina (Maui), 4
Lahainaluna Seminary (Maui), 4, 6
Lakeview (Oregon), 27
Lamb, Charles, 355n28
Lamb, Dana, 290
Lanai (Hawaii), 262
Land reclamation projects, 122, 150, 160, 186–87
Lane, Franklin, 159–60
Lang, Thomas, 193
La Paz (Baja California), 296, 297
Larkspur, 278
Lascaux (France), 202
Lasell, Edward, 14
Lasell Seminary for Young Women (Massachusetts), 14, 19, 40
Lassen Buttes, 16
Last Chance Canyon (Kern County, California), 215, 216
Last Chance Range (eastern Sierra Nevada), 245, 246
Latouche Island (Prince William Sound, Alaska), 102, 105–6
Lawson, Andrew, 29

Lead Canyon (eastern Sierra Nevada), 245, 247
Leone (American Samoa), 269, 271
Letters to Martha, xv; of affection, 16–17; on African safari, 36, 43–44; on aging, 240; on Alaskan expeditions, 58–59, 60, 69, 70, 72–73; on Alaskan magazine proposal, 181; archived collection of, 322n3; on auto camps, 231; on Bay Area home, 214, 221; on botanical specimens, 227; on Joseph Grinnell's death, 253; on Hawaiian rats, 267; on health problems, 18, 19, 220; on Katherine Hittell, 53; on Louise Kellogg, 110; on marriage, 19–21; on Mauna Loa, 263; on John Merriam, 33–34, 109–10, 114, 166, 168, 171, 343n1, 343n7; on Miocene Barstow fossils, 188; on morphological analysis, 146; on MVZ's new building, 175; on paleontology studies, 22, 29; on parents' deaths, 47, 182; on patriarchal restrictions, 34, 327n27; on personal traits, 18, 19, 29, 51, 69, 149, 185, 189, 222, 238; on politics, 184; on private collection, 55; on Shasta County expeditions, 29–30, 32, 33; on skeleton preparation, 48–49; on stock investments, 89, 334n3; on Suisun Marsh, 121, 122, 132; on Mary Wilson, 26, 27; on Yosemite, 153
Life-Histories of African Game Mammals (Roosevelt and Heller), 163
Life Sciences Building (UC Berkeley), 87
Life zones theory, 286, 354n22
Littlejohn, Chase, 66, 69, 73, 74, 331n10
Little Qualcium River (Vancouver Island), 140
Livingstone (Victoria Falls region), 46, 328n18
Livingstone, David, 35
London (United Kingdom), 47, 194, 196

London, Jack, 13
London Geological Society, 195
Long-tailed pocket mouse *(Perognathus formosus)*, 234
Lookout (California), 16
Loreto (Baja California), 294
Los Angeles County Museum of Natural History, 157
Los Angeles Times, 250–51
Los Gatos (California), 23
"The Lost Leader" (Browning), 167, 343n3
Louis XIV (king of France), 197
Lourdes (France), 200
Lower California. *See* Baja California expeditions
Lower Klamath Lake (Oregon), 27
Lower Pliocene, 216
Lower Warm Springs (Saline Valley, California), 246
Luscomb (Grizzly Island, California, resident), 123
Luxor, Temple of (Egypt), 208

Maasai people, 41
Magdalenian culture, 25 fig., 199, 201, 202
Makawao (Maui), 7
Makawao Cemetery (Maui), 312
Malaria, 6, 9, 18, 37, 326n17
Malaspina glacier (Alaska), 61
Mallow, 278
Mamelukes, tombs of, 207
Mammal collections: of Ralph Ellis, 259; of C. Hart Merriam, 59, 162, 163–64, 330n4; national ranking of MVZ's, 162. *See also* Vertebrate specimens
Mammals. *See* Alexander, Annie Montague, collecting trips
Mammals of Nevada (Hall), 256, 257, 351n15
Mammoths, 24, 194
Manono (Western Samoa), 270
A Manual of the Flowering Plants of California (Jepson), 282, 354n13
Mas-d'Azil cave (France), 201–2

Mason, Herbert, 242, 273, 279, 315; on Alexander's botanical collection, 280, 281; botanical determinations by, 283–84, 311; on Jepson, 282, 354n13

Mastodon, 108, 109

Matapao (Zimbabwe), 45

Matson, William, 11

Matson Navigation, 11

Matthew, William Diller, 172, 173, 303, 344n14

Maui: Alexander's grave on, 312; collecting activities on, 4–5; family homestead on, 262–63; Hamakua ditch project on, 9; northcentral landscape of, 7. *See also* Hawaii

Maui Railroad & Steamship Company, 11

Mauna Kea (Hawaii), 272

Mauna Loa volcano (Hawaii), 185, 263–64, 272

McDonald, Edna Wemple. *See* Wemple, Edna

McKinney, Mary Ann (Alexander's grandmother), 3, 8 fig.

Memler, Viola, 357nn20,22

Menna's tomb (Egypt), 209

Men of the Old Stone Age (Osborn), 190

Merriam, C. Hart, 48, 65, 329n4, 336n3; Alaskan expedition advice from, 59, 60–61, 63; on Alexander's bear collection, 91; Alexander's introduction to, 53, 54; Alexander's museum proposal to, 54–55, 56–57; bear specimen collection of, 59, 162, 330n4; donations to MVZ by, 163–64; formal recognition of, 57, 329–30n22; life zones theory of, 286, 354n22; ornithology pursuits of, 52; Smithsonian tour with, 84; on taxa appellations, 309

Merriam, Frank (governor of California), 250

Merriam, John C., xvi, 153, 336n13; Alaskan expedition proposal of, 52; Alexander's benefaction to, 23–24, 29–30, 65, 107, 327n17, 336n3; conflicting personal agenda of, 112–13, 114, 165–66, 168, 170–72, 343n1; departure from UC by, 166–67, 181, 343n3; fossil expedition proposals by, 24, 26, 30, 326n7; national prominence of, 165–66; Paleontology Department appointment to, 111; Rancho La Brea focus of, 107–8, 336n4; respectful of Alexander, 34, 112, 300; and Robert Sproul, 345n30; Thousand Creek expedition of, 109, 110; UCMP's friction with, 169, 170–72; UC paleontology courses of, 22; on vertebrate zoology museum, 79–80, 333n11

Mexico City trip, 290–91

Microtus coronarius (vole), 339n7

Middle Eastern tour, 203–12

Middle Triassic, 50–51

Miller, Alden H., 89, 259, 266, 304, 344n12; on Alexander's contributions, 179–80; at Alexander's home, 308; on "Analysis of Functions," 255; and E. R. Hall, 257–58; personal traits of, 255; wartime budgeting by, 261

Miller, Loye Holmes, 255, 350n9, 352n13

Miller, Virginia, 305

Milton, *Paradise Lost*, 15

Mink specimens, 123

Mint Canyon (southern California), 186–87

Miocene fossils, 23, 108, 165, 186–88

Missionaries: impact on flora of, 275; as naturalists, 4–5

Mitchell Bay (Admiralty Island, Alaska), 72

Moa bones, 194

Mojave Desert (California) fieldwork, 186, 243, 251. *See also* Alexander, Annie Montague, collecting trips; Saline Valley expedition

Mokapu Peninsula (Oahu), 267

Moku o Loe Island (Kaneohe Bay, Oahu), 267

Mole, semiaquatic. *See* Desman
Mole Harbor (Admiralty Island, Alaska), 67 fig., 69–70
Mombasa (Kenya), 35, 36
Monkeys, colobus, 44–45
Montague, Juliette (Alexander's grandmother), 5
Montague Island (Prince William Sound, Alaska), 102, 105–6
Montezuma Slough (Grizzly Island, California), 123, 125, 126, 132–33; ferry across, 136–37, 339n22; salinity of, 316
Moorea, 271
Morgan, Julia, 13
Mountain beaver *(Aplodontia)*, 146, 303
Mud hens, 125
Muir, John, 53, 244
Mulegé (Baja California), 294
Mumias (British East Africa), 36 fig., 42
Munro, George C., 264–65
"The Museum Conscience" (Grinnell), 168
Museum of Vertebrate Zoology (MVZ): academic freedom of, 156; Alexander's benefaction to, 78, 85–86, 158, 159, 160–62, 179–80; Alexander's specimen donations to, 78, 91, 222–23, 233, 240, 310, 357n7; "Analysis of Functions" of, 254–55, 256; bird specimen donations to, 158–59, 160, 259, 265, 266, 342nn9,13, 352nn11,13; cataloguing method at, 83–84; choosing Grinnell's successor at, 255–57, 350n9; conservation role of, 158, 159–60, 299–300, 356n4; construction plans for, 84–85; Ralph Ellis's specimen donations to, 259; fieldwork policies of, 97–98; financial management of, 88–90; first accession of, 75, 84, 332n22; Morris Green's specimen donations to, 241; Grinnell-Alexander relations at, 88, 93–95, 236; Grinnell's specimen donations to, 158–59, 342n9; C. Hart Merriam's research materials donation to, 163–64; Alden Miller's directorship of, 257–58, 344n12; motivations for founding of, 3, 42, 48, 55–57, 76–77; proposals on founding of, 54–55, 56, 75, 78–79, 80–82, 332n22, 333n11; public exhibitions at, 86–87, 158, 334n28; rankings of, in collections, 162, 342n20; relocation of, 174–75; research function of, 78–80, 87–88, 116, 142, 299; research prominence of, 155–56, 157–58, 313; Theodore Roosevelt's visit to, 162–63, 343n24; site locations for, 77–78; stipulations on control at, 80–82, 94–95, 117, 158, 335n19; teaching function of, 117, 299; women's admittance to, 301–2; women's exclusion from, 99; World War II's impact on, 261
Myogalea ([Galemys pyrenaicus] desman), 200

Nairoba, 37, 41
Naivasha, Lake (British East Africa), 38
Nakuru (British East Africa), 35, 36 fig., 37, 39, 40
Nanaimo (Vancouver Island), 139
National Park system, 152–53
National Research Council, 166
Nelson, Edward W., 52, 290
Ne-ne (native Hawaiian goose), 266
Neolithic artifacts, 196, 197, 201
Newcastle upon Tyne (United Kingdom), 193
New Hebrides (Vanuatu, South Pacific), 55
New York Buttes (eastern Sierra Nevada), 245
Niaux, grotte de (France), 201
Normal School (Massachusetts), 6
North, Arthur, 289
Nucheck (Hinchinbrook Island, Prince William Sound, Alaska), 98
Nyeri Desert (Taru Desert) (British East Africa), 37

Oahu (Hawaii), 23
Oakland: Alexander and Kellogg's homes in, 120; Samuel Alexander's move to, 9–10; Alexander's unhappiness in, 12, 18–19, 21, 148; attractions of, 13–14
Oakland Hills (California), 23
Oakland Public Museum, 101
O'Brien, William, 115–16
Ogilvie, Florence, 357n22
Olinda (Maui), 7, 262
Olindita (Oahu), 262
O-o bird, 266
Oraibi (Arizona), 241–42
Organic Act, 81
The Origin and Evolution of Life (Osborn), 190
Ornithology. *See* Bird specimens
Osborn, Henry Fairfield, 189, 190–91
Osgood, Wilfred, 64–65
Osmont, Vance, 30

Paia (Maui), 7
Paia Plantation (Maui), 9
Paint brush, 278
Painted Desert (Arizona), 185–186
Paleolithic cave paintings, 197, 199–200, 201
Paleontology. *See* UC Department of Paleontology
Palestine, 210–11
Palm Springs Desert (California) expedition, 303–4
Papeete museum (Tahiti), 271–72
Paradise Lost (Milton), 15
Paris (France), 15, 17, 196–97
Parker Dam (Colorado River), 276, 277
Parksville (Vancouver Island), 139, 140
Patterson, Mount (Sweetwater Mountains, California), 279
PCR (Polymerase chain reaction), 348n1
Pearl Harbor (Oahu), 23, 261
Pennask Lake (British Columbia), 226, 240

Penstemon Palmeri, 278
Peril Strait (Baranof Island, Alaska), 74
Permits, collecting, 54, 59, 61
Perognathus (pocket mouse), 225, 239–40, 304, 349n31; *formosus* (long-tailed), 234
Perturbed habitats: from automobiles/roads, 153–54; from California's growth, 55–56, 276; in desert areas, 149–50; fieldwork's importance to, 76–77, 224–25, 239, 348n1; in Hawaii, 275–76; from land reclamation projects, 122, 150, 160
Petrified Forest (Arizona), 185, 186
Phenacomys (heather vole), 233
Philadelphia Academy of Natural Sciences, 88
Photography, 38, 41, 43, 46, 51
Piedmont community (Bay Area), 214, 347–48n2
Piette collection, 197
Pika, 229
Piltdown fossils, 194–95, 197, 347n9
Pine Canyon (Utah), 225
Pine Forest Mountains (Nevada), 108, 109
Pine Grove Hills (California), 288
Pine Valley Range (Utah), 229
Pleistocene sites, 165
Pliocene fossils, 23, 109, 186, 216
Pliocene rhinoceros *(Titanotherium)*, 109
Pocket gophers: in Excelsiors, 227; Joseph Grinnell's study of, 152, 218–19, 221–22, 251, 302; in Mojave Desert, 243, 251; named for Alexander, 242; in Tehama/Colusa counties, 145
Pocket mouse *(Perognathus)*, 225, 234, 239–40, 304, 349n31
Polymerase chain reaction (PCR), 348n1
Popoi'a Islet (Kailua Bay, Oahu), 266–67
Port Florence (British East Africa), 35, 42, 43

Port Frederick (Chichagof Island, Alaska), 67 fig., 73, 74
Power Wagon. *See* Dodge Power Wagon
Prack, Frances, 357n20
Prattville (California), 16
Pre-Chellean culture, 25 fig., 199
Prince William Sound (Alaska), 96, 102, 103–4, 105
Provo Canyon (Utah), 232
Ptarmigan, 71, 103, 332n17
Public exhibitions at museums, 86–87, 334n28
Puente Viesgo (Spain), 198, 199–200
Punahou School (Honolulu), 5, 12
Pyramids of Egypt, 205, 206, 207–8

Quaternary fossils, 24, 25 fig., 109
Quincy (California), 16
Quinn River Crossing (Nevada), 108

Ramses II, 208, 210
Ranch. *See* Innisfail Ranch
Rancho La Brea excavations (Los Angeles), 107–8, 165, 336n4
Rats: Hawaiian, 266–67; kangaroo, 227, 238–39, 246, 305, 349n29; wood, 233, 246
Rattlesnake hazards, 234, 237, 293
Ray, F. S., 32
Reciprocity Treaty (1876), 9, 10
Reclamation Act (1902), 122
Red-backed vole *(Clethrionomys)*, 233
Red Bluff Bay (Baranof Island, Alaska), 67 fig., 72
Redox, Marie, 357n22
Red Rock Canyon (California), 215
Redwood Canyon (California), 23
Redwoods preservation, 153
Regillus building (Oakland), 214–15
Reinhart, Roy, 301
"Report upon mammals and birds found in portions of Trinity, Siskiyou and Shasta counties" (Kellogg), 147, 340n19
Revolutionary Reform Party, 10
Rhodes, Cecil, 45

Rhodesian skull, 194, 197
Richardson, Bill, 249, 250
Richardson, Charles, 97, 98, 109
Rift Valley (British East Africa), 38, 44
Ripon Falls (British East Africa), 44
River otters, 123, 338n6
Rock ptarmigan *(Lagopus mutus dixoni)*, 332n17
Rodman Bay (Baranof Island, Alaska), 67 fig., 74
Romantic friendships, xv–xvi, 323n6
Roosevelt, Franklin D., 185
Roosevelt, Theodore, 162–63
Roosevelt Memorial for Excellence, 57, 329–30n22
Rothchild, Charles, 334n2
Rothchild Collection at Tring, 88, 264, 334n2
Rowe, John, 132, 190
Rowley, John, 86–87, 90
Rusick, Jean, 357n22
Russell, Ward, 151, 183, 249–50
Russian Gulch State Park (Mendocino County, California), 305

Sacramento Bee, 250–51
Sacramento River Delta, 120–22, 125, 315
Saiga antelope, 200
Saline Valley (California) expedition: rescue from, 249–51; road travel on, 245, 246–47; vertebrate specimens from, 245–46, 251; weather/rations on, 248
Samford Courtney (United Kingdom), 193
Samoan trip, 269–71
Sandalwood deforestation, 275
San Diego Museum of Natural History, 334n28
Sandwich Islands. *See* Hawaii
San Francisco, 10, 13, 24
San Francisco Chronicle, 41, 117
San Ignacio (Baja California), 294
San Jacinto Mountains (California), 98
San Joaquin River (California), 120–22, 125–26, 316

San José del Cabo (Baja California), 296
Santa Clara County (California), 23
Santa Cruz Mountains (California), 23
Santa Rosalía (Baja California), 294, 315
Santillana (Spain), 198–99
Saqqâra (Egypt), 206, 207–8
Sather, Jane, 13, 114, 115
Sather, Peder, 13, 115
Sauer, Carl, 290
Saurian Hill site (American Canyon, Nevada), 50–51
Savaii (Western Samoa), 269
Science, 91, 158
Seale, Alvin, 61, 330n9
Second Mesa at Shipaulovi (Arizona), 241–42
Seldovia (Kenai Peninsula, Alaska), 62
Sequoia National Park (California), 149, 182
Setchell, William Albert, 280–81
Seton, Ernest Thompson, 38, 40
Shasta, Mount (California), 16
Shasta County (California) expeditions, 24, 30–33
Shasta Dam (California), 316
Shastasaurus alexandrae, 31
Shastasaurus expeditions, 30–33
Shipaulovi (Arizona), 241–42
Shrew *(Sorex)*, 234, 242, 310, 349n19
Shrike, 198
Sibley, Charles, 251–52
Sierra de la Laguna (Baja California), 296–97
Sierra de la Giganta (Baja California), 294
Simpson, George Gayton, 311
Siracusa (Sicily), 205
Sitkan region (Alaska), 138
Skeleton preparation, 48–49, 143–44
Skilak Lake (Alaska), 62
Skunk: Arizona striped, 233; hog-nosed, 237; spotted, 231–32
Sloughs, on Innisfail Ranch, 125, 126
Smith, James Perrin, 30–31, 50
Smith, Natasha, 303

Smithsonian Institution, 84, 88, 330n9, 338n6
Society for Vertebrate Paleontology, 93, 335n15
Sorex (shrew), 234, 242, 349n19; *grahamensis*, 310
South Pacific trip, 271–72, 353n26
Sparrows, 139–40, 265–66
Speciation: Grinnell's view of, 76, 83; in perturbed habitats, 224–25, 348n1
Specimen collections: cataloguing systems for, 83–84; for evolutionary studies, 56, 76, 83; John Merriam on value of, 79–80. *See also* Botanical specimens; Fossil specimens; preservative for, 235; Topotype specimens; Vertebrate specimens
Sphaeralcea, 295
Spiraea, 140
Spoonbill, 266, 352n12
Spotted skunk, 231–32
Spreckels, Claus, 10, 11
Sproul, Robert Gordon, 94–95, 153, 304; and Alexander's honorary degree, 176; and MVZ, 254–55, 256–57; and paleontology programs, 174, 176–77, 178, 345n25; UCMP endowment proposal to, 175, 344n20; UC presidency appointment to, 173, 345n30
Sprue disease, 220
Stanford University, 64–65, 77
Stanley, Henry, 35
Stauffacher (Hurlburt's assistant), 44, 45, 117
St-Christophe cliffs (France), 202
Stebbins, Robert, 342n20
Steinbeck, John, 294
Stephens, Frank, 69, 70, 71, 74, 75; Alaskan expedition of, 84; Alexander's benefaction to, 90, 98; career/reputation of, 63
Stephens, Kate, 63–64, 69, 74
Stevenson, Robert Louis, 270
Stewart Valley (Nevada), 219
St-Germain-en-Laye (France), 196–97

Stirton, Ruben, 259, 288, 308, 311, 351n22
Stock, Chester, 171, 183, 187, 203, 206, 344n10
Stocks: Alexander's investments in, 88–89, 334n3; benefaction's ties to, 89–90, 157, 159; endowment's earnings from, 158, 160–61, 279, 345nn32,33
Stonehenge, 192
Sugar: benefaction's ties to, 89–90, 157, 159; market competition in, 10–12
Suisun Marsh (California), 121–22. *See also* Innisfail Ranch
Suisun Marsh Preservation Act (1977), 316
Suisun Resource Conservation District, 316
Surprise Valley (California), 287
Swain Ranch (Vancouver Island), 140
Swallen, Jason, 311
Swarth, Harry S., 98, 301; Alexander's benefaction to, 90, 109; at MVZ, 97, 157; Vancouver expedition of, 138–39, 140; on vole species name, 141, 339n7
Sweetwater Mountains (Mono County, California), 276, 279–80, 289

Table Mountains (Virgin Valley, Nevada), 108
Tahiti, 271–72
Tantalus, Mount (Oahu), 262
Taxa appellations: of *Acacia* species, 355n11; Alexander's disassociation from, 141, 309–10, 339n7, 357n4; of fossils, 31, 33, 309, 327n25; honoring Alexander, 31, 33, 71, 242, 309, 317–19, 327n25, 332n17; honoring Kellogg, 319, 355n11; of pocket gopher, 242; of ptarmigan, 71
Taylor, Walter P., 90, 97, 98, 109, 313
Temple Tour, 203–12
Tempsky, Armine von, 40
Tents, Egyptian, 206, 207, 212, 215–16

Thalattosaurus alexandrae, 33, 327n25
Thomas Bay (Alaska), 67 fig., 74
Thompson Seton, Ernest, 38, 40
Thompson's Peak (Trinity Alps, California), 146
Thousand Creek Formation (Nevada), 109, 183–84
Throop Polytechnic Institute (now Cal Tech, Pasadena), 64, 65, 82
Times (London), 203
Titanotherium (Pliocene rhinoceros), 109
Tombs of the Queens (Egypt), 208
Tongass National Forest (southeastern Alaska), 331n8
Topotype specimens: of black-tailed prarie dog, 239; of cliff chipmunk, 237; from Columbia River region, 226; defined, 224; from Excelsior Mountains, 226–27; of "leucogenys" group, 242; from Mount Baldy, Arizona, 237; number of, to MVZ, 233, 240; from Palm Springs Desert, 304; from perturbed areas, 224–25, 239; of spotted skunk, 231–32; from Utah, 228–30, 231–32, 233, 234, 235
Toulouse museum (France), 200
Tractor, fat-wheeled, 125–26, 338n10
Traps, animal, 143
Triassic fossils, 30, 186
Tring Zoological Museum (Hertfordshire, United Kingdom), 88, 264, 334n2
Trinity Alps (California) expedition, 141–47; and Kellogg's fauna manuscript, 142, 147, 340n19
Tubbs Cordage (San Francisco), 100–101
Tule fog, 23
Tule Lake (California), 16
Tundra vole, 104
Tutankhamen's tomb, 203, 204, 208, 209, 347n2
Tutuila Island (American Samoa), 269, 270

Twentynine Palms (San Bernandino County, California), 243, 279
Type specimen, defined, 327n25

Uasin Gishu District (British East Africa), 36 fig., 42
UC College of Agriculture, 132
UC Department of Botany, 280–81. *See also* Botanical specimens
UC Department of Geology, 22, 111, 167, 170, 176–77
UC Department of Paleontology: Alexander's benefaction to, 107, 111–13, 167, 172–73, 336nn1,3; Alexander's course work in, 21–22, 29, 56; creation of, 111; John Merriam's departure from, 166–67, 343n3; John Merriam's personal agenda for, 112–13, 165–66, 343n1; research schism within, 111; UCMP's friction with, 170, 174, 176–77, 178; women's presence in, 302–3. *See also* Fossil specimens; UC Museum of Paleontology
UC Department of Zoology, 299
UC farm at Davis, 131
UC Herbarium: Alexander's specimens to, 242, 275, 309, 353n1; W. L. Jepson's friction with, 282–83, 354n14; Louise Kellogg's work at, 315; origins of, 280–81; specimen determinations by, 281–82, 283–84, 354n12. *See also* Botanical specimens
UC Museum of Paleontology (UCMP): Alexander's endowment of, 172–73, 175, 176–79, 344n20, 345nn23,32; Alexander's proposal for, 169; Alexander's specimens to, 357n7; Barstow Formation fossils to, 188–89; first appointments to, 170, 344n10; William Matthew's appointment to, 172–73, 344n14; John Merriam's interference with, 170–72; new building for, 174–75; women biologists at, 303

UC Museum of Vertebrate Zoology. *See* Museum of Vertebrate Zoology
University of California at Berkeley: Alexander's benefaction to, 23–24, 29–30, 32–33, 78, 85–86, 345n33, 358n12; Barrows's presidency of, 167–68; Campbell's presidency of, 170–72; earthquake's impact on, 116; Geological Survey's collections to, 326n1; Kellogg's presidency of, 13, 114–15; John Merriam's courses at, 22; MVZ proposal to, 78–79, 80–82; and MVZ salaries, 159, 337n6; as MVZ site, 77, 78; national rankings of, in 1893, 114; paleontology funding proposal to, 107, 111–13, 336nn1,3; Theodore Roosevelt's visit to, 162–63, 343n24; Sproul's presidency of, 173; UCMP proposal to, 169; Wheeler's presidency of, 32, 35, 78–82, 111–13, 115–16, 156, 163, 166, 167, 223, 337n3; women benefactors of, 115–16. *See also* Museum of Vertebrate Zoology; UC Museum of Paleontology
University of Kansas, Museum of Natural History, 258, 259–60, 356n4
Upolu Island (Western Samoa), 269, 270
U.S. Biological Survey, 52, 53, 54, 56, 63, 84, 156, 162, 163–64, 329n12
Ussat (French resort), 200–201
Utah collection expedition, 228–30, 231–32, 233, 234

Vailima (Western Samoa), 270
Valley of the Kings (Egypt), 208–9
Vancouver Island expedition, 138–41
Vassar Folklore Foundation, 118
Vertebrate specimens: from Alaskan expeditions, 59, 62, 70, 103–4, 105, 106; to Bishop Museum, by Alexander, 352n9; diminished pursuit of, 274, 276–77; from Egypt, 205, 207,

210, 332n22, 347n6; evolutionary importance of, 56, 76, 79, 146–47, 152; funds for purchase of, 160, 342n13; from Hawaii, 264–67, 352nn12,13; hunters' damage to, 144–45; loan policy on, 78–79; museum proposed for, 54–55, 56; to MVZ, by Alexander, 75, 78, 84, 91, 222–23, 310, 333n22, 357n7; named for Alexander, 71, 141, 242, 317–19, 332n17, 339n7; national ranking of MVZ's, 162, 342n20; from obscured sites, 154; permits to collect, 54, 59, 61; from perturbed areas, 76–77, 149–50, 224–25, 239, 348n1; preservation role of, 152–53; preservative for, 235; from Prince William Sound, 96, 103–6; recording of, in field catalog, 219; from Saline Valley expedition, 245–46, 251; from Samoa, 271; skeleton preparation of, 48–49, 68, 143–44, 151–52; from Trinity Alps, 143–45, 146; from Vancouver Island, 138–40, 141. *See also* Specimen collections; Topotype specimens

Vézère River district (France), 202

Victoria, Lake (British East Africa), 44

Victoria Falls, 45–46

Virgin Valley expedition (Humboldt County, Nevada), 108–9

Voles: heather, 233; named for Alexander, 141, 339n7; red-backed, 233; tundra, 104

Wagner, Warren Herb, 273, 301, 353n28

Waihee (Maui), 6

Waoili (Kauai), 4, 272

War God Spring (Utah), 242

Warm Springs (Saline Valley, California), 246

Warner Mountains (California), 286–87

Wasatch Mountains (Utah), 234

Wassuk Range (California), 287

Waterhouse, Elizabeth (Alexander's niece), 8 fig., 184

Waterhouse, Jack (Alexander's nephew), 8 fig., 266; on trips with Alexander, 269–72, 353n26

Waterhouse, John (Alexander's brother-in-law), 8 fig., 23, 49, 262, 268

Waterhouse, Martha Mabel (née Alexander, Alexander's sister), 3, 15, 226, 266–67; in Alexander's will, 312; birth of, 6; death of children of, 50, 74, 184, 229, 262; and Kellogg, 268; marriage/family of, 8 fig., 23, 49; Mount Tantalus home of, 262

Waterhouse, Montague (Alexander's nephew), 8 fig., 262

Waterhouse, Pattie (Alexander's niece), 8 fig., 269, 353n21

Waterhouse, Samuel (Alexander's nephew), 8 fig., 74

Waterhouse, Wallace (Alexander's nephew), 8 fig., 229

Waucoba Springs (eastern Sierra Nevada), 246–47

Weaverville (Trinity Alps, California), 144

Wemple, Edna, xvii; Alexander's expeditions with, 32, 50–51, 61, 62, 330n9; Alexander's letters to, 209, 210, 224–25, 231, 232, 234, 237–38, 303, 323n10; education of, 303; marriage of, 63; taxa appellations by, 309

West Indies, 21

Wheeler, Benjamin Ide, 153; Joseph Grinnell on, 155–56; and John Merriam, 166; MVZ proposal to, 78–79, 80–82, 333n11; paleontology funding proposal to, 107, 111–12, 113, 336n1; research commitment of, 32–33; retirement of, 167; university's growth under, 115, 337n3

Wildcat Peak (Trinity Alps, California), 146

Wildlife: disappearance of, 55–56, 76–77, 153–54; for food rations, 39, 40, 42–43; photographic documentation of, 38, 40, 41; sport hunting of, 35, 36–37, 44–45
Wilhelm Meister's Apprenticeship (Goethe), 225
Wilkie, Wendell, 185
Willcox (Arizona), 239, 240
Williams College (Massachusetts), 6, 14
Willow Creek (eastern Sierra Nevada), 245
Willow ptarmigan *(Lagopus lagopus alexandrae)*, 332n17
Wilson, Dick, 316
Wilson, Don, 134–35, 136, 277, 287, 316
Wilson, Mary, 26, 27, 326–27n9, 353n1
Windfall Harbor (Admiralty Island, Alaska), 66, 68
Women benefactors, 115–16
Women biologists: Alexander's benefaction to, 304–7, 357n26; fieldwork by, 303–7, 357nn20,22; Joseph Grinnell on, 99, 100, 301, 302; E. R. Hall on, 302; MVZ's admittance of, 301–2; MVZ's exclusion of, 99; in paleontology department, 302–3; in UCMP, 303
Wood rats, 233, 246
Woodward, A. Smith, 194, 195
World War I, impact of, xvi, 130, 158, 181, 182
World War II, impact of, 261, 272–73, 301, 306

Yakutat (Alaska), 102–3
A Yosemite Flora (Hall and Hall), 277
Yosemite National Park (California), 153, 159–60, 277
Yreka (California), 26
Yucca, 278

Zaikof Bay (Montague Island, Prince William Sound, Alaska), 105
Zambesi Railway Bridge, 45–46
Zinc specimen cases, 78, 85, 261
Zorilla *(Ictonyx libyca)*, 207
Zuni Well (Arizona), 186

Compositor: Integrated Composition Systems
Text: 10/13 Aldus
Display: Aldus
Printer and binder: Friesens